Ecological Monitoring and Assessment of Freshwater Ecosystems: New Trends and Future Challenges

Ecological Monitoring and Assessment of Freshwater Ecosystems: New Trends and Future Challenges

Eva Papastergiadou
Kostas Stefanidis

MDPI

Basel • Beijing • Wuhan • Barcelona • Belgrade • Novi Sad • Cluj • Manchester

Eva Papastergiadou
Department of Biology
University of Patras
Patras
Greece

Kostas Stefanidis
Institute of Marine Biological
Resources and Inland Waters
Hellenic Centre for Marine Research
Anavyssos
Greece

Editorial Office
MDPI AG
Grosspeteranlage 5
4052 Basel, Switzerland

This is a reprint of articles from the Special Issue published online in the open access journal *Water* (ISSN 2073-4441) (available at: www.mdpi.com/journal/water/special_issues/water_Ecosystems).

For citation purposes, cite each article independently as indicated on the article page online and using the guide below:

Lastname, A.A.; Lastname, B.B. Article Title. *Journal Name* **Year**, *Volume Number*, Page Range.

ISBN 978-3-7258-1724-5 (Hbk)
ISBN 978-3-7258-1723-8 (PDF)
https://doi.org/10.3390/books978-3-7258-1723-8

Contents

About the Editors

Eva Papastergiadou

Eva Papastergiadou is a professor of Ecology at the University of Patras, with expertise in ecosystem-scale dynamics in freshwaters, functional diversity patterns, and processes of aquatic macrophyte assemblages. She has experience as key partner and national representative of Greece and Cyprus in the WFD Intercalibration process, in many national and international projects on biodiversity conservation, shallow lakes functioning; monitoring and management of protected areas; the organization of workshops and coordination of scientific working teams (e.g., Implementation of the Natura 2000 network, ecological quality assessment of inland waters, WFD 2000/60). Also, she is a member of several professional committees, editorial board member of eight international journals, and guest editor of three Special Issues. Her research interests are focused on Freshwater Ecology, Conservation, and Management; functional aspects of lake and river communities; environmental and climate changes on aquatic ecosystems functioning; the evaluation of ecological quality; the dynamics of aquatic and riparian vegetation; patterns and processes structuring macrophyte assemblages at various spatial scales; biodiversity monitoring; impacts of human activities; and the effects of landscape structure and habitat quality on freshwater biodiversity. Currently, she is a director of the Environmental Sciences Lab of the School of Natural Sciences, University of Patras and director of the post graduate program of the Department of Biology for Master Thesis, named "Applied Ecology and Environmental Management".

Kostas Stefanidis

Konstantinos Stefanidis is a post-doctorate researcher at the Institute of Marine Biological Resources and Inland Waters of the Hellenic Centre for Marine Research. He has expertise in freshwater ecology with emphasis placed on freshwater plant diversity, ecological processes of shallow lakes, and climate change impacts on freshwater ecosystems. He has participated in several national and international projects on the monitoring and management of freshwater ecosystems, biodiversity conservation, and ecosystem functioning. He is an editorial board member of three international journals and has served as a guest editor of three Special Issues. His research interests include functional and taxonomic diversity patterns of freshwater plants, ecological assessment of freshwater ecosystems, multiple stressor effects on freshwater ecosystems, hydromorphological alterations in rivers, habitat heterogeneity, and ecosystem functioning.

Preface

Freshwater ecosystems are of worldwide importance for maintaining biodiversity and ecosystem services; at the same time, they are highly vulnerable to increasing anthropogenic activities and global climate change. Due to the multifunctionality of freshwater ecosystems, conservation management and restoration measures could lead to an improvement in biodiversity, ecological quality, and the supply of clean water and other ecosystem services to humans. In recent decades, extensive national and international regulations have been adopted to protect water resources. Biological monitoring/assessment methods and classification systems were greatly improved by the EU Water Framework Directive (WFD 2000/60) through monitoring programmes based on species composition and abundance. This Special Issue focuses on current research, and challenges, related to aquatic diversity that explores important issues on the monitoring of ecological quality, experimental studies, modelling, and the decline in species and contributes to conservation and preventing future loss of freshwater biodiversity. Within the broad framework of "Ecological Monitoring and Assessment of Freshwaters", this Special Issue aims to highlight new research findings and significant advances concerning all aspects of bio-assessment and aquatic ecosystem processes. Emphasis was given to contributions that explore the dynamics and functioning of freshwater ecosystems, develop new methods for monitoring and assessing ecological quality, investigate the use of biotic metrics or indices, consider environmental DNA methods, include experimental studies, and promote the use of modelling approaches. We believe that this reprint of "Ecological Monitoring and Assessment of Freshwater Ecosystems: New Trends and Future Challenges" will offer readers comprehensive and in-depth knowledge, thereby promoting further research in the field of freshwater monitoring and assessment of ecological quality. This collection of twelve original articles, accepted for publication from four continents of the world (Europe, North America, Africa, and Asia), expands our current knowledge on various topics, including the use of freshwater biota as indicators of environmental change, the application of models for predicting biological parameters, and the use of eDNA methods for monitoring invasive species. Open access to all articles increases the opportunity to share knowledge between researchers, managers and the public worldwide. Finally, we would like to express our gratitude to the authors, the reviewers, and the editors for their time and expertise to help this reprint evolve to its present edition.

Eva Papastergiadou and Kostas Stefanidis
Editors

water

MDPI

Editorial

Ecological Monitoring and Assessment of Freshwater Ecosystems: New Trends and Future Challenges

Konstantinos Stefanidis [1,2,*] and Eva Papastergiadou [1,*]

1 Department of Biology, School of Natural Sciences, University of Patras, University Campus Rio, 26500 Patras, Greece
2 Hellenic Centre for Marine Research (HCMR), Institute of Marine Biological Resources and Inland Waters, 46.7 km of Athens—Sounio Ave., 19013 Attiki, Greece
* Correspondence: kstefani@upatras.gr (K.S.); evapap@upatras.gr (E.P.); Tel.: +30-2291076439 (K.S.); +30-261096925 (E.P.)

Citation: Stefanidis, K.; Papastergiadou, E. Ecological Monitoring and Assessment of Freshwater Ecosystems: New Trends and Future Challenges. *Water* **2024**, *16*, 1460. https://doi.org/10.3390/w16111460

Received: 26 April 2024
Accepted: 17 May 2024
Published: 21 May 2024

1. Introduction

Freshwater ecosystems, particularly rivers and lakes, are under severe pressure due to increasing anthropogenic activities, such as water extraction, flow regulation, pollution, and habitat fragmentation [1–3]. Local, regional, and global drivers of environmental change (e.g., land cover transformation, pollution, the introduction of invasive species, and climate change) are responsible for the loss of many freshwater biota and ecosystem functions all around the world [4,5]. The global biodiversity crisis is more acute in freshwater ecosystems than in any other ecosystem. The current rate of wetland loss is three times that of forest loss and almost 27% of freshwater species are threatened with extinction [5]. Furthermore, human-driven changes greatly impact the delivery of ecosystem services, affecting the well-being of humans. Hence, introducing new conservation management and restoration measures is mandatory to improve biodiversity, ecological quality, and the supply of clean water and other ecosystem services to humans. Over recent decades, extensive national and international regulations have been adopted to protect water resources. In Europe, the biological monitoring, assessment methods, and classification systems in use have been greatly improved by the EU Water Framework Directive (WFD 2000/60) through monitoring programs based on species composition and abundance [6,7]. The goal of the Water Framework Directive is to restore or maintain the ecological state of the freshwater systems that are present across all of the EU member states. Thus, the WFD provides very detailed guidelines for the implementation of ecological monitoring and the assessment of all European inland and coastal waters [6,7].

Ecological monitoring is essential for understanding an ecosystem's functions and dynamics. The collection of biological and environmental data has greatly improved our capacity to understand the impact of many anthropogenic activities on biotic communities and the overall health of an ecosystem. Traditionally, ecological monitoring is based on extensive field surveys to acquire information about the diversity and composition of species' communities and their relationships with the environment. Significant advancements in monitoring have created novel methods and have delivered new tools that further increase the efficiency of data collection and reduce the associated costs. The use of wireless sensor network arrays (e.g., camera traps, acoustic sensors) promotes real-time monitoring with high-frequency measurements over large spatial scales [8]. The analysis of satellite images is another source of high-quality biotic and abiotic data with wide applications in ecological studies including freshwater ecosystems [9]. Recently, function-based assessments (species traits) and molecular methods (eDNA-based bioassessment) have been proposed to complement or even replace current monitoring methods [10–14].

This Editorial refers to the Special Issue "Ecological Monitoring and Assessment of Freshwater Ecosystems: New Trends and Future Challenges". The wide scope of the Special

Issue aims to highlight new research findings and significant advances concerning all aspects of bioassessment and the processes that occur within aquatic ecosystems. Emphasis was placed on contributions that explored the dynamics and functioning of freshwater ecosystems, developed new methods for monitoring and assessing ecological quality, studies that included the use of biotic metrics or indices, environmental DNA methods, experimental studies, and those that promoted the use of modelling approaches.

Twelve articles were finally accepted for publication from a total of twenty submissions that were considered for the SI. The twelve selected studies cover four continents of the world (Europe, North America, Africa, and Asia) and present results within the scope of the SI that expand our current knowledge on various topics, including the use of freshwater biota as indicators of environmental change, the application of models for predicting biological parameters, and the use of eDNA methods for monitoring invasive species (Table 1).

Table 1. Analysis of the contributions published in this Special Issue.

No of Contribution	Focus on Biotic Component/Ecosystem Process	Type of Research	Spatial Scale/Geographical Coverage
1	Chlorophyll-a concentration/eutrophication	Modelling	Ecosystem/Greece
2	Aquatic macrophtes/responses to salinization and increased irradiance	Experiment/Ecophysiology	Ecosystem/Habitat/Greece
3	Riverine habitats	Assessment method/index development	Catchment/China
4	Fish/predator effects	Experiment	South Africa
5	Phosphorus and nutrient losses at catchment scale	Modelling	Multiple catchments/Europe
6	Zebra and quagga mussels/eDNA assessment	eDNA assessment	Ecosystem/the USA
7	Aquatic macrophytes/ecological quality index	Assessment method/index development	Multiple ecosystems/Greece
8	Meiofauna/responses to salinization	Field study	Catchment/Vietnam
9	Cyanobacteria/study on akinetes	Biochemical/molecular study	Organism/South Korea
10	Water quality and biological quality indices	Field study	Catchment/South Korea
11	Macroinvertebrates and fish/responses to pollution	Field study	Multiple ecosystems/South Africa
12	Freshwater sponges/responses to pollution	Field study	Multiple ecosystems/Serbia

2. Main Messages of the Special Issue and the Book

The current Special Issue and the contributions within it discuss many themes related to aquatic diversity that focus on the monitoring of ecological quality, experimental studies, modelling, and the decline in species and also contribute to conservation and preventing future loss of freshwater biodiversity. The first article, by Hadjizolomou et al., examines the potential to optimize artificial neural networks (ANN) for predicting the chlorophyll-a concentration in lakes with limited field data. The authors found that the ANN's performance is greater when the leave-one-out (LOO) cross validation method is used and increases with the k-fold number. They also investigated the effects of the input parameters on the prediction of Chl-a concentration by conducting a sensitivity analysis, and they found that changes in conductivity and water temperature caused a higher % of changes in the predicted outcome. Based on their results, the authors concluded that ANN models can be a useful tool for predicting chlorophyll-a, and potentially other lake variables, even when data scarcity is an issue.

The next contribution was written by Malea et al., who experimentally investigated the physiological responses of the submerged plant *Stuckenia pectinata* to high levels of salinity and irradiance. Their results highlighted the plant's significant photo-acclimation potential, which could be used to regulate the number and size of its reaction centres and photosynthetic electron transport chain through the dissipation of excess energy into heat. They also found that the interaction between salinity and irradiance had a significant effect on the plant's Chl-a, b contents, which may indicate its potential ability to acclimatize by adjusting the Chl a, b contents. However, they did not report significant impacts on the relative growth rate, which could mean that the plant may become acclimatized by reallocating resources to compensate for growth. Thus, the authors conclude that the regulation of photosynthetic pigment content and photosystem II performance comprised the plant's primary growth strategy within the high salinity and irradiance conditions that are likely to occur due to eutrophication and future climatic change.

The third article, by Yu et al., deals with the use of satellite imagery for assessing the habitat quality of mountain streams in the Chishui basin in China. This study employs a series of metrics based on water environmental status, river morphology, riparian zone, and human disturbance, combined with stream order, elevation, slope, and sinuosity, to classify habitats into types. The habitat assessment was conducted with the use of the habitat quality index (CHQI) and the results indicated that the headwaters of three rivers (Tongmin, Tongzi, and Xishui) have been impacted by anthropogenic activity. The authors conclude that their habitat assessment method can be used for further biomonitoring in other mountain river systems as well.

The fourth contribution, by Munyai et al., comes from South Africa and investigates the effect of two predator fish, *Oreochromis mossambicus* and *Enteromius paludinosus*, on Chironomidae prey. The study uses a comparative functional response approach to assess the interactions of the two predator species when they are feeding on a readily consumed prey within multiple predator scenarios. The findings of this article reported that each species displayed a significant Type II FR, characterized by high feeding rates at low prey densities. *Oreochromis mossambicus* had a steeper (initial slope, i.e., higher attack rate) and higher (asymptote of curve, i.e., shorter handling time and higher maximum feeding rate) FR, whereas *E. paludinosus* exhibited lower-magnitude FRs (i.e., lower attack rate, longer handling time, and lower feeding rate). In multiple predator scenarios, the feeding rates were predicted well by using those of single predators, both in conspecific and interspecific pairs, and thus the authors did not find evidence for antagonistic or synergistic multiple predator effects (MPEs). The results from this study, although experimental, improve our current knowledge about how trophic interactions among conspecific or interspecific fish species in Austral tropical wetlands might influence their aquatic prey species.

The article by Wade et al. is the fifth contribution to the SI and examines the relative importance of the climate and land use drivers in nutrient loss in nine study catchments in Europe and a neighbouring country (Turkey). Catchment-scale biophysical models were applied within a common framework to quantify the integrated effects of projected changes in the climate, land use (including wastewater inputs), N deposition, and water use on the quantity and quality of river and lake water in the mid-21st century. The proposed land use changes were derived from catchment stakeholder workshops, and the assessment quantified changes in mean annual N and P concentrations and loads. The main finding of this study was that, at most of the sites, the projected effects of climate change alone on nutrient concentrations and loads were small, whilst land use changes had a larger effect and were of sufficient magnitude that, overall, a move to more environmentally focused farming achieved a reduction in N and P concentrations and loads despite the projected effects climate change. However, at Beyşehir lake in Turkey, increased temperatures and lower precipitation reduced water flows considerably, making climate change, rather than more intensive nutrient usage, the greatest threat to the freshwater ecosystem. Individual site responses did, however, vary, and were dependent on the balance of diffuse and point source inputs. The simulated changes in the chlorophyll-a content in the lake were not generally proportional to changes in nutrient loading. Thus, further work is required to accurately simulate the extremes in flow and water quality and determine how reductions in freshwater N and P translate into an aquatic ecosystem response.

The next article by Marshall et al. presents and evaluates new eDNA assays that target the extended repeat sections of zebra and quagga mussels. These assays lower the limit of detection of genomic DNA by 100-fold for zebra mussels and 10-fold for quagga mussels. Additionally, these newly developed assays facilitated longer durations of detection during degradation mesocosm experiments and a greater sensitivity for the detection of eDNA in water samples collected across western Lake Erie compared to standard assays that target mitochondrial genes. Finally, this study illustrates how important it is to understand the complete genomic structure of an organism in order to improve eDNA analyses.

The seventh contribution, by Stefanidis et al., presents a methodological approach for the implementation of a WFD-compliant macrophyte index in the riverine systems of Greece and the results from the pilot application of the index. The study analyses the methodological framework for defining the stressor gradients and the least disturbed sites along with the reference conditions that are required for the derivation of the ecological quality classes. It also includes the classification of the river reaches into five quality classes that were derived from the application of the Macrophyte Biological Index IBMR for Greek rivers ($IBMR_{GR}$). The main findings showed that hydromorphological modifications were the main environmental stressors and that they were strongly with the correlated $IBMR_{GR}$, while physicochemical stressors were of lesser importance. In addition, the ecological assessment showed that almost 60% of the sites failed the WFD target of a "Good" ecological quality class, which agrees with classification assessments that were based on other BQEs for Greece and many other Mediterranean countries. Overall, this work provides the first assessment of the ecological classification of Greek rivers using the BQE of aquatic macrophytes, which has significant implications for ecological monitoring and decision making within the framework of the implementation of the WFD.

Nguyen et al. contributed to the SI with an article that assesses the impact of salinity variations on riverine ecosystems with a particular focus on the responses of meiofauna to salinization along the Van Uc River continuum in Vietnam. The main findings of this study were that the meiofaunal richness indices were higher in the estuary and slightly decreased upriver. Nematoda was the most dominant taxon at the salty stations, while Rotifera was more abundant at the less salty ones. The results from a multiple variate analysis indicated a strong interplay among salinity, nutrients, and pore water conductivity, which shaped the meiofaunal distribution. The inclusion of pore water salinity, nutrients, and meiofaunal community structure indicated that there was a greater extent of the saline ecosystem in the estuary than previously thought, posing a greater risk to freshwater salinization. Hence, the contribution by Nguyen et al. highlights the potential role of meiofauna as a bioindicator but also calls for a reformation of salinity assessment for better freshwater conservation and management.

The next contribution, by Kim et al., analyses the akinete structure, as well as akinete-specific proteins and their amino acid sequences, of the cyanobacterium *Dolichospermum circinale*. The akinetes were produced from vegetative cells isolated from the North Han River, Korea, while the akinete protein was obtained using electrophoresis, and its antibody-binding reaction potential (ig-score) was quantified. The authors found that the homology of the *D. circinale* akinete-specific protein was very low (9.8%) compared to that of *Anabaena variabilis*, indicating that its composition was substantially different, even among phylogenetically close taxa. Overall, this article represents the first known report on the *D. circinale* akinete protein and its amino acid sequence and offers insights into their practical application for detecting akinetes in freshwater systems.

The tenth article is by Lee et al., also from South Korea, and investigates the effects of land use on stream water quality and biological conditions in sub-watersheds and micro-watersheds across the Han River. By employing random forest models, the authors found that water quality and biological indicators were significantly affected by forest areas at both scales, and the sub-watershed models performed better than the micro-watershed models. The effects on the water quality and biological indicators were similar regardless of the scales, although the relationship between land use and stream conditions was slightly more sensitive in the micro-watersheds than in the sub-watersheds. In addition, their results showed that urban and agricultural areas showed a lower proportion of water quality and biological condition variability in the micro-watersheds than in the sub-watersheds, while the forests showed the opposite results. Hence, the authors concluded that the spatial scale matters when developing effective watershed management strategies to maintain stream ecosystems.

The next study, by Munyai et al., examined the relationships between environmental parameters and biotic communities, with an emphasis on the effects of water and sediment quality parameters on the macroinvertebrate and fish communities in three subtropical reservoirs in South Africa. The results of their redundancy analysis and two-way ANOVA showed that there were significant differences among the reservoirs, and they identified four water variables (water temperature, oxidation–reduction potential, pH and conductivity) and one sediment metal (Mg) as the most important parameters driving the structure of the fish community. In addition, high concentrations of metals in the sediment of one of the three reservoirs (Nandoni) suggested that anthropogenic activities have significantly influenced the sediment quality, with severe implications for the ecological conditions. Finally, the authors propose the need to adopt measures that improve the conservation of these ecosystems.

The final contribution, by Andjus et al., assess the frequency of spicule malformations in freshwater sponges in relation to selected environmental parameters of the streams and the presence of river pollutants. The authors conducted a morphological analysis using light and scanning electron microscopy, and recorded the number of anomalies (spicules with bulbous enlargements that are sharply bent, bifurcated, scissor- and cross-like, and t-shaped). The results reported single- and double-bent spicules as the main types of anomalies. An important finding was that the authors found statistically significant differences in the concentrations of ammonia, orthophosphates, sodium, chloride, manganese, and lead between the sites with the lowest and the highest numbers of anomalies. Thus, this study indicates that several pollutants could be responsible for the occurrence of spicule anomalies.

This Special Issue is devoted to articles that present new and original research on topics focused on the ecological monitoring and assessment of freshwater ecosystems. More than half of the articles examine the relationships between environment and freshwater biota at various scales, from the organism level to the community and ecosystem levels (Table 1). These studies focused on various biotic groups, including aquatic macrophytes (contributions No 2 and 7), freshwater sponges (contribution No 12), fish (contribution No 4 and 11), and invertebrates (contributions No 8 and 11). In addition, some articles explored new methods and research techniques for ecological monitoring, such as contribution No 6 which assesses novel eDNA assays for zebra and quagga mussels, contribution No 9 which deals with the use of akinete structure of cyanobacteria as indicator of environmental change, and contribution No 1 which focuses on using ANN models to predict Chl-a concentration in shallow lakes. The spatial scale of the studies varied from the local (ecosystem scale) to the regional (catchment scale) and transregional (multiple catchments) scales. Six of the articles include studies conducted at the local scale (one ecosystem or catchment), three of the articles concern studies at a wider scale (multiple ecosystems or catchments), and two of the articles were based on experimental research. Finally, one study involved the biophysical modelling of nine catchments from Europe and Turkey to investigate the effects of climate change and land use transformation on nutrient losses under different future scenarios.

3. Conclusions

Freshwater ecosystems cover less than 1% of the Earth's surface and yet are among the most diverse and threatened systems in the world [5]. The importance of freshwater ecosystems in maintaining biodiversity and ecosystem services is becoming increasingly clear, but, at the same time, freshwater ecosystems are highly vulnerable to human impacts such as climate change and land use changes. The collection of articles in this SI show that regional (e.g., land cover transformation, pollution, hydromorphological alterations, and invasive species) and global (e.g., climate change) environmental changes are responsible for the loss of many aquatic biota and ecosystem functions worldwide. Furthermore, there is a growing number of studies, and EU policies such as the WFD 2000/60 and the habitats directive, that highlight the need to improve current monitoring schemes and undertake conservation and restoration actions within inland waters.

In this context, future research will prioritize gaining a better understanding of how these changes can affect species, communities, functions, and ecosystem services by employing new methods and tools.

In the current SI, several topics were discussed which provide significant insights into the monitoring, conservation, and management of freshwater ecosystems. Catchment models, machine learning, eDNA assessments, and remote sensing are all tools that are gaining ground in ecological monitoring (see contributions 1,3,5,6). In addition, ecological studies based on field monitoring have begun to focus on less explored biotic groups, such as sponges and meiofauna, and ecophysiological responses, producing promising results with further implications for the ecological assessment of freshwater ecosystems (contributions 2,4,8,9,12). Finally, the development and testing of new ecological indices remain top priorities for freshwater ecologists (contributions 3 and 7) but also for managers who are called in to assess the impact of anthropogenic stressors on freshwater ecosystems.

Author Contributions: Conceptualization, E.P. and K.S.; writing—original draft preparation E.P. and K.S.; writing—review and editing, E.P. and K.S. All authors have read and agreed to the published version of the manuscript.

Conflicts of Interest: The authors declare no conflicts of interest.

List of Contributions

1. Hadjisolomou, E.; Stefanidis, K.; Herodotou, H.; Michaelides, M.; Papatheodorou, G.; Papastergiadou, E. Modelling Freshwater Eutrophication with Limited Limnological Data Using Artificial Neural Networks. *Water* **2021**, *13*, 1590. https://doi.org/10.3390/w13111590.
2. Malea, L.; Nakou, K.; Papadimitriou, A.; Exadactylos, A.; Orfanidis, S. Physiological Responses of the Submerged Macrophyte Stuckenia pectinata to High Salinity and Irradiance Stress to Assess Eutrophication Management and Climatic Effects: An Integrative Approach. *Water* **2021**, *13*, 1706. https://doi.org/10.3390/w13121706.
3. Yu, F.; Liu, F.; Xia, Z.; Lin, P.; Xu, C.; Wang, J.; Hou, M.; Zou, X. Classification and Assessment Methods for Mountain Channel Habitats in the Chishui River Basin, China. *Water* **2022**, *14*, 515. https://doi.org/10.3390/w14040515.
4. Munyai, L.; Dalu, T.; Wasserman, R.; Mugwedi, L.; Dondofema, F.; O'Brien, G.; Cuthbert, R. Functional Responses and Additive Multiple Predator Effects of Two Common Wetland Fish. *Water* **2022**, *14*, 699. https://doi.org/10.3390/w14050699.
5. Wade, A.; Skeffington, R.; Couture, R.; Erlandsson Lampa, M.; Groot, S.; Halliday, S.; Harezlak, V.; Hejzlar, J.; Jackson-Blake, L.; Lepistö, A.; Papastergiadou, E.; Riera, J.; Rankinen, K.; Shahgedanova, M.; Trolle, D.; Whitehead, P.; Psaltopoulos, D.; Skuras, D. Land Use Change to Reduce Freshwater Nitrogen and Phosphorus will Be Effective Even with Projected Climate Change. *Water* **2022**, *14*, 829. https://doi.org/10.3390/w14050829.

6. Marshall, N.; Vanderploeg, H.; Chaganti, S. Improving Environmental DNA Sensitivity for Dreissenid Mussels by Targeting Tandem Repeat Regions of the Mitochondrial Genome. *Water* **2022**, *14*, 2069. https://doi.org/10.3390/w14132069.
7. Stefanidis, K.; Dimitrellos, G.; Sarika, M.; Tsoukalas, D.; Papastergiadou, E. Ecological Quality Assessment of Greek Lowland Rivers with Aquatic Macrophytes in Compliance with the EU Water Framework Directive. *Water* **2022**, *14*, 2771. https://doi.org/10.3390/w14182771.
8. Nguyen, H.; Gourdon, L.; Bui, H.; Dao, D.; Mai, H.; Do, H.; Nguyen, T.; Ouillon, S. Ecological Responses of Meiofauna to a Saltier World—A Case Study in the Van Uc River Continuum (Vietnam) in the Dry Season. *Water* **2023**, *15*, 1278. https://doi.org/10.3390/w15071278.
9. Kim, K.; Park, C.; Hwang, S. Analyzing the Akinete Protein of the Harmful Freshwater Cyanobacterium, Dolichospermum circinale. *Water* **2023**, *15*, 2746. https://doi.org/10.3390/w15152746.
10. Lee, J.; Park, S.; Lee, S. Effect of Land Use on Stream Water Quality and Biological Conditions in Multi-Scale Watersheds. *Water* **2023**, *15*, 4210. https://doi.org/10.3390/w15244210.
11. Munyai, L.; Liphadzi, T.; Mutshekwa, T.; Mutoti, M.; Mofu, L.; Murungweni, F. Water and Sediment Chemistry as Drivers of Macroinvertebrates and Fish Assemblages in Littoral Zones of Subtropical Reservoirs. *Water* **2024**, *16*, 42. https://di.org/10.3390/w16010042.
12. Andjus, S.; Tubić, B.; Vasiljević, B.; Nikolić, V.; Paunović, M. Anomalies of Sponge Spicules: Exploring Links to Environmental Pollution. *Water* **2024**, *16*, 332. https://doi.org/10.3390/w16020332.

References

1. Best, J. Anthropogenic Stresses on the World's Big Rivers. *Nat. Geosci.* **2019**, *12*, 7–21. [CrossRef]
2. Woolway, R.I.; Albergel, C.; Frölicher, T.L.; Perroud, M. Severe Lake Heatwaves Attributable to Human-Induced Global Warming. *Geophys. Res. Lett.* **2022**, *49*, e2021GL097031. [CrossRef]
3. Grill, G.; Lehner, B.; Thieme, M.; Geenen, B.; Tickner, D.; Antonelli, F.; Babu, S.; Borrelli, P.; Cheng, L.; Crochetiere, H.; et al. Mapping the World's Free-Flowing Rivers. *Nature* **2019**, *569*, 215–221. [CrossRef]
4. Williams-Subiza, E.A.; Epele, L.B. Drivers of Biodiversity Loss in Freshwater Environments: A Bibliometric Analysis of the Recent Literature. *Aquat. Conserv. Mar. Freshw. Ecosyst.* **2021**, *31*, 2469–2480. [CrossRef]
5. Tickner, D.; Opperman, J.J.; Abell, R.; Acreman, M.; Arthington, A.H.; Bunn, S.E.; Cooke, S.J.; Dalton, J.; Darwall, W.; Edwards, G.; et al. Bending the Curve of Global Freshwater Biodiversity Loss: An Emergency Recovery Plan. *Bioscience* **2020**, *70*, 330–342. [CrossRef]
6. Quevauviller, P.; Barceló, D.; Beniston, M.; Djordjevic, S.; Harding, R.J.; Iglesias, A.; Ludwig, R.; Navarra, A.; Navarro Ortega, A.; Mark, O.; et al. Integration of Research Advances in Modelling and Monitoring in Support of WFD River Basin Management Planning in the Context of Climate Change. *Sci. Total Environ.* **2012**, *440*, 167–177. [CrossRef] [PubMed]
7. Maia, R. The WFD Implementation in the European Member States. *Water Resour. Manag.* **2017**, *31*, 3043–3060. [CrossRef]
8. Besson, M.; Alison, J.; Bjerge, K.; Gorochowski, T.E.; Høye, T.T.; Jucker, T.; Mann, H.M.R.; Clements, C.F. Towards the Fully Automated Monitoring of Ecological Communities. *Ecol. Lett.* **2022**, *25*, 2753–2775. [CrossRef] [PubMed]
9. Carrea, L.; Crétaux, J.F.; Liu, X.; Wu, Y.; Calmettes, B.; Duguay, C.R.; Merchant, C.J.; Selmes, N.; Simis, S.G.H.; Warren, M.; et al. Satellite-Derived Multivariate World-Wide Lake Physical Variable Timeseries for Climate Studies. *Sci. Data* **2023**, *10*, 30. [CrossRef] [PubMed]
10. Van Driessche, C.; Everts, T.; Neyrinck, S.; Halfmaerten, D.; Haegeman, A.; Ruttink, T.; Bonte, D.; Brys, R. Using Environmental DNA Metabarcoding to Monitor Fish Communities in Small Rivers and Large Brooks: Insights on the Spatial Scale of Information. *Environ. Res.* **2023**, *228*, 115857. [CrossRef] [PubMed]
11. Blancher, P.; Lefrançois, E.; Rimet, F.; Vasselon, V.; Argillier, C.; Arle, J.; Beja, P.; Boets, P.; Boughaba, J.; Chauvin, C.; et al. A Strategy for Successful Integration of DNA-Based Methods in Aquatic Monitoring. *Metabarcoding Metagenomics* **2022**, *6*, 215–226. [CrossRef]
12. Magni, P.; Vesal, S.E.; Giampaoletti, J.; Como, S.; Gravina, M.F. Joint Use of Biological Traits, Diversity and Biotic Indices to Assess the Ecological Quality Status of a Mediterranean Transitional System. *Ecol. Indic.* **2023**, *147*, 109939. [CrossRef]

13. Bonilla-Valencia, L.; Castillo-Aguero, S.; Zavala-Hurtado, J.A.; García, F.J.E.; Lindig-Cisneros, R.; Martínez-Orea, Y. Linking Functional Diversity to Ecological Indicators: A Tool to Predict Anthropogenic Effects on Ecosystem Functioning. *Environ. Rev.* **2022**, *30*, 175–183. [CrossRef]
14. Stefanidis, K.; Oikonomou, A.; Dimitrellos, G.; Tsoukalas, D.; Papastergiadou, E. Relationships between Environmental Factors and Functional Traits of Macrophyte Assemblages in Running Waters of Greece. *Diversity* **2023**, *15*, 949. [CrossRef]

water

MDPI

Article

Modelling Freshwater Eutrophication with Limited Limnological Data Using Artificial Neural Networks

Ekaterini Hadjisolomou [1,*], Konstantinos Stefanidis [2,3], Herodotos Herodotou [1], Michalis Michaelides [1], George Papatheodorou [4] and Eva Papastergiadou [2]

1 Department of Electrical Engineering, Computer Engineering and Informatics, Cyprus University of Technology, 30 Arch. Kyprianos Street, Limassol 3036, Cyprus; herodotos.herodotou@cut.ac.cy (H.H.); michalis.michaelides@cut.ac.cy (M.M.)
2 Department of Biology, School of Natural Sciences, University of Patras, University Campus Rio, 26500 Patra, Greece; kstefanidis@hcmr.gr (K.S.); evapap@upatras.gr (E.P.)
3 Hellenic Centre for Marine Research (HCMR), Institute of Marine Biological Resources and Inland Waters, 46.7 km of Athens—Sounio Ave., 19013 Anavyssos, Greece
4 Laboratory of Marine Geology and Physical Oceanography, Department of Geology, University of Patras, 26504 Patra, Greece; George.Papatheodorou@upatras.gr
* Correspondence: e.hadjisolomou@cut.ac.cy

Abstract: Artificial Neural Networks (ANNs) have wide applications in aquatic ecology and specifically in modelling water quality and biotic responses to environmental predictors. However, data scarcity is a common problem that raises the need to optimize modelling approaches to overcome data limitations. With this paper, we investigate the optimal *k*-fold cross validation in building an ANN using a small water-quality data set. The ANN was created to model the chlorophyll-*a* levels of a shallow eutrophic lake (Mikri Prespa) located in N. Greece. The typical water quality parameters serving as the ANN's inputs are pH, dissolved oxygen, water temperature, phosphorus, nitrogen, electric conductivity, and Secchi disk depth. The available data set was small, containing only 89 data samples. For that reason, *k*-fold cross validation was used for training the ANN. To find the optimal *k* value for the *k*-fold cross validation, several values of *k* were tested (ranging from 3 to 30). Additionally, the leave-one-out (LOO) cross validation, which is an extreme case of the *k*-fold cross validation, was also applied. The ANN's performance indices showed a clear trend to be improved as the *k* number was increased, while the best results were calculated for the LOO cross validation as expected. The computational times were calculated for each *k* value, where it was found the computational time is relatively low when applying the more expensive LOO cross validation; therefore, the LOO is recommended. Finally, a sensitivity analysis was examined using the ANN to investigate the interactions of the input parameters with the Chlorophyll-*a*, and hence examining the potential use of the ANN as a water management tool for nutrient control.

Keywords: data scarcity; *k*-fold cross validation; artificial neural network; eutrophication

Citation: Hadjisolomou, E.; Stefanidis, K.; Herodotou, H.; Michaelides, M.; Papatheodorou, G.; Papastergiadou, E. Modelling Freshwater Eutrophication with Limited Limnological Data Using Artificial Neural Networks. *Water* **2021**, *13*, 1590. https://doi.org/10.3390/w13111590

Academic Editor: Peter Goethals

Received: 22 April 2021
Accepted: 1 June 2021
Published: 4 June 2021

1. Introduction

Data-driven models are systematically used in the field of water resources management because they are less demanding in terms of data acquisition and quantity than the empirical and the physical-based hydrological models [1]. Artificial Neural Networks (ANNs) are such data-driven models with a wide application in water resources and aquatic sciences [2]. In particular, ANNs are used for modelling several domains of aquatic ecology such as benthic macroinvertebrates, planktonic communities, fish assemblages, and biomanipulation assessment [3]. During the last two decades, several ANN water quality modelling studies were performed with very good modelling results [4–6]. Their main advantages include their ability to model complex and nonlinear processes and the fact that they do not require assumptions about the distribution of the data or the relationships between input and output variables [2]. In addition, they offer the user the flexibility to successfully model environmental relationships with limited knowledge of the problem [7]. Thus, ANNs are considered ideal for modelling aquatic ecosystems, which are characterized by complex dynamics and nonlinear analytics [3].

Water quality monitoring programmes in particular produce a large amount of data with complex structures [8]. Unfortunately, there are also many cases where insufficient data quantity and quality can be a problem in ecological quality assessment and management [9]. Several factors such as bad weather conditions during monitoring, lack of funding, and problematic sensors are responsible for the lack of data. Data scarcity of water resources is an issue highlighted in the study of Cigizoglu and Kisi [10]; the authors address this issue successfully by applying *k*-fold cross validation into the ANN's training data set. Generally, the phenomenon of data scarcity/small data sets when modelling with ANNs is an issue found in many scientific fields and the method of *k*-fold cross validation is widely used [11]. The problem of having a small data set when applying *k*-fold cross validation is discussed in a water quality modelling study of Goethals et al. [5], where it is stated that lower *k*-values may produce more robust ANN models, but with lower performance, and therefore a high *k*-value is recommended.

In this article, we address the issue of training an ANN with a small dataset of water quality parameters collected from a eutrophic lake in Greece. Eutrophication is one of the most significant problems today that is responsible for the degradation of water quality in many freshwater, coastal, and marine ecosystems worldwide [12,13]. Because freshwater lakes are major providers of ecosystem services (e.g., water supply for potable use and irrigation), eutrophication has severe socioeconomic implications that threaten human well-being, particularly in areas of the world where water is scarce. Therefore, scientists are developing new tools and methods for efficient and improved monitoring of water quality. In addition, they employ advanced modelling techniques and algorithms for creating novel forecasting schemes of eutrophication, since a lake's ecosystem is very complex and sensitive and even a mild external pressure (such as tourist activity) with small inflows of biogenic elements can promote eutrophication [14]. Not surprisingly, ANNs have been used before in modelling eutrophication processes in lakes [15]. Linear regression methods and decision tree methods are some alternative methodologies used for eutrophication modelling. These methodologies have the merits of needing few parameters to adjust and being simple; however, they might not perform well when the data sample is small or does not follow certain assumptions about linear or Gaussian distributions [16]. In contrast, ANN models are not affected by the non-linearity effect, which is often observed between the environmental parameters associated with eutrophication [17].

The purpose of this modelling study is two-fold. First, we trained an ANN by implementing an approach of *k*-fold cross validation. Then, we compared the obtained modelling results between models built with different *k* values. Additionally, the factor of the needed computational time was also taken into consideration as a criterion for choosing the optimal model, with respect to the ANN performance and complexity in terms of computation time. Second, we examined the explanatory power of the optimal model to identify whether it can act as a water management tool. We show that the ANN managed to predict the chlorophyll-*a* levels of Lake Mikri Prespa with high accuracy. Based on this model, the contribution of each environmental parameter was evaluated with the use of a sensitivity analysis algorithm. The results of the sensitivity analysis produced useful conclusions about the role of each parameter with emphasis on the role of nutrients levels changes on the trophic status of the lake.

2. Materials and Methods

2.1. Study Area and Data Collection

The used dataset contains environmental variables obtained from Lake Mikri Prespa, a shallow eutrophic lake located in northwestern Greece. The lake is an ecosystem of great ecological importance as it belongs to the Prespa National park, is a Ramsar wetland site, an important bird area, and a Natura 2000 site [18]. The lake lies at an elevation of 853 m above sea level [19] and is located in the wider transboundary Prespa area shared by Greece, Albania, and North Macedonia (Figure 1). The climate of the area is characterized as sub-Mediterranean with continental influences, with frequent snowfall in the winter and summer rain drops [20]. The lake is characterized as a shallow lake, with approximately 48 km^2 of surface area, a maximum water depth of 8 m [21] and mean depth of 4.1 m [19]. The trophic status of the lake is characterized as eutrophic. As a result, prolonged cyanobacterial blooms have been recorded, which may start in spring and persist until December [22].

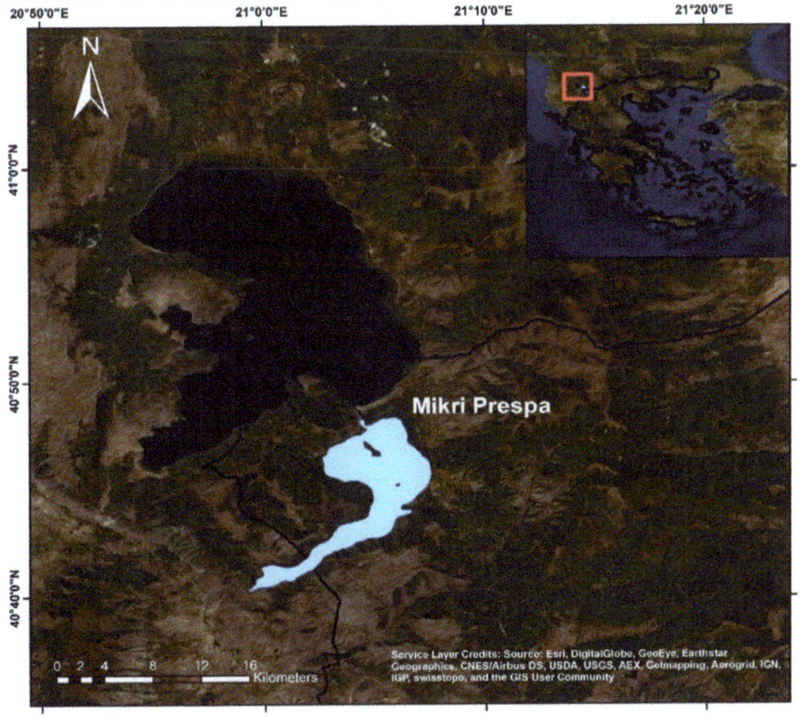

Figure 1. Map of the transboundary Mikri Prespa Lake located in northwestern Greece.

Chlorophyll-*a* concentration and environmental data were collected from fifteen sampling sites on a seasonal basis from 2006 to 2008 (see Hadjisolomou et al. [19]). We used the following environmental parameters as predictors of chlorophyll-*a* (Chl-*a*): pH, surface dissolved oxygen concentration (DO), electrical conductivity (EC), Secchi disk depth (SD), water depth (WD), surface water temperature (WT), total phosphorus concentration (TP) and dissolved inorganic nitrogen concentration (DIN).

The basic statistical properties of the environmental parameters are shown with the use of box plots in Figure 2.

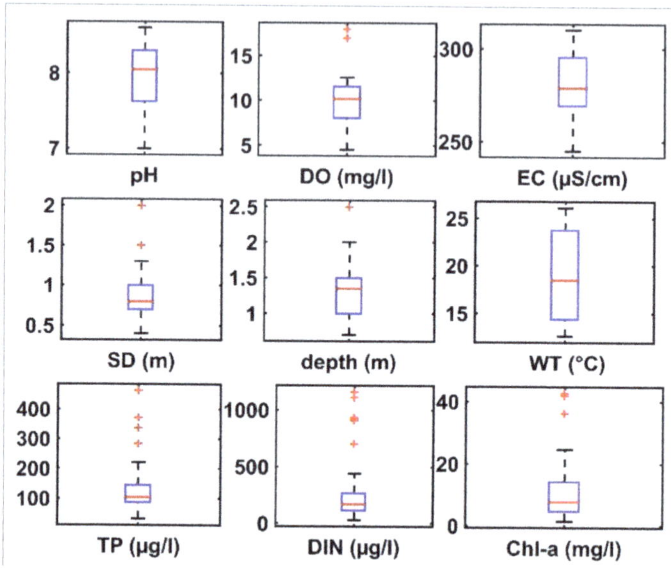

Figure 2. Box plot of the monitored environmental parameters (red horizontal line: median value; box: 25–75% percentile; whiskers: valid range; red marks: extreme values referred as "outliers").

2.2. Preliminaries on ANN and Model Construction

ANNs mimic the way biological neurons learn and make logical conclusions. A very popular ANN is the multi-layer feedforward (MLF) network trained by the back-propagation algorithm [23]. Feedforward networks are considered suitable for function approximation problems. The MLF networks are divided into at least three layers and each layer consists of neurons. The first layer is the input layer, followed by at least one intermediate hidden layer and the output layer that produces the final output. Each neuron of a layer relates to all the neurons of the next layer using synaptic weights. Every neuron performs aggregation on its weighted inputs and yields an output through an activation function. The most commonly used activation functions are the linear, the logistic and the hyperbolic tangent activation function [24]. The output value of the j-th neuron (o_j) is given by the equations as described by Dedecker et al. [25]:

$$o_j = f(u_j) \tag{1}$$

$$u_j = \sum_{\forall i} w_{ij} x_i + z_j \tag{2}$$

where f is the activation function, x_i is the input from i-th neuron belonging to the immediate previous layer, w_{ij} is the synaptic weight that connects x_i with the j-th neuron and z_j a bias term. The output of each neuron is computed and propagated through the next layer until the last layer, producing a network output that compares with the given output [26]. The learning procedure is repeated several times with the use of a training algorithm and each time the synaptic weights are adjusted until they minimize an error function, usually taken as the mean square difference between the predicted and the given output [27]. The Levenberg–Marquardt (LM) algorithm has the fastest convergence among the existing variations of backpropagation algorithms when it comes to ANNs with up to a few hundred parameters [23].

The topology of an ANN that determines the number of hidden layers and the number of neurons in each layer can be determined after a trial and error procedure [28,29]. The topology of a 3-layer ANN can be presented as *L1-H1-L2* where *L1* is the number of neurons in the input layer, *H1* the number of neurons in the hidden layer and *L2* the number of neurons in the output layer. In order to avoid overfitting, the maximum number of the hidden layer neurons can be computed according to the following the rule of thumb proposed by Maier et al. [30]:

$$N^H \leq 2N^I + 1 \tag{3}$$

$$N^H \leq 2N^{TR} / \left(N^I + 1\right) \tag{4}$$

where N^H is the number of hidden layer neurons, N^I the number of inputs, and N^{TR} the number of training samples. The maximum N^H must be the smallest number found by those two rules.

Data normalization is a common procedure since the ANN's performance is improved [25,31] and after network training the data is set back to its initial form. Dimensionality reduction of the measured variables is also recommended in order to achieve a reduced size of the original data set of variables [32]. The variable/parameter's dimension reduction is performed not only because it reduces the model computational complexity, but also eliminates the possibility for model's misconvergence and poor accuracy [4]. ANNs are evaluated based on several performance indices for their test set [33]. The Root Mean Square Error (RMSE) and Mean Absolute Error (MAE) are some commonly used performance indices and have the following mathematical formulas, respectively [34,35]:

$$RMSE = \sqrt{\frac{\sum_{i=1}^{n} (o_i - s_i)^2}{n}} \tag{5}$$

$$MAE = \frac{1}{n} \sum_{i=1}^{n} |o_i - s_i| \tag{6}$$

where for the above set of equations the parameter o_i is the observed value, s_i the simulated/predicted value, and n the number of observations.

The sensitivity analysis is performed based on the 'Perturb' method. The 'Perturb' method is a sensitivity analysis methodology, which is computing the perturbation effect of the input variables regarding the output variable. The 'Perturb' method is examining the effect that a small change of an input variable has on the ANN's output, therefore the input variables can be classified by an order of importance [36]. The mathematical formula that describes the 'Perturb' method of sensitivity analysis is explained by Lee et al. [37] as follows:

$$\text{Sensitivity } (\%) = \frac{1}{n} \sum_{i=1}^{n} \left(\frac{\text{change in output } (\%)}{\text{change in input } (\%)} \right)_i \times 100 \tag{7}$$

where n represents the number of observations.

The k-fold cross validation method is used in order to avoid overfitting and the data set is divided into k equally sized folds/subsamples [38]. For each k-fold of the data set, an ANN model is trained on the other $k - 1$ folds of the data set and validated for the k-th fold. The cross-validation procedure is repeated k times (hence, each fold is used exactly once as the validation data set) and finally the average of the k calculated validation performance indices is computed [39,40]. According to Goethals et al. [5], the optimal k-value is found based on an evaluation procedure, where the robustness and reliability of the developed ANN models is assessed. A high value of k is recommended when the dataset is small and has very few observations. An extreme case of the k-fold cross validation method is the case of leave-one-out (LOO) cross validation, when the arithmetic value of k equals the number of measured samples (n) in the dataset.

3. Results

The monitored variables (pH, DO, EC, SD, depth, WT, TP, DIN, and Chl-*a*) were analyzed for collinearity by calculating the Pearson correlation coefficient (r). As stated by Gebler et al. [41], the deletion of the collinear parameters helps the ANN to avoid unnecessary/superfluous information and simplifies the model's structure. The results of the correlation analysis (Table 1) revealed a strong correlation between the parameters SD depth and water depth ($r = 0.722$). Therefore, the parameter water depth is eliminated from the ANN's inputs. The Matlab (R2018b) software was used for data analysis and ANNs' development for the needs of this study.

Table 1. Results of the Pearson correlation coefficient (r) for the monitored environmental parameters.

	pH	DO	EC	SD	Depth	WT	TP	DIN	Chl-*a*
pH	1.000								
DO	0.089	1.000							
EC	0.260	−0.169	1.000						
SD	−0.385	−0.229	0.094	1.000					
depth	−0.153	−0.186	0.051	**0.722**	1.000				
WT	0.442	−0.077	0.602	−0.278	−0.006	1.000			
TP	−0.106	−0.164	0.048	0.299	0.216	−0.066	1.000		
DIN	−0.131	0.251	0.097	0.059	−0.093	−0.179	0.198	1.000	
Chl-*a*	0.234	−0.101	0.267	−0.124	−0.163	0.362	−0.014	−0.172	1.000

DO (dissolved oxygen, mg/L); EC (electric conductivity μS/cm); SD (Secchi disk, m) depth; depth (water depth, m); WT (water temperature, °C); TP (total phosphorus, μg/L); DIN (dissolved inorganic nitrogen, μg/L); Chl-*a* (Chlorophyll-*a*, mg/L). For strong correlations ($r > 0.7$), the associated value is shown with bold characters.

Several topologies were examined using a typical trial-and-error methodology, as recommended by Ozesmi et al. [29] and Palani et al. [42]; while a more analytical procedure for finding the ANN's optimal topology is described in the study of Tuhtan et al. [43]. The 7-8-1 topology was found to be the optimal based on the calculated performance indices, which in our case are the Root Mean Square Error (RMSE) and the Mean Absolute Error (MAE). The k-fold cross validation method is applied in this modeling study, since the data sample for the ANN learning and training is small ($n = 89$, corresponding to sample points without any missing values). The optimal k-value was calculated among a candidate set for k values ranging between [3,30] and the LOO cross validation case (where $k = 89$). For each k-value, k ANNs (with the same topology 7-8-1) were created and their performance indices for the test sets were averaged (see Figures 3 and 4). Based on RMSE and MAE results for each k-value, it was calculated that the optimal k-value was $k = 89$ since it produced the lowest RMSE and MAE. The calculated RMSE and MAE values for $k = 89$ are always equal because there is only one observation ($n = 1$) per test data set (recall Equations (5) and (6)). It must also be noted that $k = 89$ exhibits the maximum number of outliers when calculating the RMSE and MAE performance indices. However, the ratio of outlier number per k-fold number (when $k = 89$) is lower than the calculated ratio for the majority of $k = 3:30$.

The computational time needed to train and test each ANN model for the different given k values was recorded (Figure 5). Based on the different time values that are needed for each k-fold, the comparison for the needed time and the ANN's performance for the given k values is enabled (by observing Figures 3–5). The LOO case had the highest computational time ($t = 314$ s), as expected. Even though the LOO case had the highest computational time, this time value is relatively low because the data set is small and consists of only 89 samples. The RMSE and MAE values are following a clear pattern, where they decrease as the k number increases. For the $k = 89$ case, the RMSE and MAE are equal to 0.36; however, for $k = 30$ the RMSE = 2.33 and MAE = 1.42. Therefore, for the $k = 89$ value there is a significant improvement of the ANN's modelling performance compared with that of $k = 30$ with low computational overhead. So, by considering both the computational times and the performance indices, it is decided that the $k = 89$ (LOO) is the best value for the modelling needs of this study.

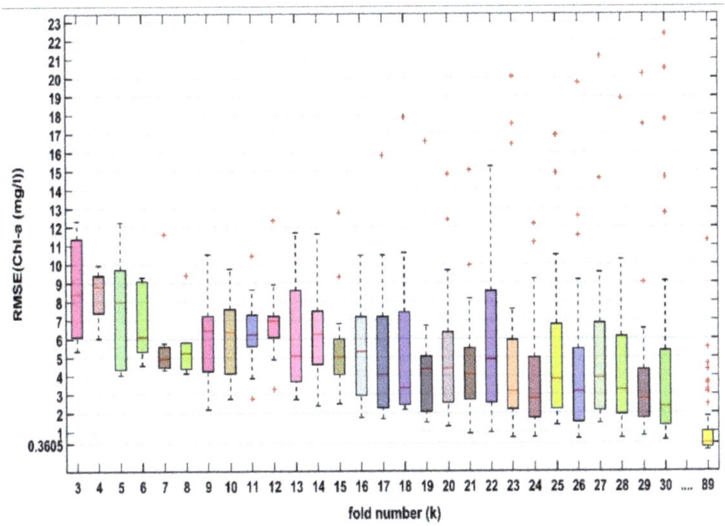

Figure 3. Comparison of the ANN's Root Mean Square Error (RMSE) performance index for k = 3:30 and k = 89 (leave-one-out case) values. k = 89 achieves the lowest RMSE = 0.3605. Red marks are extreme values referred as "outliers".

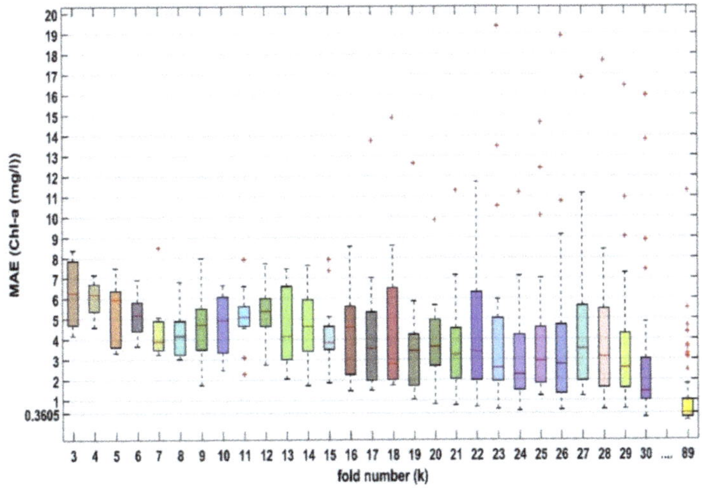

Figure 4. Comparison of the ANN's Mean Absolute Error (MAE) performance index for k = 3:30 and k = 89 (leave-one-out case) values. k = 89 achieves the lowest MAE = 0.3605. Red marks are extreme values referred as "outliers".

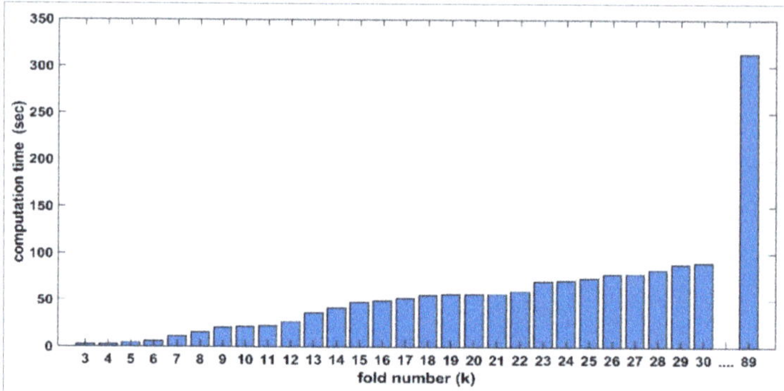

Figure 5. Comparison of the ANN's computational times for k = 3:30 and k = 89 (leave-one-out case) values.

The associated results for predicting Lake's Mikri Prespa Chl-*a* production are based on the averaged values for the LOO case. The ANN model's sensitivity analysis is also expressed based on these averaged values.

The created ANN model managed to predict with good accuracy the Chl-*a* levels, as it was demonstrated from the RMSE and MAE indices. The measured values of the Chl-*a* parameter are plotted against those predicted by the ANN (Figure 6) for both the test dataset and the averaged predicted values of the entire dataset. It can be observed that these three graphical plots are almost identical. Therefore, the ANN model can be characterized as a reliable predictor for the Chl-*a* parameter. The biggest differences between the real and the averaged predicted data are observed for some elevated values of the Chl-*a* parameter. However, the ANN typically produces output results that match/describe the lake's tendency towards a eutrophic or hypereutrophic status for these elevated values of Chl-*a* input data. Additionally, the test data points also match the real data points closely, with the exception of a few instances.

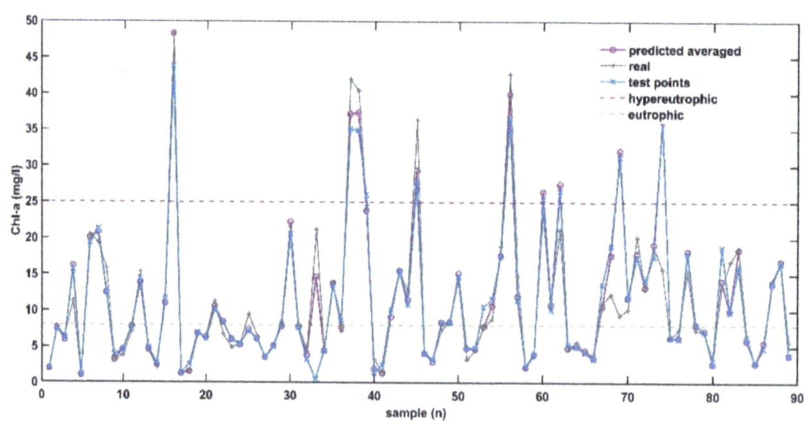

Figure 6. Graphical representation of the monitored (real) chlorophyll-*a* (Chl-*a*) values against the averaged predicted values and the test data. The green and red dotted horizontal lines show the Chl-*a* limits for when a lake becomes eutrophic and hypereutrophic [44], respectively.

The ANN's sensitivity analysis algorithm allowed the calculation of each input environmental parameter's impact on the Chl-*a* parameter. The input parameters were perturbated by −10%; −5%; +5%; and +10% for each of the 89 inputs and the average results are presented in Figure 7. These input's fluctuations correspond to different modelling scenarios that examine the Chl-*a* level's changes in regard with the perturbation of the ANN's inputs.

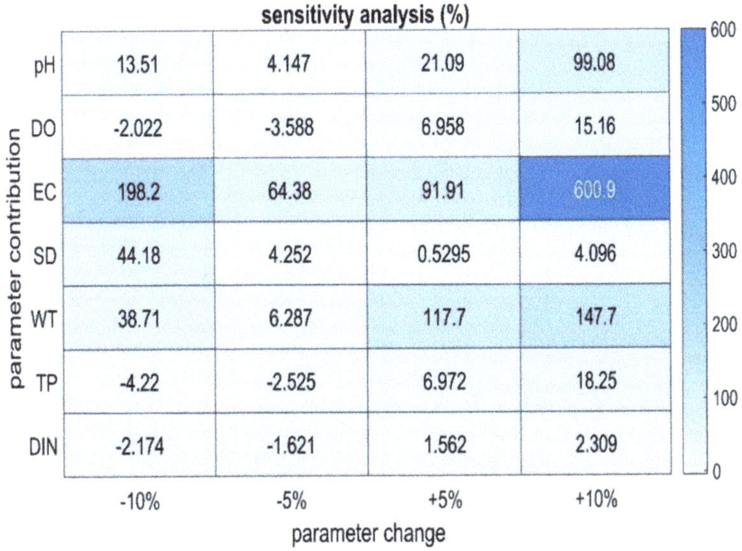

Figure 7. The sensitivity analysis calculating parameter contribution (%) after a change. Where ANN's input parameters: pH; dissolved oxygen (DO); electric conductivity (EC); Secchi disk (SD) depth; water temperature (WT); total phosphorus (TP); and dissolved inorganic nitrogen (DIN).

The pH parameter perturbations between −10% and +5% produced no clear results about the interactions between pH and Chl-*a*. However, for increased pH values (+10% change) a significant change of Chl-*a* levels is observed (+99.08%).

The DO parameter perturbations seem to follow a pattern, where the Chl-*a* levels are increased when the DO parameter is increased and vice versa. EC had the greatest contribution to the Chl-*a* production (600.9% when the EC is increased by 10% and 198.2% when decreased by 10%). Interestingly, smaller increases or decreases of EC lead to smaller increases to Chl-*a*.

Regarding SD parameter's negative or positive fluctuations, the ANN also did not reveal a clear pattern. Although algal production is not the only mechanism affecting the SD, lower water transparency (−10%) was related with an increase of Chl-*a* by approximately 44%.

The WT fluctuations were associated with increase in Chl-*a* levels. However, in the cases of WT increases, the Chl-*a* increased drastically, showing the strong impact of WT on Chl-*a*. Both TP and DIN are related to the Chl-*a* parameter. It is clearly shown that Chl-*a* changes follow the changes in TP and DIN.

4. Discussion

Data scarcity is a computational issue which lake modelers must address very often. As it is highlighted by Jeong et al. [45], the unavailability of suitable/enough freshwater quality data is a problem found when applying machine learning techniques. Generally, ANNs must be trained with a big enough data number in order to produce good results. When more data are available during the training process, more accurate results are given by the ANNs [46]. As mentioned by Kavzoglu [47], a small sample size might result in poor performance of the ANN. Under these conditions, *k*-fold cross validation is considered a good option [48]. 10-fold cross validation is widely used in ANN water quality modelling studies (e.g., [39,49]), while 5-fold cross validation (e.g., [50]) is also used. In some cases, researchers are using less commonly used k values. For example, in a water quality modelling study by Chang et al. [51], the created ANN was verified with the use of 14-fold cross validation.

In our case, since the available data set was very small (n = 89), the k-fold cross validation method was applied and investigated. The k number was ranged for values between [3,30] and LOO cross validation was also investigated. It was clearly observed that the bigger values of k produced better outputs. In the case of LOO cross validation, the predicted outputs were highly reliable, since they were evaluated based on a very low RMSE = 0.3605. Additionally, the graphical plots of the real data and the predicted data demonstrate an almost perfect match, except for a few instances related with very high values of Chl-a concentrations (eutrophic/hypertrophic values). However, even in that case, the ANN managed to capture the Lake's Mikri Prespa tendency for eutrophic conditions and the predicted outputs were characterized as eutrophic instances/values.

LOO cross validation was not only chosen for the needs of this ANN modelling study because it calculated the best outputs, but the computational time factor was also taken into consideration. The computational time for k = 89 (LOO), was higher than the time needed for smaller k values. However, the computational time needed for k = 89 is still relatively low (t = 314 s) and is considered practical. Hence, choosing the LOO cross validation for the ANN modelling is not a limitation based on computational time terms. Therefore, even though the k = 1:30 values can produce faster computations, the ANN's good performance calculated with the use of LOO cross validation prevailed.

Besides the development of a reliable model based on ANN techniques adjusted for the needs of Lake Mikri Prespa, another objective of this study was the extraction of information related with the role/impact of the measured water quality parameters to algal production. The sensitivity analysis had a crucial role for investigating a parameter's contribution. The negative or positive fluctuations of the input parameters, and specifically nutrient perturbations, allowed us to create various modelling scenarios. Based on the application of sensitivity analysis, useful results about Mikri Prespa Lake algal production can be extracted and the constructed ANN can serve as a lake restoration management tool [52]. The impact of the environmental parameters on the Chl-a can be directly measured/quantified based on sensitivity analysis algorithm. ANN-based modelling studies (e.g., [17,53]) dealing with eutrophication control are of great environmental significance. ANNs are ideal for ecological modeling, since they can model phenomena with non-linear and complex data [52]. Furthermore, ANNs require no a priori assumptions about the model or the data distribution [54] and are considered advanced modelling techniques, because they can be used on a heterogeneous data set. ANNs are successfully addressing these modelling issues, which other modelling methods might fail to overcome. Therefore, the development of ANNs for limnological studies is considered a good modelling practice. Modelling scenarios based on ANNs can aid management authorities with the implementation of measures for lake water quality improvement and ecosystem restoration. For example, quantifying the ecosystem response to a simulated decrease of one or both nutrients (phosphorus and nitrogen) can provide useful insight whether managers should target only phosphorus or both phosphorus and nitrogen. With this study we showed a strong connection between Chl-a and nutrients, which corroborates many limnological studies [55–57].

Concerning the other parameters (i.e., EC, SD and WT), the ANN did not produce a clear pattern. However, as stated by Hadjislomou et al. [19], the modelling of limnological parameters is a very case-sensitive task and the underlying mechanisms controlling limnological parameters are complex and usually their interactions are not easily correlated/examined. ANNs are data-driven models and the relationships/interactions among the associated parameters are not always easy to be understood, since ANNs require no a priori assumptions about the model or the data distribution [54]. Additionally, even if the created ANN is a good predictor for the Chl-a parameter, there are many other factors that might play an important role in explaining the observed patterns of Chl-a. As stated by

Napiórkowska-Krzebietke et al. [58], the dynamics and the accumulation of cyanobacterial blooms in lakes are controlled by many factors, such as wind strength, while wind-driven sediment resuspension is a common feature in shallow lakes. Additionally, other special characteristics of the lake (e.g., location) must be taken into consideration when examining algal production. For example, the extended duration of cyanobacterial blooms in Lake Mikri Prespa is favored by the warm Mediterranean climate throughout the year [22]. Regarding the WT, the results of the sensitivity analysis are in agreement with a relevant modeling study of Lake Mikri Prespa [19], where the interactions among the environmental variables with the help of an unsupervised ANN were examined. It was concluded that the data from Lake Mikri Prespa are primarily associated with the WT. Temperature has a crucial role for the Mikri Prespa lake's functioning, and because it is a shallow lake, it is affected by the seasonality effect. The results of our modeling study agree with these findings, since in our case the WT is calculated to be the second most influencing parameter.

Another interesting finding derived from the sensitivity analysis results is the relatively small increase of Chl-*a*, which is observed for negative changes of the WT parameter and corresponds to temperatures of less hot months, might be attributed to other meteorological conditions/factors, which exist during spring and autumn. For example, wind mixing is more intense during these seasons and can lead to the release of phosphorus and nitrogen from the sediments, a process which favors eutrophication [59]. In addition, based on the sensitivity analysis results, it is concluded that the EC parameter role is not easy to be understood, since it is associated with complex processes such as wind mixing and inorganic dissolved matter inflow from the lake's catchment, e.g., surface runoff and river inflow [60]. Nevertheless, the ANN managed to associate the increased levels of EC with elevated algal production, which possibly shows the effect of increased nutrients on the algal productivity.

According to the ANN's results, the reduction of TP is associated with a reduction of Chl-*a* levels and vice versa. This modelling scenario, which is related to TP perturbations, is supported by the fact that strong relationships between increased phosphorus loadings and eutrophication have been shown in freshwater ecosystems [61]. The second scenario, which is related with DIN perturbations, also showed a reduction to Chl-*a* levels when the DIN parameter decreased. The DIN parameter has similar behavior with the TP parameter; therefore, any increase of nitrogen levels is associated with increased algal productivity [62]. The linkage between nutrient loading and eutrophication is often non-linear, because of the complex mechanisms by which hydrological and meteorological conditions interfere with the nutrients [63]. Based on this statement, the non-proportional increase/decrease for Chl-*a* levels in regard to the associated nutrient perturbations that were observed during the sensitivity analysis can be justified.

Based on the ANN sensitivity analysis results, the TP parameter has greater impact on algal production than the DIN parameter. The stronger impact of TP compared to the DIN parameter is more noticeable when TP concentration increases. Nevertheless, DIN's role should not be underestimated in eutrophication management/control, since it is observed that high DIN levels are related with continued serious eutrophication problems caused by cyanobacterial blooms [64]. Even though the role of nitrogen as a limiting factor is debated, it has an important role in shallow polymictic eutrophic lakes [65]. The ANN simulation scenarios regarding the DIN parameter clearly showed that DIN additions into the lake are promoting algal production, while DIN level decrease is related with Chl-*a* level decrease. Therefore, it is recommended that the lake's nutrient management should not only be focused on phosphorus, but on nitrogen as well. Additionally, it is documented that the simultaneous decrease of both nutrients has a bigger reduction of Chl-*a* levels and is related with the synergistic effect of DIN and TP parameters [17]. In the case of synergistic/combined perturbations of nutrients, a similar behavior with the case of only one nutrient fluctuation/perturbation is observed. Yet, the combined reduction of DIN and TP leads to even lower levels of Chl-*a* than the single reduction of DIN or TP concentration.

According to several studies eutrophication levels in lakes are linked with both nutrients (nitrogen and phosphorus) and high levels of nutrients results in high levels of Chl-*a* [13].

Even a small reduction of phosphorus and nitrogen concentration into Lake Mikri Prespa has a beneficial effect on lake trophic status, while adding nutrients into the Lake promotes eutrophication. A recent study by Verstijnen et al. [66] stated that Lake Mikri Prespa is very sensitive to nutrient increase, and that even small additions of nutrients derived from waterbirds are associated with cyanobacterial blooms. The ANN clearly captured this relationship for both nutrients (DIN and TP) and how prone Mikri Prespa Lake is to eutrophication. In conclusion, the created ANN is a reliable predictor for Chl-*a* levels and can successfully investigate different management scenarios related with nutrient control. However, since the data set is small, the generalization ability of the ANN might not be sufficient for new data related with abnormal/unusual conditions (e.g., huge increase of a nutrient). Based on this limitation, the created management scenarios were restricted to parameter's fluctuations up to $\pm$ 10. Therefore, the re-calibration of the model when more data is available is recommended in order to extend its capabilities as a management tool.

Generally, the LOO cross validation method is considered to be the best option for modelling small datasets with the use of ANNs. However, some concerns related with the effect of overtraining might exist. The LOO cross validation provides the benefit of having more data available for training, but at the same time the data set that is used for validation becomes smaller and the evaluation becomes less reliable and robust. This issue is compensated by the fact that more training and validations are applied. Of course, the LOO cross validation should not be limited to ANN-based applications, but it can be used for modelling with other machine learning methods such as linear regression and random forests.

5. Conclusions

Data scarcity is a very common issue observed when modelling limnological data sets with the use of ANNs. In the case of Lake Mikri Prespa, a relatively small number of observations ($n = 89$) was used to develop an ANN and model the trophic status of the Lake with the use of k-fold cross validation. For that purpose, several k values were examined. The LOO cross validation produced the better outputs, while the computational time that was needed was relatively low. Therefore, LOO cross validation is recommended for the needs of this eutrophication-related modelling study. Additionally, the created ANN was a good Chl-*a* parameter predictor and can serve as a water management tool. Based on sensitivity analysis, the ANN examined the scenarios when the nutrients (phosphorus and nitrogen) levels into the lake increased or decreased. In the case of nutrient increase, the model clearly showed that Lake Mikri Prespa's water quality declines even more. On the other hand, when the nutrient levels are decreased the ANN model clearly showed that algal production reduced.

Author Contributions: Conceptualization, E.H., K.S.; methodology, E.H.; software, E.H.; data analysis, E.H., H.H., M.M.; data curation, K.S., E.P.; writing—original draft preparation, E.H.; writing—review and editing, E.H., K.S., H.H., M.M., E.P., G.P.; supervision, H.H., M.M.; funding acquisition, M.M., H.H. All authors have read and agreed to the published version of the manuscript.

Funding: This work was co-funded by the European Regional Development Fund and the Republic of Cyprus through the Research and Innovation Foundation (STEAM Project: INTEGRATED/0916/0063 and MARI-Sense Project: INTEGRATED/0918/0032).

Data Availability Statement: Currently data is not publicly available.

Conflicts of Interest: The authors declare no conflict of interest.

References

1. Adnan, R.M.; Zounemat-Kermani, M.; Kuriqi, A.; Kisi, O. Machine Learning Method in Prediction Streamflow Considering Periodicity Component. In *Springer Transactions in Civil and Environmental Engineering*; Springer: Singapore, 2020; pp. 383–403.
2. Oyebode, O.; Stretch, D. Neural network modeling of hydrological systems: A review of implementation techniques. *Nat. Resour. Model.* **2019**, *32*, e12189. [CrossRef]
3. Kim, D.-K.; Park, K.; Jo, H.; Kwak, I.-S. Comparison of Water Sampling between Environmental {DNA} Metabarcoding and Conventional Microscopic Identification: A Case Study in Gwangyang Bay, South Korea. *Appl. Sci.* **2019**, *9*, 3272. [CrossRef]
4. Muttil, N.; Chau, K.W. Neural network and genetic programming for modelling coastal algal blooms. *Int. J. Environ. Pollut.* **2006**, *28*, 223. [CrossRef]
5. Goethals, P.L.M.; Dedecker, A.P.; Gabriels, W.; Lek, S.; De Pauw, N. Applications of artificial neural networks predicting macroinvertebrates in freshwaters. *Aquat. Ecol.* **2007**, *41*, 491–508. [CrossRef]
6. Zounemat-Kermani, M. Principal Component Analysis (PCA) for estimating Chlorophyll concentration using forward and generalized regression neural networks. *Appl. Artif. Intell.* **2014**, *28*, 16–29. [CrossRef]
7. Bennett, C.; Stewart, R.A.; Beal, C.D. ANN-based residential water end-use demand forecasting model. *Expert Syst. Appl.* **2013**, *40*, 1014–1023. [CrossRef]
8. Yotova, G.; Lazarova, S.; Kudłak, B.; Zlateva, B.; Mihaylova, V.; Wieczerzak, M.; Venelinov, T.; Tsakovski, S. Assessment of the Bulgarian Wastewater Treatment Plants' Impact on the Receiving Water Bodies. *Molecules* **2019**, *24*, 2274. [CrossRef]
9. Moustaka-Gouni, M.; Sommer, U.; Economou-Amilli, A.; Arhonditsis, G.B.; Katsiapi, M.; Papastergiadou, E.; Kormas, K.A.; Vardaka, E.; Karayanni, H.; Papadimitriou, T. Implementation of the Water Framework Directive: Lessons Learned and Future Perspectives for an Ecologically Meaningful Classification Based on Phytoplankton of the Status of Greek Lakes, Mediterranean Region. *Environ. Manag.* **2019**, *64*, 675–688. [CrossRef]
10. Cigizoglu, H.K.; Kisi, Ö. Flow prediction by three back propagation techniques using k-fold partitioning of neural network training data. *Hydrol. Res.* **2005**, *36*, 49–64. [CrossRef]
11. Cunningham, P.; Carney, J.; Jacob, S. Stability problems with artificial neural networks and the ensemble solution. *Artif. Intell. Med.* **2000**, *20*, 217–225. [CrossRef]
12. Smith, V.H.; Joye, S.B.; Howarth, R.W. Eutrophication of freshwater and marine ecosystems. *Limnol. Oceanogr.* **2006**, *51*, 351–355. [CrossRef]
13. Smith, V.H.; Tilman, G.D.; Nekola, J.C. Eutrophication: Impacts of excess nutrient inputs on freshwater, marine, and terrestrial ecosystems. *Environ. Pollut.* **1999**, *100*, 179–196. [CrossRef]
14. Dynowski, P.; Senetra, A.; Źróbek-Sokolnik, A.; Kozłowski, J. The Impact of Recreational Activities on Aquatic Vegetation in Alpine Lakes. *Water* **2019**, *11*, 173. [CrossRef]
15. Hadjisolomou, E.; Stefanidis, K.; Papatheodorou, G.; Papastergiadou, E. Evaluating the contributing environmental parameters associated with eutrophication in a shallow lake by applying artificial neural networks techniques. *Fresenius Environ. Bull.* **2017**, *26*, 3200–3208.
16. Brown, M.G.L.; Skakun, S.; He, T.; Liang, S. Intercomparison of Machine-Learning Methods for Estimating Surface Shortwave and Photosynthetically Active Radiation. *Remote Sens.* **2020**, *12*, 372. [CrossRef]
17. Hadjisolomou, E.; Stefanidis, K.; Papatheodorou, G.; Papastergiadou, E. Assessing the contribution of the environmental parameters to eutrophication with the use of the "PaD" and "PaD2" methods in a hypereutrophic lake. *Int. J. Environ. Res. Public Health* **2016**, *13*, 764. [CrossRef] [PubMed]
18. Stefanidis, K.; Sarika, M.; Papastegiadou, E. Exploring environmental predictors of aquatic macrophytes in water-dependent Natura 2000 sites of high conservation value: Results from a long-term study of macrophytes in Greek lakes. *Aquat. Conserv. Mar. Freshw. Ecosyst.* **2019**, *29*, 1133–1148. [CrossRef]
19. Hadjisolomou, E.; Stefanidis, K.; Papatheodorou, G.; Papastergiadou, E. Assessment of the eutrophication-related environmental parameters in two mediterranean lakes by integrating statistical techniques and self-organizing maps. *Int. J. Environ. Res. Public Health* **2018**, *15*, 547. [CrossRef]
20. Panagiotopoulos, K.; Aufgebauer, A.; Schäbitz, F.; Wagner, B. Vegetation and climate history of the Lake Prespa region since the Lateglacial. *Quat. Int.* **2013**, *293*, 157–169. [CrossRef]
21. Stefanidis, K.; Papastergiadou, E. Linkages between Macrophyte Functional Traits and Water Quality: Insights from a Study in Freshwater Lakes of Greece. *Water* **2019**, *11*, 1047. [CrossRef]
22. Vardaka, E.; Moustaka-Gouni, M.; Cook, C.M.; Lanaras, T. Cyanobacterial blooms and water quality in Greek waterbodies. *J. Appl. Phycol.* **2005**, *17*, 391–401. [CrossRef]
23. Hagan, M.T.; Demuth, H.B.; Beale, M.H.; De Jesús, O. *Neural Network Design*, 2nd ed.; Martin Hagan: Stillwater, OK, USA, 2014.
24. Chen, J.-C.; Wang, Y.-M. Comparing Activation Functions in Modeling Shoreline Variation Using Multilayer Perceptron Neural Network. *Water* **2020**, *12*, 1281. [CrossRef]
25. Dedecker, A.P.; Goethals, P.L.M.; Gabriels, W.; De Pauw, N. Optimization of Artificial Neural Network ({ANN}) model design for prediction of macroinvertebrates in the Zwalm river basin (Flanders, Belgium). *Ecol. Model.* **2004**, *174*, 161–173. [CrossRef]
26. Ghalkhani, H.; Golian, S.; Saghafian, B.; Farokhnia, A.; Shamseldin, A. Application of surrogate artificial intelligent models for real-time flood routing. *Water Environ. J.* **2012**, *27*, 535–548. [CrossRef]

27. Vilas, L.G.; Spyrakos, E.; Palenzuela, J.M.T. Neural network estimation of chlorophyll *a* from MERIS full resolution data for the coastal waters of Galician *rias* (NW Spain). *Remote Sens. Environ.* **2011**, *115*, 524–535. [CrossRef]
28. Heddam, S.; Ptak, M.; Zhu, S. Modelling of daily lake surface water temperature from air temperature: Extremely randomized trees (ERT) versus Air2Water, MARS, M5Tree, RF and MLPNN. *J. Hydrol.* **2020**, *588*, 125130. [CrossRef]
29. Özesmi, S.L.; Tan, C.O.; Özesmi, U. Methodological issues in building, training, and testing artificial neural networks in ecological applications. *Ecol. Model.* **2006**, *195*, 83–93. [CrossRef]
30. Maier, H.R.; Dandy, G.C.; Burch, M.D. Use of artificial neural networks for modelling cyanobacteria Anabaena spp. in the River Murray, South Australia. *Ecol. Model.* **1998**, *105*, 257–272. [CrossRef]
31. Scardi, M.; Harding, L.W. Developing an empirical model of phytoplankton primary production: A neural network case study. *Ecol. Model.* **1999**, *120*, 213–223. [CrossRef]
32. Herodotou, H.; Aslam, S.; Holm, H.; Theodossiou, S. Big Maritime Data Management. In *Maritime Informatics, Progress in IS*; Lind, M., Michaelides, M., Ward, R.T., Watson, R., Eds.; Springer International Publishing: Cham, Switzerland, 2021; pp. 313–334.
33. Karamoutsou, L.; Psilovikos, A. The Use of Artificial Neural Network in Water Quality Prediction in Lake Kastoria, Greece. In Proceedings of the 14th Conference of the Hellenic Hydrotechnical Association, Volos, Greece, 16–17 May 2019; pp. 882–889.
34. Gebler, D.; Kolada, A.; Pasztaleniec, A.; Szoszkiewicz, K. Modelling of ecological status of Polish lakes using deep learning techniques. *Environ. Sci. Pollut. Res.* **2020**, *28*, 5383–5397. [CrossRef]
35. Melesse, A.M.; Khosravi, K.; Tiefenbacher, J.P.; Heddam, S.; Kim, S.; Mosavi, A.; Pham, B.T. River Water Salinity Prediction Using Hybrid Machine Learning Models. *Water* **2020**, *12*, 2951. [CrossRef]
36. Gevrey, M.; Dimopoulos, I.; Lek, S. Review and comparison of methods to study the contribution of variables in artificial neural network models. *Ecol. Model.* **2003**, *160*, 249–264. [CrossRef]
37. Lee, J.H.W.; Huang, Y.; Dickman, M.; Jayawardena, A.W. Neural network modelling of coastal algal blooms. *Ecol. Model.* **2003**, *159*, 179–201. [CrossRef]
38. Zhou, T.; Wang, F.; Yang, Z. Comparative Analysis of ANN and SVM Models Combined with Wavelet Preprocess for Groundwater Depth Prediction. *Water* **2017**, *9*, 781. [CrossRef]
39. Olden, J.; Lawler, J.; Poff, N.L. Machine Learning Methods Without Tears: A Primer for Ecologists. *Q. Rev. Biol.* **2008**, *83*, 171–193. [CrossRef]
40. Mamun, M.; Kim, J.-J.; Alam, M.A.; An, K.-G. Prediction of Algal Chlorophyll-*a* and Water Clarity in Monsoon-Region Reservoir Using Machine Learning Approaches. *Water* **2019**, *12*, 30. [CrossRef]
41. Gebler, D.; Szoszkiewicz, K.; Pietruczuk, K. Modeling of the river ecological status with macrophytes using artificial neural networks. *Limnologica* **2017**, *65*, 46–54. [CrossRef]
42. Palani, S.; Liong, S.-Y.; Tkalich, P. An ANN application for water quality forecasting. *Mar. Pollut. Bull.* **2008**, *56*, 1586–1597. [CrossRef]
43. Tuhtan, J.A.; Fuentes-Perez, J.F.; Toming, G.; Kruusmaa, M. Flow velocity estimation using a fish-shaped lateral line probe with product-moment correlation features and a neural network. *Flow Meas. Instrum.* **2017**, *54*, 1–8. [CrossRef]
44. Organisation for Economic Co-operation and Development. *Eutrophication of Waters: Monitoring, Assessment and Control*; Organisation for Economic Co-operation and Development: Paris, France, 1982; ISBN 9264122982.
45. Jeong, K.-S.; Kim, D.-K.; Chon, T.-S.; Joo, G.-J. Machine Learning Application to the Korean Freshwater Ecosystems. *Korean J. Ecol.* **2005**, *28*, 405–415. [CrossRef]
46. Aria, S.H.; Asadollahfardi, G.; Heidarzadeh, N. Eutrophication modelling of Amirkabir Reservoir (Iran) using an artificial neural network approach. *Lakes Reserv. Res. Manag.* **2019**, *24*, 48–58. [CrossRef]
47. Kavzoglu, T. Increasing the accuracy of neural network classification using refined training data. *Environ. Model. Softw.* **2009**, *24*, 850–858. [CrossRef]
48. Solomatine, D.P.; Ostfeld, A. Data-driven modelling: Some past experiences and new approaches. *J. Hydroinform.* **2008**, *10*, 3–22. [CrossRef]
49. Olomukoro, J.O.; Odigie, J.O. Ecological modelling using artificial neural network for macroinvertebrate prediction in a tropical rainforest river. *Int. J. Environ. Waste Manag.* **2020**, *26*, 325–348. [CrossRef]
50. Deng, T.; Chau, K.-W.; Duan, H.-F. Machine learning based marine water quality prediction for coastal hydro-environment management. *J. Environ. Manag.* **2021**, *284*, 112051. [CrossRef]
51. Chang, F.-J.; Tsai, Y.-H.; Chen, P.-A.; Coynel, A.; Vachaud, G. Modeling water quality in an urban river using hydrological factors—Data driven approaches. *J. Environ. Manag.* **2015**, *151*, 87–96. [CrossRef] [PubMed]
52. Lek, S.; Guégan, J.F. Artificial neural networks as a tool in ecological modelling, an introduction. *Ecol. Model.* **1999**, *120*, 65–73. [CrossRef]
53. Yotova, G.; Varbanov, M.; Tcherkezova, E.; Tsakovski, S. Water quality assessment of a river catchment by the composite water quality index and self-organizing maps. *Ecol. Indic.* **2021**, *120*, 106872. [CrossRef]
54. Teles, L.O.; Vasconcelos, V.; Teles, L.O.; Pereira, E.; Saker, M.; Vasconcelos, V. Time Series Forecasting of Cyanobacteria Blooms in the Crestuma Reservoir (Douro River, Portugal) Using Artificial Neural Networks. *Environ. Manag.* **2006**, *38*, 227–237. [CrossRef]
55. Atoui, A.; Hafez, H.; Slim, K. Occurrence of toxic cyanobacterial blooms for the first time in Lake Karaoun, Lebanon. *Water Environ. J.* **2012**, *27*, 42–49. [CrossRef]

56. Jeppesen, E.; Pekcan-Hekim, Z.; Lauridsen, T.L.; Sondergaard, M.; Jensen, J.P. Habitat distribution of fish in late summer: Changes along a nutrient gradient in Danish lakes. *Ecol. Freshw. Fish* **2006**, *15*, 180–190. [CrossRef]
57. Søndergaard, M.; Larsen, S.E.; Jørgensen, T.B.; Jeppesen, E. Using chlorophyll a and cyanobacteria in the ecological classification of lakes. *Ecol. Indic.* **2011**, *11*, 1403–1412. [CrossRef]
58. Napiórkowska-Krzebietke, A.; Kalinowska, K.; Bogacka-Kapusta, E.; Stawecki, K.; Traczuk, P. Cyanobacterial Blooms and Zooplankton Structure in Lake Ecosystem under Limited Human Impact. *Water* **2020**, *12*, 1252. [CrossRef]
59. Liu, X.; Zhang, G.; Sun, G.; Wu, Y.; Chen, Y. Assessment of Lake Water Quality and Eutrophication Risk in an Agricultural Irrigation Area: A Case Study of the Chagan Lake in Northeast China. *Water* **2019**, *11*, 2380. [CrossRef]
60. Borowiak, M.; Borowiak, D.; Nowiński, K. Spatial Differentiation and Multiannual Dynamics of Water Conductivity in Lakes of the Suwałki Landscape Park. *Water* **2020**, *12*, 1277. [CrossRef]
61. Heisler, J.; Glibert, P.M.; Burkholder, J.M.; Anderson, D.M.; Cochlan, W.; Dennison, W.C.; Dortch, Q.; Gobler, C.J.; Heil, C.A.; Humphries, E.; et al. Eutrophication and harmful algal blooms: A scientific consensus. *Harmful Algae* **2008**, *8*, 3–13. [CrossRef] [PubMed]
62. Akagha, S.C.; Nwankwo, D.I.; Yin, K. Dynamics of nutrient and phytoplankton in Epe Lagoon, Nigeria: Possible causes and consequences of reoccurring cyanobacterial blooms. *Appl. Water Sci.* **2020**, *10*, 1–16. [CrossRef]
63. Paerl, H.W. Assessing and managing nutrient-enhanced eutrophication in estuarine and coastal waters: Interactive effects of human and climatic perturbations. *Ecol. Eng.* **2006**, *26*, 40–54. [CrossRef]
64. Paerl, H.W.; Xu, H.; McCarthy, M.J.; Zhu, G.; Qin, B.; Li, Y.; Gardner, W.S. Controlling harmful cyanobacterial blooms in a hyper-eutrophic lake (Lake Taihu, China): The need for a dual nutrient (N&P) management strategy. *Water Res.* **2011**, *45*, 1973–1983. [CrossRef]
65. Jeppesen, E.; Søndergaard, M.; Meerhoff, M.; Lauridsen, T.L.; Jensen, J.P. Shallow lake restoration by nutrient loading reduction—some recent findings and challenges ahead. *Hydrobiologia* **2007**, *584*, 239–252. [CrossRef]
66. Verstijnen, Y.J.M.; Maliaka, V.; Catsadorakis, G.; Lürling, M.; Smolders, A.J.P. Colonial nesting waterbirds as vectors of nutrients to Lake Lesser Prespa (Greece). *Inland Waters* **2021**, 1–17. [CrossRef]

water

MDPI

Article

Physiological Responses of the Submerged Macrophyte *Stuckenia pectinata* to High Salinity and Irradiance Stress to Assess Eutrophication Management and Climatic Effects: An Integrative Approach

Lamprini Malea [1], Konstantinia Nakou [1], Apostolos Papadimitriou [1], Athanasios Exadactylos [2] and Sotiris Orfanidis [1,*]

[1] Benthic Ecology & Technology Laboratory, Fisheries Research Institute (ELGO-DIMITRA), Nea Peramos, 640 07 Kavala, Greece; lamprini.m379@gmail.com (L.M.); nakou@inale.gr (K.N.); apostolispap@inale.gr (A.P.)
[2] Department of Ichthyology and Aquatic Environment, School of Agricultural Sciences, University of Thessaly, Fytokou str., 384 46 Volos, Greece; thanos046@gmail.com
* Correspondence: sorfanid@inale.gr

Abstract: *Stuckenia pectinata*, a submerged macrophyte of eutrophic to hyper-eutrophic fresh to brackish waters, faces management and climatic-forced increment of salinity and irradiance in Vistonis Lake (Greece) that may endanger its existence and the ecosystem functioning. A pre-acclimated clone under low irradiance and salinity conditions was treated to understand the effects of high salinity and irradiance on a suite of subcellular (chlorophyll *a* fluorescence kinetics and JIP-test, and chlorophyll content) to organismal (relative growth rate—RGR) physiological parameters. The responses to high irradiance indicated the plant's great photo-acclimation potential to regulate the number and size of the reaction centers and the photosynthetic electron transport chain by dissipation of the excess energy to heat. A statistically significant interaction ($p < 0.01$) of salinity and irradiance on Chl *a*, *b* content indicated acclimation potential through adjusting the Chl *a*, *b* contents. However, no significant ($p > 0.05$) difference was observed on Chl *a/b* ratio and the RGR, indicating the species' potential to become acclimatized by reallocating resources to compensate for growth. Thus, the regulation of photosynthetic pigment content and photosystem II performance consisted of the primary growth strategy to present and future high salinity and irradiance stressful conditions due to eutrophication management and the ongoing climatic changes.

Keywords: brackish water; factorial experiment; JIP-test; relative growth rate; repeated ANOVA; RDA

Citation: Malea, L.; Nakou, K.; Papadimitriou, A.; Exadactylos, A.; Orfanidis, S. Physiological Responses of the Submerged Macrophyte *Stuckenia pectinata* to High Salinity and Irradiance Stress to Assess Eutrophication Management and Climatic Effects: An Integrative Approach. *Water* **2021**, *13*, 1706. https://doi.org/10.3390/w13121706

Academic Editor: Eva Papastergiadou

Received: 21 May 2021
Accepted: 17 June 2021
Published: 20 June 2021

1. Introduction

Submerged macrophytes are aquatic plants of remarkable phenotypic plasticity [1] that grow across contrasting conditions, from pristine to degraded lakes, estuaries, and coastal lagoons [2,3]. They provide food and habitat for invertebrates, larvae and juvenile fishes [4]. They also play a crucial role in water quality improvement by stabilizing the sediments [2]. In addition, they take up excess nitrogen and phosphorus and excrete allelopathic substances [5], contributing to reductions in nuisance algal blooms [6]. Therefore, they are valuable components maintaining the functioning of ecosystems [7,8].

Among the submerged macrophytes, *Stuckenia pectinata* (L.) Börner (syn. *Potamogeton pectinatus* L.) is a common species of several eutrophic to hyper-eutrophic fresh to brackish waters worldwide [9–11]. Because the plant is food for waterfowl [12], it spreads over long distances [13] and in different environments, where it survives through acclimatization by means of plastic phenotypic responses [1,14,15] and genetic differentiation [16,17].

Stuckenia pectinata is a shade-adapted plant [18] with a relatively low irradiance optimum (50–60 μmol photons m^{-2} s^{-1}) for photosynthesis [1,18]. Under highly turbid eutrophic conditions, it lengthens its bundles in the upper water layer without exposing

them to full sunlight [10,19] to avoid excess light. Thus, this species has been indicated as a bioindicator of moderately turbid-degraded conditions in European lakes [20].

There are conflicting results regarding the effects of salinity on *S. pectinata* physiology, especially regarding the upper tolerance limit, which is reported from 4.2 to 20.1 [21] or less than 30 [22]. Rodríguez-Gallego et al. [23] reported a salinity tolerance range from 0 to 16.5, while a salinity tolerance range from ca. 5 to 18 has been also reported [24]. van Wijck et al. [25] showed successful plant growth at salinity 0 to 6, while at salinity 9, a decrease in biomass production was observed. Borgnis and Boyer [3] showed that *S. pectinata* showed four times the biomass production and ten times the number of shoots at salinity values between 0 and 5. While at salinity 10, the plant had double biomass production, at a salinity of 15, it could survive, but with obvious aging. At salinities 0 and 5, it showed the maximum growth and reproduction rate, while at salinities 10 and 15, there was no flowering.

In a recent study, Hu et al. [26] indicated that salinity 12 and 18 restricted light conversion efficiency at high irradiance (340 μmol photons m^{-2} s^{-1}), and reduced chlorophyll *a* (Chl *a*) content, while increasing heat dissipation in *S. pectinata*. However, there is limited information regarding the effects of high irradiance on the growth and photosynthesis of *S. pectinata*, especially under fluctuating salinity conditions.

Vistonis is one of the largest Greek shallow lakes (2 m mean depth, ca. 45 km^2 area) suffering occasionally from high turbidity (10 July 2014: light attenuation coefficient-k = 1.05–6.77 m^{-1}) and eutrophic to hypertrophic waters (10 July 2014: Chl *a* = 9.1—240.5 μg/L, total dissolved inorganic nitrogen = 3.35—32.18 μmol/L, soluble reactive phosphorus = 0.16—2.15 μmol/L). Such conditions often favor cyanobacterial blooms, leading to anoxia and massive fish kills [27,28]. Under such conditions, a strong irradiance gradient formed, such that the 1% level of incident photosynthetically active radiation (PAR), often taken as the limit for macrophyte growth, penetrated from 2.19 to 0.61 m deep [29]. For water quality management, the Lake of Vistonis is connected through short channels to nearby lagoons and by a long channel with the Vistonikos Gulf (Northern Aegean Sea). So, its salinity fluctuates between 0 (north part) and 5.9, sometimes reaching 14 [30] and, occasionally, at deep levels, as much as 34 (July 2020; south part), with an increasing tendency. Climatic-forced, sea-level rise may increase the inflow of clear seawater in the lake, increasing the salinity and light availability and leading to benthic vegetation changes that will endanger the existing ecosystem services [31].

Chlorophyll *a* fluorescence analysis consists of an easy, non-destructive plant method which allows the estimation of photosynthetic physiology under different environmental stress conditions [32]. The OJIP polyphasic Chl *a* fluorescence rise kinetics plotted on logarithmic time-scale (JIP-test) is a tool to assess the photochemical quantum yield of PSII photochemistry and electron transport activity [33–35]. Some of the environmental stressors that caused malfunctions in the plant's physiology are irradiance and salinity, and JIP-test has been used to assess the plant's photosynthetic performance [36,37]. For example, high irradiance stress has been referred to as increasing the absorption flux (ABS), the trapping flux (TR), and the dissipation energy flux (DI) per reaction center (RC) in *Lemna minor* L. [38]. On the other hand, under high salinity conditions, ions enter the cell, disrupting the electron transport chain at the donor and acceptor sides of PSII [39].

Energy generated by photosynthesis is allocated to growth, reproduction, and defense [40]. However, as plants experience stressful conditions, to ensure their survival, they may be acclimatized by reallocating resources toward increased growth or alterations in morphology [41]. There is a need for an integrated (i.e., at different biological levels) study to assess plant response to stress [42].

This study aimed to investigate *S. pectinata*'s responses under medium (MI) and high irradiance (HI) and medium (MS) and high salinity (HS) stress, representative of existing and future water conditions of the brackish-south part of Vistonis Lake (Greece). The temperature was regulated to early-summer conditions and the nutrient levels to eutrophic conditions. A clone pre-acclimated for two years under low irradiance and salinity laboratory conditions was used to avoid seasonality or acute effects on species physiology. We applied a suite of physiological parameters representative of subcellular (Chl *a* fluorescence and Chl content) to organismal (growth) processes. The results will be valuable (a) in gaining an insight into on the photosynthetic and growth responses that allow the species to inhabit the brackish and clear water depths, (b) in contributing a management strategy to protect the freshwater habitats of the lake under the ongoing climatic changes.

2. Materials and Methods

2.1. Plant Material

A *S. pectinata* strain isolated from Vistonis Lake (Greece) was cultivated for two years in a PVC tank with tap water in the laboratory at 15–18 °C and 30 μmol photons m^{-2} s^{-1}, 14 h light per day. Twenty-four apical shoots of the plant were randomly chosen for the experiment and pre-acclimated for 14 days at increased salinity from 4 to 19 (1 salinity degree per day), in 2 L aquariums at 21–22 °C and 60 μmol photons m^{-2} s^{-1}, 14 h light per day. The eutrophic medium was renewed once a week, and it consisted of artificial aerated medium (60 μmol/L N-NO_3^- and 2 μmol/L P-PO_4^-) produced by Münster sea salt (Meersalz) diluted in resin-filtered tap water (<15 μS cm^{-1}).

2.2. Experimental Design and Treatments

A factorial experiment (n = 6) was carried out in culture chamber at 24–26 °C and lasted ten days in which combined effects of salinity (two levels: 9 and 19) and irradiance (two levels: 100 and 400 μmol photons m^{-2} s^{-1}, MI and HI, respectively) on the pre-acclimated *S. pectinata* were tested. There were four experimental conditions. C1: salinity 9 and irradiance 100 μmol photons m^{-2} s^{-1}, C2: salinity 19 and irradiance 100 μmol photons m^{-2} s^{-1}, C3: salinity 9 and irradiance 400 μmol photons m^{-2} s^{-1}, C4: salinity 19 and irradiance 400 μmol photons m^{-2} s^{-1}. Irradiance was provided for 14 h per day by LED Fyto-Panels (81 × 27 cm; Photon Systems Instruments, Drasov, Czech Republic). Two hundred milliliters of the medium was renewed every day inside small glass jars covered by transparent glass (2 mm) to avoid evaporation. All glass jars used for the experiment were placed on shakers to avoid medium stratification.

2.3. Chlorophyll-a Fluorescence Measurements

Chlorophyll *a* fluorescence measurements were carried out on experimental days 1, 5, and 8. The polyphasic Chl *a* fluorescence kinetics (OJIP) were measured using a continuous excitation plant efficiency analyzer (Handy PEA; Hansatech Instruments Ltd., Norfolk, UK). Before the measurement of each specimen, a dark adaptation of 15 min was conducted. Plant leaves were illuminated with continuous red light (wavelength in peak 650 nm) from three diodes of 3.000 μmol photons m^{-2} s^{-1}. The fast fluorescence rise kinetics was recorded from 10 μs to 1 s. The fluorescence intensity at 20 μs, 50 μs, 100 μs 300 μs, 2 ms, and 30 ms was measured, and the maximum fluorescence was extracted. Several basic parameters were calculated from the extracted data, and the basic parameters derived by the JIP-test models included nine biophysical parameters (Table 1).

Table 1. Glossary, definition of terms and formulae of the JIP-test used for the analysis of the Chl *a* fluorescence transients OJIP [35,43].

DATA EXTRACTED FROM THE RECORDED FLUORESCENCE TRANSIENT OJIP	
F_t (or, simply, F)	fluorescence at time t after onset of actinic illumination
$F_{20\,\mu s}$	minimal reliable recorded fluorescence, at 20 μs
$F_{50\,\mu s}$	fluorescence at 50 μs (for the calculation of the slopes)
$F_{100\,\mu s}$	fluorescence at 100 μs
$F_{300\,\mu s}$	fluorescence at 300 μs
$F_J \equiv F_{2ms}$	fluorescence at the J-step (2 ms) of OJIP
$F_I \equiv F_{30ms}$	fluorescence at the I-step (30 ms) of OJIP
F_P	maximal recorded fluorescence, at the peak P of OJIP
BASIC PARAMETERS CALCULATED FROM THE EXTRACTED DATA	
$F_0 \cong F_{20\,\mu s}$	fluorescence when all PSII RCs are open (≅ to the minimal reliable recorded fluorescence)
F_M (= F_P)	maximal fluorescence, when all PSII RCs are closed (= F_P when the actinic light intensity is above 500 [μmol(photon) m^{-2} s^{-1}])
$V_t \equiv (F_t - F_0)/(F_M - F_0)$	relative variable fluorescence at time t
M_0 and $M_0{}' \equiv [(\Delta F/\Delta t)_0]/(F_M - F_0)$	approximated initial slopes (in ms^{-1}) of the V_t = f(t) kinetics
$M_0 \equiv 4 \times [(F_{300\,\mu s} - F_{50\,\mu s})/(F_M - F_0)]/(\Delta t)_0$	with $(\Delta t)_0$ = (300 − 50) μs = 0.25 ms
$M_0{}' \equiv 20 \times [(F_{100\,\mu s} - F_{50\,\mu s})/(F_M - F_0)]/(\Delta t)_0$	with $(\Delta t)_0$ = (100 − 50) μs = 0.05 ms
BIOPHYSICAL PARAMETERS DERIVED FROM THE BASIC PARAMETERS BY THE JIP-TEST	
Quantum yields and efficiencies/probabilities	
$\varphi_{Pt} \equiv TR_t/ABS = [1 - (F_t/F_M)] = \Delta F_t/F_M$	quantum yield for primary photochemistry, leading to Q_A reduction (i.e. trapped energy flux TR per absorption flux ABS), at any time t
$\varphi_{P0} \equiv TR_0/ABS = [1 - (F_0/F_M)]$	maximum quantum yield for primary photochemistry
$\varphi_{E0} \equiv ET_0/ABS = [1 - (F_0/F_M)] \times (1 - V_J)$	quantum yield for electron transport (ET) further than $Q_A{}^-$
$\varphi_{R0} \equiv RE_0/ABS = [1 - (F_0/F_M)] \times (1 - V_I)$	quantum yield for reduction of end electron acceptors (RE) at the PSI acceptor side
$\psi_{E0} \equiv ET_0/TR_0 = (1 - V_J)$	efficiency/probability that an electron moves further than $Q_A{}^-$
$\delta_{R0} \equiv RE_0/ET_0 = (1 - V_I)/(1 - V_J)$	efficiency/probability that an electron from the intersystem electron carriers is transferred to reduce end electron acceptors at the PSI acceptor side
Specific energy flux (per active, i.e., per Q_A-reducing PSII reaction centre - RC), in ms^{-1}	
$TR_0/RC = M_0 \times (1/V_J)$	trapped energy flux, per RC
$DI_0/RC = ABS/RC - TR_0/RC$	specific energy flux for dissipation per RC
Density of active RCs	
$RC/ABS = (TR_0/ABS) \times (TR_0/RC)^{-1}$	RCs per PSII antenna Chl *a*
$ABS/RC = M_0 \times (1/V_J) \times (1/\varphi_{P0})$	absorption flux (exciting PSII antenna Chl *a* molecules) per RC (also used as a unit-less measure of PSII apparent antenna size)
Energetic connectivity of PSII units	
$M_0/M_0{}'$	grouping or connectivity among PSII units (the higher is the ratio indicate the less of connectivity)

2.4. Chlorophyll (a, b) Content

Chlorophyll *a* and Chl *b* content was determined at the end of the experiment (day 10th), according to [44]. The chlorophyll concentration was expressed as mg/g wet biomass (wb) of a leaf. The extraction was carried out in a dark room using 10 mL of 90% acetone and clear sand and the centrifugation lasted 10 min at 1500 rpm. Specimens' absorbance was measured at the wavelengths of 647, 664, and 750 nm using a Shimadzu UV-1800 spectrophotometer, Kyoto, Japan. The chlorophyll content was calculated according to the following equations: Chl *a* (μg mL^{-1}) = $11.93E_{664} - 1.93E_{647}$, Chl *b* (μg mL^{-1}) = $20.3E_{647} - 4.68E_{664}$, where E = corrected absorbency (absorbency at the wavelength − absorbency at 750 nm). The final chlorophyll content was expressed as mg Chl g^{-1} wb using the volume of acetone used to extract the pigments, divide by the wb of the leaf.

2.5. *Relative Growth Rate*

The relative growth rate (RGR) was calculated as RGR (day^{-1}) = ($LnWB_{t2} - LnWB_{t1}$)/($t_2 - t_1$), where WB_{t2} was the wet biomass on the experimental day 10th, and WB_{t1} was the wet biomass on the experimental day 1st.

3. Statistical Analysis

All statistical analyses were conducted using the software STATISTICA for Windows (version 7.1; StatSoft and TIBCO Software Inc., Palo Alto, CA, USA). Data were presented as mean (n = 6) ± standard error (SE) for each sample. Normality was tested by the Shapiro–Wilk's W test, while Levene's test tested the homogeneity. The effects of salinity (fixed factor, two levels: 9 and 19) and irradiance (fixed factor, two levels: 100 and 400 μmol photons m^{-2} s^{-1}) on JIP-test parameters were analyzed with the parametric repeated measures analysis of variance (two-way repeated-measures ANOVA) and on chlorophyll content and RGR with parametric factorial analysis of variance (two-way ANOVA). The multiple comparisons were tested by Duncan's post hoc test. In the present study, samples did not have normal distribution and homogeneity of variance; thus, the significance level (*p*-value) was defined as 0.01 [45]. Redundancy analysis (RDA) was used to quantify the irradiance and salinity effects on the variation of chlorophyll fluorescence (day 8th), chlorophyll content (day 8th), and RGR (day 10th) parameters using Canoco 5.1 (Microcomputer Power, Ithaca, NY, USA). Statistical tests were run using the Monte Carlo permutation procedure.

4. Results

4.1. *Chlorophyll a Fluorescence Measurements*

(a) Normalizations and Subtractions

The chlorophyll *a* fluorescence polyphasic O-J-I-P kinetics of dark-adapted samples grown under C1, C2, C3, and C4, for 1, 5, and 8 days are evident in Figure 1. Each kinetic is the average of six replicates, after each of the kinetics Ft = f(t) was normalized on its initial value; i.e., the averages are plotted as F_t/F_0 = f(t). This normalization was permitted as the minor variations observed among the raw $F_{20\mu s}$ values were random and not depending on the treatment. Hence, the experimental $F_{20\mu s}$ was reasonably taken as the real F_0. The kinetics demonstrated clearly that high light levels (C3 and C4) result in a pronounced suppression of the variable fluorescence ($F_t - F_0$) throughout the entire time course of the 1 s.

(b) The JIP-Test Parameters

The trapped energy flux per RC (TR_0/RC) was not affected by irradiance or salinity and their interaction (Figure 2A, Table S1), but the effect of irradiance was significant on RCs per PSII antenna Chl *a* (RC/ABS; $p < 0.001$, Table S1). Mean RC/ABS values (Figure 2B) were higher at 100 μmol photons m^{-2} s^{-1} (C1, C2) than at 400 μmol photons m^{-2} s^{-1} (C3, C4). The maximum RC/ABS values were observed at C2 and at C1 on the 5th day, while the minimum values were observed at C3 on the 5th day and at C4 on the 1st day. Post hoc comparisons (Table S2) showed that RC/ABS values were significantly lower in C3 and C4 than in C1 and C2.

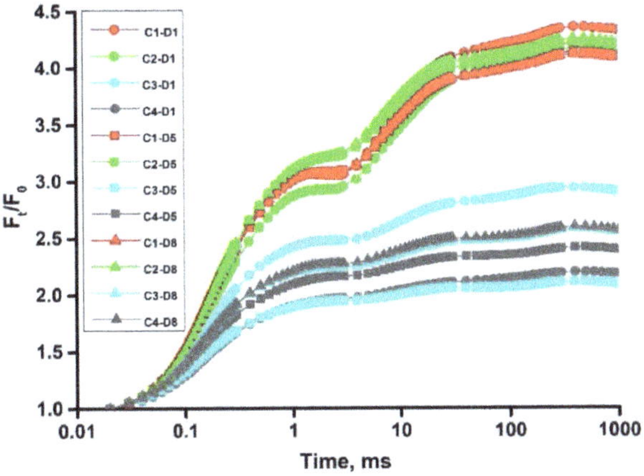

Figure 1. OJIP fluorescence transients expressed as F_t/F_0 = f(t), induced by strong red actinic light (3.000 µmol photons m^{-2} s^{-1}; peak at 650 nm) and measured with a plant efficiency analyzer (Handy PEA; Hansatech Instruments Ltd., Norfolk, UK) in *S. pectinata* leaves and dark-adapted for 30 min. Induction curves plotted on logarithmic scale from 20 µs to 1 s (JIP-time). Experimental conditions C1: salinity 9 and irradiance 100 µmol photons m^{-2} s^{-1}, C2: salinity 19 and irradiance 100 µmol photons m^{-2} s^{-1}, C3: salinity 9 and irradiance 400 µmol photons m^{-2} s^{-1}, C4: salinity 19 and irradiance 400 µmol photons m^{-2} s^{-1}. Days: Day 1 (D1), Day 5 (D5), Day 8 (D8).

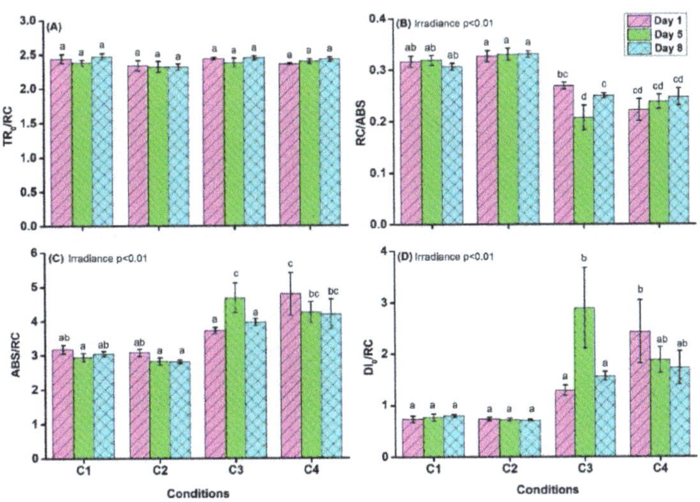

Figure 2. Effects of salinity and irradiance on (**A**) TR_0/RC, (**B**) RC/ABS, (**C**) ABS/RC, and (**D**) DI_0/RC. Values are means ± standard error, n = 6. Statistical differences from the post hoc test ($p < 0.01$) are represented using different Latin letters. For further information see Table 1.

The effect of irradiance was significant on the light absorption flux of antenna chlorophyll molecules per RC (also a measure of PSII apparent antenna size-ABS/RC; $p < 0.001$, Table S1). Mean ABS/RC values (Figure 2C) were higher at 400 µmol photons m^{-2} s^{-1} (C3, C4) than at 100 µmol photons m^{-2} s^{-1} (C1, C2). The maximum values of ABS/RC were observed at C3 on the 5th day and at C4 on the 1st day, while the minimum values were observed at C1 on the 5th day and at C2 on the 8th day. Post hoc comparisons (Table S3) showed that ABS/RC values were significantly lower in C1 and C2 than in C3 and C4 conditions. There was a significant interaction between salinity and treatment time on ABS/RC ($p < 0.01$, Table S1). At salinity 19 (C2, C4), no significant differences were observed. At 400 μmol photons m^{-2} s^{-1} (C3), ABS/RC values increased from the 1st to the 5th day, but they decreased on the 8th day.

The effect of irradiance was significant on the dissipated energy flux per RC (DI_0/RC; $p < 0.001$, Table S1). Mean DI_0/RC values were higher at 400 μmol photons m^{-2} s^{-1} (C3, C4) than at 100 μmol photons m^{-2} s^{-1} (C1, C2). The maximum of DI_0/RC were observed at C3 on the 5th day and at C4 on the 1st day, while the minimum values were observed at C2 on the 8th day and at C2 on the 1st day. Post hoc comparisons (Table S4) showed that DI_0/RC values were significantly lower in C1 and C2 than in C3 and C4 conditions.

The effect of irradiance was significant on φ_{P0}, i.e., the efficiency by which PSII transforms excitation energy to oxido-reduction energy (reduction of the primary electron quinone acceptor Q_A to Q_A^-) ($p < 0.01$, Table S1). Mean φ_{P0} values (Figure 3A) were higher at 100 μmol photons m^{-2} s^{-1} (C1, C2) than at 400 μmol photons m^{-2} s^{-1} (C3, C4). The maximum φ_{P0} values were observed at C1 on the 1st day and at C2 on the 8th day, while the minimum values were observed at C3 on the 5th day and at C4 on the 1st day. Post hoc comparisons (Table S5) showed that φ_{P0} values were significantly lower in C3 and C4 than in C1 and C2. There was also a significant interaction between salinity, irradiance and treatment time on φ_{P0} ($p < 0.01$, Table S1). At salinity 9 and irradiance 400 μmol photons m^{-2} s^{-1} (C3), φ_{P0} values decreased on the 5th day, but they partly recovered on the 8th day. In the other conditions (C1, C3, C4), φ_{P0} did not exert significant changes over time.

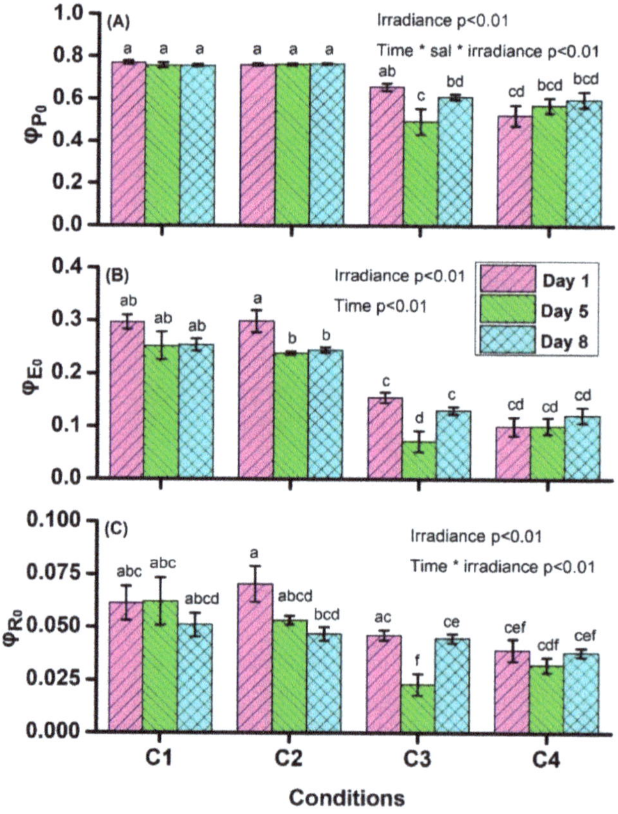

Figure 3. Effects of salinity and irradiance on (**A**) φ_{P0}, (**B**) φ_{E0}, and (**C**) φ_{R0}. Values are means ± standard error, n = 6. * = statistically significant interaction. For further information see Table 1 and Figure 2.

The effect of irradiance was significant on the quantum yields for electron transport to the plastoquinone (PQ) pool (φ_{E0}; $p < 0.01$, Table S1). Mean φ_{E0} values (Figure 3B) were higher at 100 μmol photons m^{-2} s^{-1} (C1, C2) than at 400 μmol photons m^{-2} s^{-1} (C3, C4). The maximum φ_{E0} values were observed at C2 and at C1 on the 1st day, while the minimum values were observed at C3 on the 5th day and at C4 on the 1st day. Post hoc comparisons (Table S6) showed that φ_{E0} values were significantly lower in C3 and C4 than in C1 and C2. The effect of treatment time was significant on φ_{E0} ($p < 0.01$, Table S1). φ_{E0} values in C3 decreased on the 5th day, but they partly recovered on the 8th day.

The effect of irradiance was significant on the quantum yields for electron transport to the electron acceptors at the PSI acceptor side (φ_{R0}; $p < 0.01$, Table S1). Mean φ_{R0} values (Figure 3C) were higher at 100 μmol photons m^{-2} s^{-1} (C1, C2) than at 400 μmol photons m^{-2} s^{-1} (C3, C4). The maximum φ_{R0} values were observed at C2 on the 1st day and at C2 on the 5th day, while the minimum values were observed at C3 and at C4 on the 5th day. Post hoc comparisons (Table S7) showed that φ_{R0} values were significantly lower in C3 and C4 than in C1 and C2. There was also a significant interaction between irradiance and treatment time on φ_{R0} ($p < 0.01$, Table S1). At irradiance 400 μmol photons m^{-2} s^{-1} (C3, C4), φ_{R0} decreased from the 1st to the 5th day, but they partly recovered on the 8th day, while at 100 μmol photons m^{-2} s^{-1} (C1, C2), φ_{R0} decreased over time.

The effect of irradiance was significant on the efficiency/probability that an electron moves further than Q_A^- (ψ_{E0}; $p < 0.01$, Table S1). Mean ψ_{E0} values (Figure 4A) were higher at 100 μmol photons m^{-2} s^{-1} (C1, C2) than at 400 μmol photons m^{-2} s^{-1} (C3, C4). The maximum ψ_{E0} values were observed at C1 on the 1st day and at C2 on the 1st day, while the minimum values were observed at C3 on the 5th day and at C4 on the 5th day. Post hoc comparisons (Table S8) showed that ψ_{E0} values were significantly lower in C3 and C4 than in C1 and C2. The effect of treatment time was significant on ψ_{E0} ($p < 0.01$, Table S1). Ψ_{E0} values decreased on the 5th day, but they partly recovered on the 8th day.

The effect of irradiance was significant on the efficiency/probability that an electron from the intersystem electron carriers is transferred to reduce end electron acceptors at the PSI acceptor side (δ_{R0}; $p < 0.01$, Table S1). Mean δ_{R0} values (Figure 4B) were higher at 400 μmol photons m^{-2} s^{-1} (C3, C4) than at 100 μmol photons m^{-2} s^{-1} (C1, C2). The maximum δ_{R0} values were observed at C4 on the 1st day and at C3 on the 5th day, while the minimum values were observed at C2 and at C1 on the 8th day. Post hoc comparisons (Table S9) showed that δ_{R0} values were significantly higher in C3 and C4 than in C1 and C2.

The effect of irradiance was significant on the connectivity of PSII photosynthetic units (M_0'/M_0; $p < 0.01$, Table S1). Mean M_0'/M_0 values (Figure 4C) were higher at 400 μmol photons m^{-2} s^{-1} (C3, C4) than at 100 μmol photons m^{-2} s^{-1} (C1, C2). The maximum M_0'/M_0 values were observed at C3 and at C4 on the 5th day, while the minimum values were observed at C1 on the 1st day and at C2 on the 8th day. Post hoc comparisons (Table S10) showed that M_0'/M_0 values were significantly lower in C3 and C4 than in C1 and C2. Differences due to salinity and time are found not to be statistically significant.

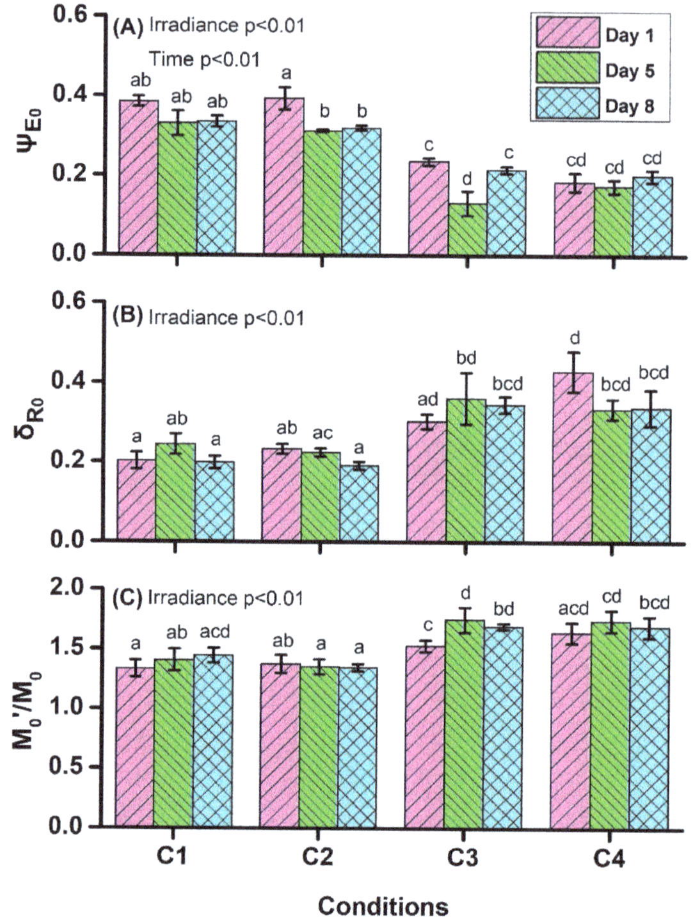

Figure 4. Effects of salinity and irradiance on (**A**) ψ_{E0}, (**B**) δ_{R0}, and (**C**) M_0'/M_0. Values are means ± standard error, n = 6. For further information see Table 1 and Figure 2.

4.2. Chlorophyll Content

There was a significant interaction between irradiance and salinity on leaf Chl *a* and Chl *b* content ($p < 0.001$, Table S11). Mean Chl *a* and Chl *b* values are shown in Figure 5A,B and were maximal at C3 and minimal at C1. Post hoc comparisons showed that Chl *a* (Table S12) and Chl *b* (Table S13) values under C1 and C4 were significantly lower than those under C3. There was not any significant effect of irradiance and salinity on Chl *a/b* ratio (Table S2). Mean values of the Chl *a/b* ratio are shown in Figure 5C. The maximum mean Chl *a/b* values were at C2 and the minimum values were at C4.

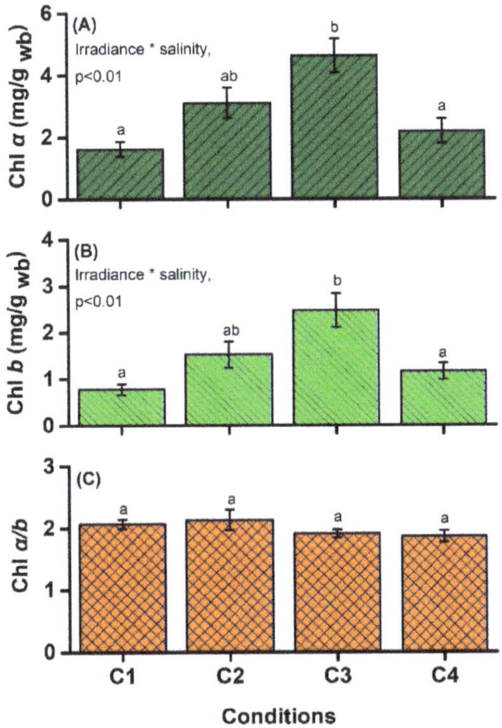

Figure 5. Effects of salinity and irradiance on (**A**) Chlorophyll *a*, (**B**) Chlorophyll *b*, and (**C**) Chlorophyll *a/b* ratio. Values are means ± standard error and measured on the day 10th of the experiment, n = 6. Wb = wet biomass. For further information see Figures 2 and 3.

4.3. Relative Growth Rate

There was no significant effect of salinity and irradiance on RGR (Table S14). Mean RGR values (Figure 6) were maximal at C3 and minimal at C2.

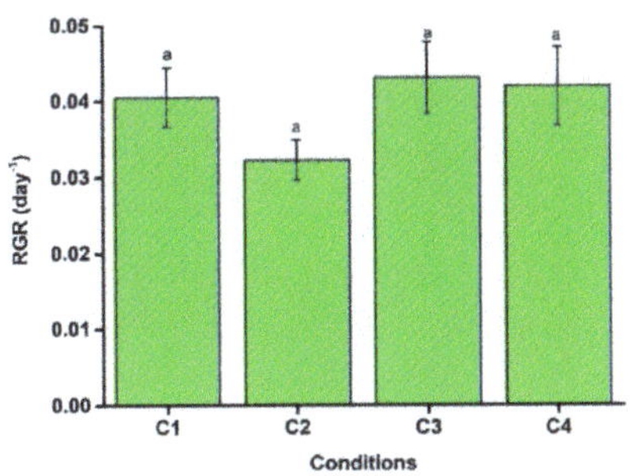

Figure 6. Effects of salinity and irradiance on relative growth rate (RGR). Values are means ± standard error, n = 6. For further information see Figure 2.

4.4. Redundancy Analysis (RDA)

Analysis of the correlations between the physiological parameters and stress factors (irradiance and salinity) was performed using an RDA forward selection model (Figure 7). According to the RDA results, irradiance was the only statistically significant factor that added 98.2% (Monte Carlo permutation test: pseudo F = 17.4, $p = 0.001$) to the explanatory power. The first axis accounted for 44.2% of the variation (Monte Carlo permutation test: F = 0.88) and was correlated with the irradiance, while the second axis accounted for 0.38% of the variation. The third axis accounted for 9.2% of the variation, while the fourth axis accounted for 3.9% of the total variation. While DI_0/RC, ABS/RC, δ_{R0} and M_0'/M_0 were positively correlated with Axis 1, ψ_{E0}, φ_{E0}, φ_{P0}, and RC/ABS were negatively correlated (Table 2). The experimental points were distributed rather homogeneously on the PC1/PC2 plane, with PC1 to clearly separate low light from high light conditions. This indicates that the irradiance was by far the most important factor affecting the studied physiological parameters of *S. pectinata*.

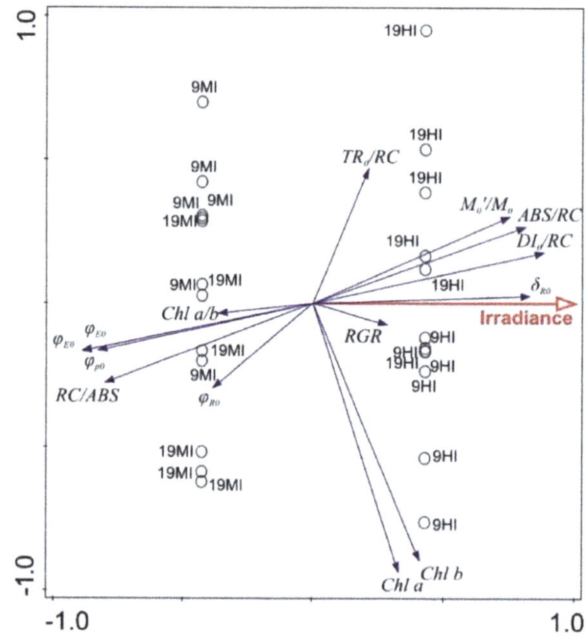

Figure 7. Triplot from the redundancy analysis (RDA) that shows the relationships between the JIP-test parameters, chlorophyll content and relative growth rate (RGR) parameters of *Stuckenia pectinata*. The length of the arrows indicates the strength of representation and contribution of each parameter to the PC axes. MI = moderate irradiance, HI = high irradiance, 9 and 19 correspond to salinity values. For further information, see Tables 1 and 2. The first axis accounted for 44.2% of the variation, while the second axis accounted for 0.38%.

Table 2. Loading scores of JIP-test, chlorophyll content and relative growth rate (RGR) parameters in the redundancy analysis (RDA) in temperate and subtropical forests. The correlation coefficients (ρ) are shown in the matrix. For further information, see Table 1. Axis = RDA axis.

	Axis 1	Axis 2	Axis 3	Axis 4
φ_{P0}	**−0.829**	−0.165	−0.521	0.011
TR_0/RC	0.214	0.468	0.025	0.220
DI_0/RC	**0.881**	0.177	0.423	0.023
RC/ABS	**−0.803**	−0.279	−0.471	−0.052
ABS/RC	**0.811**	0.266	0.492	0.044
M_0'/M_0	**0.750**	0.301	0.398	0.198
φ_{R0}	−0.386	−0.297	−0.235	**−0.699**
φ_{E0}	**−0.889**	−0.167	−0.398	−0.066
ψ_{E0}	**−0.892**	−0.168	−0.334	−0.101
δ_{R0}	**0.827**	0.026	0.350	−0.345
Chl *a*	0.331	**−0.936**	0.042	0.021
Chl *b*	0.410	**−0.896**	0.135	0.026
Chl *a*/*b*	−0.365	−0.037	−0.379	−0.021
RGR	0.285	−0.076	0.136	**−0.768**

5. Discussion

5.1. Irradiance Effects

The irradiance significantly affected the photosynthetic performance from the two factors tested, but not the chlorophyll content and RGR of *S. pectinata* (Figure 7). Acclimation of plants to HI includes regulation of (a) the number of the reaction centers and the size of their light-harvesting antennae, (b) the photosynthetic electron transport chain by dissipation of the excess energy to heat, and (c) the stoichiometry of photosynthetic components such as chlorophylls [46–48].

Under HI conditions (C3, C4), the density of active RCs (RC/ABS) per PSII, i.e., the number of Q_A-reducing RCs per PSII antenna Chl, as a result of partial inactivity of RCs [49], decreased. This result indicates that some active RCs were converted into heat sinks, enhancing the energy dissipation, as also indicated by DI_0/RC. DI_0/RC represents the total dissipation of untrapped excitation energy from all RCs concerning the number of active RCs [35]. As the number of inactive centers under HI increased (Figure 2B), the DI_0/RC ratio also increased (Figure 2D) because the inactive centers could not trap the photon, so the number of untrapped photons increased. Thus, the heat dissipation of excessive excitation energy slowed the overreduction of the photosynthetic electron transfer chain and minimized the potential photooxidative damage. On the other hand, an increase in the absorbance flux per reaction center (ABS/RC) value, as a measure for the average PSII antennae size [49], indicated overloading of PSII RC with electrons [38]. Although, in general, a smaller photosystems antenna size is reported under high light [50], we speculate that this is related to a Chl content increase, at least under the moderate salinity condition (C3).

HI affected negatively the parameters related to yields (φ_{P0}, φ_{E0}, and φ_{R0}) and efficiency/probability (ψ_{E0}) ($p < 0.01$), indicating that photoinhibition on PSII activity occurred of *S. pectinata* that considered as a positive acclimation for the downregulation of photosynthetic excitation pressure [51,52]. HI was also observed to reduce the maximum quantum yield in other plants [53,54]. Hu et al. [26] also showed that HI induced the decrease in effective quantum yield of PSII in *S. pectinata*. In agreement, TR_0/RC represents the maximal rate by which an exciton is trapped by the RC, resulting in the reduction of Q_A^- [55], which remained unaffected with all the treatments (Figure 2A). These results confirm that *S. pectinata* seemed to exhibit a photo acclimation rather than the photodamage of the reaction center complexes when exposed to HI.

Among the numerous biophysical parameters derived by the JIP-test, φ_{R0} is theoretically related to the energy supply from PSI towards the Calvin cycle, which is by definition the quantum yield of this supply. Indeed, there was a weak correlation φ_{R0} with RGR (Table 2, RDA axis 4th). However, the actual energy supply to the Calvin cycle is affected by several other factors in addition to φ_{R0}, and RGR does not depend only on the activity of the Calvin cycle. Thus, in this study, HI did not impair the RGR. As a result, the plants could relocate resources to compensate for any potential costs associated with photoinhibition, repair of photosystem II, and processes of high-light acclimation.

On the other hand, δ_{R0} increased when *S. pectinata* was exposed to HI, suggesting an improvement in the efficiency of electron transport, and therefore, energy, from plastoquinol (PQH_2) via cytochrome b_6/f to the PSI end electron acceptors (plastocyanin, ferredoxin) [56]. Thus, overexcitation energy pressure in PSII can be alleviated despite not increasing the sum of energy flux into PSI acceptors. Indeed, δ_{R0} values increased during the first steps of stress, and only severe stress caused depletion of the efficiency of electron transport up to end electron acceptors of PSI [57]. The result was consistent with those in the stems of *Hexinia polydichotoma*. It suggests that the electron transport pattern related to PSI can play an important role in the photoprotection of PSII in adverse conditions [58].

Acclimation to HI was also expressed by a decrease in the energetic connectivity of PSII photosynthetic units compared to moderate irradiance treated samples (C1, C2). It is worth remembering that, in separate units, the initial fluorescence rise is exponential and that the bigger the degree of connectivity, the more the curves deviate towards sigmoidal [35]. Though none of the kinetics in the present study appear sigmoidal, the deviation from exponential could not be excluded, and it was evaluated by comparing M_0' with M_0 (see Table 1) since the higher is their ratio, the more towards the exponential curve is the fluorescence rise.

5.2. Interactive Effects between Irradiance and Salinity

The chlorophyll content is associated with acclimation to different irradiance levels. HI or sun-grown plants contain lower chlorophyll than low light-grown or shade plants, which is considered a long-term regulation mechanism that controls light absorption capacity [59]. The results reveal that Chl *a* and Chl *b* content in *S. pectinata* increased significantly ($p < 0.01$) under HI and low salinity (Figure 5, Table S2), inconsistent with the general trend. Hu et al. [26] and Pilon and Santamaría [1] showed that the total chlorophyll concentration of *S. pectinata* was higher at low irradiance, which resulted in higher photosynthetic performance. Obviously, the HI condition used in this study was not too high to make the plants unable to produce more chlorophyll. This result is in line with a previous study focusing on *Zostera marina* (L) responses to light stress applied within controlled laboratory conditions [60]. Hu et al. [26] also noticed that HS and HI reduced ($p < 0.001$) the Chl *a* content in *S. pectinata*, in agreement with the results of this study, where Chl *a* content at HI and HS was lower ($p < 0.01$) than HI and MS.

The Chl *a/b* ratio indicates the degree of sun/shade acclimation and the structure of the photosynthetic apparatus to improve light capture [61,62]. Chl *a/b* ratio decreases under low irradiance conditions because the plants develop bigger light-harvesting complexes (LHCs). Both Chl *a* and Chl *b* are found in LHCs, yet only Chl *a* is found in the reaction centers. As a result, bigger LHCs as a response to low light levels lead to increments of both Chl *a* and Chl *b*; still, the overall Chl *a/b* ratio drops as there is no increase in the number of Chl *a* molecules forming the reaction centers of the photosystems. This pattern was not observed in our study, where the Chl *a/b* ratio remained constant under all treatments, as in the studies of [38,63] regarding the species of *Lemna*. Accordingly, the dissipation of energy became extremely important, reducing the risk of photoinhibition (see above). In any case, plants had a light-harvesting antenna size (mean Chl *a/b* ratios < 6.5) not susceptible to HI [64]. The absence of significant changes in the Chl *a/b* ratio under irradiance stress also revealed that the irradiance levels were adequate for the plant's growth throughout the experiment [65]. It is well-known that under stressful conditions, the plant may reallocate resources toward increased growth or alterations in morphology at the expense of defensive compounds and reproductive structures [41].

5.3. Interactive Effects with Time

Photo-acclimation corresponds to different processes involving many cellular components and occurring over a broad range of time scales, from seconds to days. This study focused on long-term photo acclimation (days) monitored by changes in the number and size of the reaction centers and the photosynthetic electron transport chain. Changes within the eight days of the experiment duration were noticed in all measured parameters but statistical changes in the parameters related to yields (φ_{P0}, φ_{E0}, and φ_{R0}) and in DI_0/RC. Besides regulation of irradiance as discussed above, φ_{P0} and DI_0/RC acclimation is also affected by salinity, indicating that the photosynthetic apparatus could cope with salinity stress, and the plant was tolerant within the range from 9 to 19. The great range of tolerance of *S. pectinata* to salinity (9–18) has been reported by other studies [3,23,24]. Hu et al. [26] also reported that PSII units and electron flow were not influenced under salinity stress.

5.4. Stuckenia pectinata in Vistonis Lake

Ruppia maritima L. is a submerged macrophyte that survives in brackish estuarine and coastal lagoon waters near Vistonis Lake. Following the scenario of the replacement of *S. pectinata* by *R. maritima* due to sea level and salinity rise, it seems the lower biomass which would be available to the overwintering waterfowl at Vistonis will impact the structure and function of higher trophic levels [66]. However, based on this study, that is a not near-future scenario as the Vistonis *S. pectinata* clone tolerated salinity of at least up to 19 and can occasionally survive salinity increases up to 34. High tolerance was also confirmed by the RGR but evidenced a tendency of a decrease at a salinity of 19 (Figure 6). This is a topic that should be further investigated in the future through long-term experiments.

6. Conclusions

This study investigates the correlations among parameters representing processes of two different biological levels, from subcellular (photosynthetic) to organismal (growth). It is now clear that the submerged macrophyte *S. pectinata* had great acclimation potential to present and future stressful irradiance and salinity conditions in hypertrophic Vistonis Lake. Such physiology might also be of great importance for other environmental disturbances, such as heavy metals. However, further research on the role of other abiotic factors, such as temperature and different nutrient forms, is needed to better determine how climate change will influence *S. pectinata*'s abundance. In addition, other species aspects, such as morphology and reproduction, should be further investigated in the future in long-term experiments.

Supplementary Materials: The following are available online at https://www.mdpi.com/article/10.3390/w13121706/s1. Table S1. Two-way repeated measures of variance of JIP-test in *Stuckenia pectinata*. Table S2. Duncan's post hoc test for RC/ABS in *Stuckenia pectinata*. Table S3. Duncan's post hoc test for ABS/RC in *Stuckenia pectinata*. Table S4. Duncan's post hoc test for DI_0/RC in *Stuckenia pectinata*. Table S5. Duncan's post hoc test for φ_{P0} in *Stuckenia pectinata*. Table S6. Duncan's post hoc test for φ_{E0} in *Stuckenia pectinata*. Table S7. Duncan's post hoc test for φ_{R0} in *Stuckenia pectinata*. Table S8. Duncan's post hoc test for ψ_{E0} in *Stuckenia pectinata*. Table S9. Duncan's post hoc test for δ_{R0} in *Stuckenia pectinata*. Table S10. Duncan's post hoc test for M_0'/M_0 in *Stuckenia pectinata*. Table S11. Two-way measures of variance of chlorophyll content and ratio in *Stuckenia pectinata*. Table S12. Duncan's post hoc test for Chl *a* in *Stuckenia pectinata*. Table S13. Duncan's post hoc test for Chl *b* in *Stuckenia pectinata*. Table S14. Two-way measures of variance of RGR in *Stuckenia pectinata*.

Author Contributions: This work has been conducted at the Fisheries Research Institute (ELGO DIMITRA) as a Master's dissertation of L.M. at the University of Thessaly with supervisors A.E. and S.O., S.O. conceived and designed research, analyzed and interpreted data, wrote the article;

L.M. conducted experiments, analyzed data and wrote the article; K.N. and A.P. pre-acclimated the material and helped in the experiments; A.E. contributed to the writing and review of the manuscript. All authors have read and agreed to the published version of the manuscript.

Funding: This research received no external funding.

Institutional Review Board Statement: Not applicable.

Informed Consent Statement: Not applicable.

Acknowledgments: We thank Merope Tsimilli-Michael for her clarifications concerning the concept of the JIP-test and the interpretation of the derived parameters.

Conflicts of Interest: The authors declare no conflict of interest.

References

1. Pilon, J.; Santamaría, L. Clonal variation in morphological and physiological responses to irradiance and photoperiod for the aquatic angiosperm *Potamogeton pectinatus*. *J. Ecol.* **2002**, *90*, 859–870. [CrossRef]
2. Scheffer, M.; de Redelijkheid, M.R.; Noppert, F. Distribution and dynamics of submerged vegetation in a chain of shallow eutrophic lakes. *Aquat. Bot.* **1992**, *42*, 199–216. [CrossRef]
3. Borgnis, E.; Boyer, K.E. Salinity tolerance and competition drive distributions of native and invasive submerged aquatic vegetation in the upper San Francisco Estuary. *Estuaries Coasts* **2016**, *39*, 707–717. [CrossRef]
4. Jeppesen, E.; Søndergaard, M.; Søndergaard, M.; Kirsten, C. Alternative Stable states. In *The Structering Role of Submerged Macrophytes in Lakes*; Jeppesen, E., Søndergaard, M., Christoffersen, K., Eds.; Springer: Berlin/Heidelberg, Germany, 1998; p. 427.
5. Gao, Y.N.; Dong, J.; Fu, Q.Q.; Wang, Y.P.; Chen, C.; Li, J.H.; Li, R.; Zhou, C.J. Allelopathic effects of submerged macrophytes on phytoplankton. *Allelopath. J.* **2017**, *40*, 1–22. [CrossRef]
6. Chen, J.; Zhang, H.; Han, Z.; Ye, J.; Liu, Z. The influence of aquatic macrophytes on *Microcystis aeruginosa* growth. *Ecol. Eng.* **2012**, *42*, 130–133. [CrossRef]
7. Engelhardt, K.A.M.; Ritchie, M.E. Effects of macrophyte species richness on wetland ecosystem functioning and services. *Nature* **2001**, *411*, 687–689. [CrossRef]
8. Hilt, S.; Brothers, S.; Jeppesen, E.; Veraart, A.J.; Kosten, S. Translating regime shifts in shallow lakes into changes in ecosystem functions and services. *Bioscience* **2017**, *67*, 928–936. [CrossRef]
9. Albertoni, E.F.; Palma-Silva, C.; Trindade, C.R.T.; Furlanetto, L.M. Field evidence of the influence of aquatic macrophytes on water quality in a shallow eutrophic lake over a 13-year period. *Acta Limnol. Bras.* **2014**, *26*, 176–185. [CrossRef]
10. Bakker, E.S.; Van Donk, E.; Declerck, S.A.J.; Helmsing, N.R.; Hidding, B.; Nolet, B.A. Effect of macrophyte community composition and nutrient enrichment on plant biomass and algal blooms. *Basic Appl. Ecol.* **2010**, *11*, 432–439. [CrossRef]
11. Hidding, B.; Brederveld, R.J.; Nolet, B.A. How a bottom-dweller beats the canopy: Inhibition of an aquatic weed (*Potamogeton pectinatus*) by macroalgae (*Chara* spp.). *Freshw. Biol.* **2010**, *55*, 1758–1768. [CrossRef]
12. Triest, L.; Tran Thi, V.; Le Thi, D.; Sierens, T.; van Geert, A. Genetic differentiation of submerged plant populations and taxa between habitats. *Hydrobiologia* **2010**, *656*, 15–27. [CrossRef]
13. Abbasi, S.; Afsharzadeh, S.; Saeidi, H.; Triest, L. Strong genetic differentiation of submerged plant populations across mountain ranges: Evidence from *Potamogeton pectinatus* in Iran. *PLoS ONE* **2016**, *11*, 1–20. [CrossRef] [PubMed]
14. Ganie, A.H.; Reshi, Z.A.; Wafai, B.A. Reproductive ecology of *Potamogeton pectinatus* L. (= *Stuckenia pectinata* (L.) Börner) in relation to its spread and abundance in freshwater ecosystems of the Kashmir Valley, India. *Trop. Ecol.* **2016**, *57*, 787–803.
15. Pilon, J.; Santamaría, J. Seasonal acclimation in the photosynthetic and respiratory temperature responses of three submerged freshwater macrophyte species. *New Phytol.* **2001**, *151*, 659–670. [CrossRef]
16. Sandsten, H.; Beklioglu, M.; Ince, O. Effects of waterfowl, large fish and periphyton on the spring growth of *Potamogeton pectinatus* L. in Lake Mogan, Turkey. *Hydrobiologia* **2005**, *537*, 239–248. [CrossRef]
17. Abbasi, S.; Afsharzadeh, S.; Saeidi, H. Genetic diversity of *Potamogeton pectinatus* L. in Iran as revealed by ISSR markers. *Acta Bot. Croat.* **2017**, *76*, 177–182. [CrossRef]
18. Hootsmans, M.J.M.; Drovandi, A.A.; Soto Perez, N.; Wiegman, F. Photosynthetic plasticity in *Potamogeton pectinatus* L. from Argentina: Strategies to survive adverse light conditions. *Hydrobiologia* **1996**, *340*, 1–5. [CrossRef]
19. Shabnam, N.; Sharmila, P.; Sharma, A.; Strasser, R.J.; Govindjee; Pardha-Saradhi, P. Mitochondrial electron transport protects floating leaves of long leaf pondweed (*Potamogeton nodosus* Poir) against photoinhibition: Comparison with submerged leaves. *Photosynth. Res.* **2015**, *125*, 305–319. [CrossRef] [PubMed]
20. Poikane, S.; Portielje, R.; Denys, L.; Elferts, D.; Kelly, M.; Kolada, A.; Mäemets, H.; Phillips, G.; Søndergaard, M.; Willby, N.; et al. Macrophyte assessment in European lakes: Diverse approaches but convergent views of 'good' ecological status. *Ecol. Indic.* **2018**, *94*, 185–197. [CrossRef]
21. Kondo, K.; Kawabata, H.; Ueda, S.; Hasegawa, H.; Inaba, J.; Mitamura, O.; Seike, Y.; Ohmomo, Y. Distribution of aquatic plants and absorption of radionuclides by plants through the leaf surface in brackish Lake Obuchi, Japan, bordered by nuclear fuel cycle facilities. *J. Radioanal. Nucl. Chem.* **2003**, *257*, 305–312. [CrossRef]

22. Kantrud, H.A. Sago pondweed (*Potamogeton pectinatus* L.): A literature review. *Resour. Publ. US Fish Wildl. Serv.* **1990**, *176*, 23.
23. Rodríguez-Gallego, L.; Sabaj, V.; Masciadri, S.; Kruk, C.; Arocena, R.; Conde, D. Salinity as a major driver for submerged aquatic vegetation in coastal lagoons: A Multi-Year Analysis in the Subtropical Laguna de Rocha. *Estuaries Coasts* **2014**, *38*, 451–465. [CrossRef]
24. Dhir, B. Status of aquatic macrophytes in changing climate: A perspective. *J. Environ. Sci. Technol.* **2015**, *8*, 139–148. [CrossRef]
25. van Wijck, C.; Grillas, P.; de Groot, C.J.; Ham, L.T. A comparison between the biomass production of *Potamogeton pectinatus* L. and *Myriophyllum spicatum* L. in the Camargue (southern France) in relation to salinity and sediment characteristics. *Vegetatio* **1994**, *113*, 171–180. [CrossRef]
26. Hu, Q.; Turnbull, M.; Hawes, I. Salinity restricts light conversion efficiency during photo-acclimation to high irradiance in *Stuckenia pectinata*. *Environ. Exp. Bot.* **2019**, *165*, 83–91. [CrossRef]
27. Moustaka-Gouni, M.; Hiskia, A.; Genitsaris, S.; Katsiapi, M.; Manolidi, K.; Zervou, S.K.; Christophoridis, C.; Triantis, T.M.; Kaloudis, T.; Orfanidis, S. First report of *Aphanizomenon favaloroi* occurrence in Europe associated with saxitoxins and a massive fish kill in Lake Vistonis, Greece. *Mar. Freshw. Res.* **2017**, *68*, 793–800. [CrossRef]
28. Vardaka, E.; Moustaka-Gouni, M.; Cook, C.M.; Lanaras, T. Cyanobacterial blooms and water quality in Greek waterbodies. *J. Appl. Phycol.* **2005**, *17*, 391–401. [CrossRef]
29. Lüning, K. *Seaweeds: Their Environment, Biogeography, and Ecophysiology*; Wiley: Hoboken, NJ, USA, 1990.
30. Markou, D.A.; Sylaios, G.K.; Tsihrintzis, V.A.; Gikas, G.D.; Haralambidou, K. Water quality of Vistonis Lagoon, Northern Greece: Seasonal variation and impact of bottom sediments. *Desalination* **2007**, *210*, 83–97. [CrossRef]
31. Dolbeth, M.; Crespo, D.; Leston, S.; Solan, M. Realistic scenarios of environmental disturbance lead to functionally important changes in benthic species-environment interactions. *Mar. Environ. Res.* **2019**, *150*, 104770. [CrossRef]
32. Maxwell, K.; Johnson, G.N. Chlorophyll fluorescence—A practical guide Kate. *J. Exp. Bot.* **2000**, *51*, 659–668. [CrossRef]
33. Strasser, R.J.; Tsimilli-Michael, M.; Dangre, D.; Rai, M. Biophysical phenomics reveals functional building blocks of plants systems biology: A case study for the evaluation of the impact of mycorrhization with *Piriformospora indica*. *Adv. Tech. Soil Microbiol.* **2007**, *11*, 319–341. [CrossRef]
34. Strasser, R.J.; Srivastava, A.; Tsimilli-Michael, M. The fluorescence transient as a tool to characterize and screen photosynthetic samples. In *Probing Photosynthesis: Mechanism, Regulation & Adaptation*; CRC: Boca Raton, FL, USA, 2000; pp. 443–480.
35. Tsimilli-Michael, M. Revisiting JIP-test: An educative review on concepts, assumptions, approximations, definitions and terminology. *Photosynthetica* **2020**, *58*, 275–292. [CrossRef]
36. Wang, G.; Chen, L.; Hao, Z.; Li, X.; Liu, Y. Effects of salinity stress on the photosynthesis of *Wolffia arrhiza* as probed by the OJIP test. *Fresenius Environ. Bull.* **2011**, *20*, 432–438.
37. Xu, C.; Wang, M.T.; Yang, Z.Q.; Zheng, Q.T. Low temperature and low irradiation induced irreversible damage of strawberry seedlings. *Photosynthetica* **2020**, *58*, 156–164. [CrossRef]
38. Lepeduš, H.; Vidaković-Cifrek, Ž.; Šebalj, I.; Antunović Dunić, J.; Cesar, V. Effects of low and high irradiation levels on growth and PSII efficiency in *Lemna minor* L. *Acta Bot. Croat.* **2020**, *79*, 185–192. [CrossRef]
39. Parihar, P.; Singh, S.; Singh, R.; Singh, V.P.; Prasad, S.M. Effect of salinity stress on plants and its tolerance strategies: A review. *Environ. Sci. Pollut. Res.* **2015**, *22*, 4056–4075. [CrossRef] [PubMed]
40. Tuller, J.; Marquis, R.J.; Andrade, S.M.M.; Monteiro, A.B.; Faria, L.D.B. Trade-offs between growth, reproduction and defense in response to resource availability manipulations. *PLoS ONE* **2018**, *13*, 1–12. [CrossRef]
41. Gleeson, S.K.; Tilman, D. Plant allocation and the multiple limitation hypothesis. *Am. Nat.* **1992**, *139*, 1322–1343. [CrossRef]
42. Schubert, N.; Freitas, C.; Silva, A.; Costa, M.M.; Barrote, I.; Horta, P.A.; Rodrigues, A.C.; Santos, R.; Silva, J. Photoacclimation strategies in northeastern Atlantic seagrasses: Integrating responses across plant organizational levels. *Sci. Rep.* **2018**, *8*, 1–14. [CrossRef]
43. Tsimilli-Michael, M.; Strasser, R.J. In vivo assessment of stress impact on plant's vitality: Applications in detecting and evaluating the beneficial role of mycorrhization on host plants. *Mycorrhiza* **2008**, 679–703. [CrossRef]
44. Granger, S.; Lizumi, H. Water quality measurement methods for seagrass habitat. *Glob. Seagrass Res. Methods* **2001**, 393–406. [CrossRef]
45. Underwood, A.J. *Experiments in Ecology*; Cambridge University Press: Cambridge, UK, 1996.
46. Owens, T.G.; Falkowski, P.G.; Whitledge, T.E. Diel periodicity in cellular chlorophyll content in marine diatoms. *Mar. Biol.* **1980**, *59*, 71–77. [CrossRef]
47. Ruban, A.V.; Berera, R.; Ilioaia, C.; Van Stokkum, I.H.M.; Kennis, J.T.M.; Pascal, A.A.; Van Amerongen, H.; Robert, B.; Horton, P.; Van Grondelle, R. Identification of a mechanism of photoprotective energy dissipation in higher plants. *Nature* **2007**, *450*, 575–578. [CrossRef] [PubMed]
48. Dietz, K.J. Efficient high light acclimation involves rapid processes at multiple mechanistic levels. *J. Exp. Bot.* **2015**, *66*, 2401–2414. [CrossRef]
49. Stirbet, A. Govindjee On the relation between the Kautsky effect (chlorophyll a fluorescence induction) and Photosystem II: Basics and applications of the OJIP fluorescence transient. *J. Photochem. Photobiol. B Biol.* **2011**, *104*, 236–257. [CrossRef] [PubMed]
50. Wientjes, E.; Van Amerongen, H.; Croce, R. LHCII is an antenna of both photosystems after long-term acclimation. *Biochim. Biophys. Acta Bioenerg.* **2013**, *1827*, 420–426. [CrossRef] [PubMed]
51. Raven, J.A. The cost of photoinhibition. *Physiol. Plant.* **2011**, *142*, 87–104. [CrossRef] [PubMed]

52. Jiang, H.X.; Chen, L.S.; Zheng, J.G.; Han, S.; Tang, N.; Smith, B.R. Aluminum-induced effects on Photosystem II photochemistry in *Citrus* leaves assessed by the chlorophyll a fluorescence transient. *Tree Physiol.* **2008**, *28*, 1863–1871. [CrossRef]
53. Broetto, F.; Monteiro Duarte, H.; Lüttge, U. Responses of chlorophyll fluorescence parameters of the facultative halophyte and C_3-CAM intermediate species *Mesembryanthemum crystallinum* to salinity and high irradiance stress. *J. Plant Physiol.* **2007**, *164*, 904–912. [CrossRef]
54. Hazrati, S.; Tahmasebi-Sarvestani, Z.; Modarres-Sanavy, S.A.M.; Mokhtassi-Bidgoli, A.; Nicola, S. Effects of water stress and light intensity on chlorophyll fluorescence parameters and pigments of *Aloe vera* L. *Plant Physiol. Biochem.* **2016**, *106*, 141–148. [CrossRef]
55. Stirbet, A.D.; Strasser, R.J. Numerical simulation of the in vivo fluorescence in plants. *Math. Comput. Simul.* **1996**, *42*, 245–253. [CrossRef]
56. Yan, K.; Chen, P.; Shao, H.; Shao, C.; Zhao, S.; Brestic, M. Dissection of photosynthetic electron transport process in Sweet Sorghum under Heat Stress. *PLoS ONE* **2013**, *8*, 1–6. [CrossRef]
57. Goltsev, V.; Zaharieva, I.; Chernev, P.; Kouzmanova, M.; Kalaji, H.M.; Yordanov, I.; Krasteva, V.; Alexandrov, V.; Stefanov, D.; Allakhverdiev, S.I.; et al. Drought-induced modifications of photosynthetic electron transport in intact leaves: Analysis and use of neural networks as a tool for a rapid non-invasive estimation. *Biochim. Biophys. Acta Bioenerg.* **2012**, *1817*, 1490–1498. [CrossRef] [PubMed]
58. Li, L.; Zhou, Z.; Liang, J.; Lv, R. In vivo evaluation of the high-irradiance effects on PSII activity in photosynthetic stems of *Hexinia polydichotoma*. *Photosynthetica* **2015**, *53*, 621–624. [CrossRef]
59. Ruban, A.V. Plants in light. *Commun. Integr. Biol.* **2009**, *2*, 50–55. [CrossRef]
60. Bertelli, C.M.; Unsworth, R.K.F. Light stress responses by the eelgrass, *Zostera marina* (L). *Front. Environ. Sci.* **2018**, *6*, 1–13. [CrossRef]
61. Esteban, R.; Barrutia, O.; Artetxe, U.; Fernández-Marín, B.; Hernández, A.; García-Plazaola, J.I. Internal and external factors affecting photosynthetic pigment composition in plants: A meta-analytical approach. *New Phytol.* **2015**, *206*, 268–280. [CrossRef]
62. Collier, C.J.; Waycott, M.; Ospina, A.G. Responses of four Indo-West Pacific seagrass species to shading. *Mar. Pollut. Bull.* **2012**, *65*, 342–354. [CrossRef] [PubMed]
63. Paolacci, S.; Harrison, S.; Jansen, M.A.K. The invasive duckweed *Lemna minuta* Kunth displays a different light utilisation strategy than native *Lemna minor* Linnaeus. *Aquat. Bot.* **2018**, *146*, 8–14. [CrossRef]
64. Wu, G.; Ma, L.; Sayre, R.T.; Lee, C.H. Identification of the optimal light harvesting antenna size for high-light stress mitigation in plants. *Front. Plant Sci.* **2020**, *11*, 1–11. [CrossRef]
65. Zhu, S.; Qin, L.; Feng, P.; Shang, C.; Wang, Z.; Yuan, Z. Treatment of low C/N ratio wastewater and biomass production using co-culture of *Chlorella vulgaris* and activated sludge in a batch photobioreactor. *Bioresour. Technol.* **2019**, *274*, 313–320. [CrossRef] [PubMed]
66. Casagranda, C.; Boudouresque, C.F. Biomass of *Ruppia cirrhosa* and *Potamogeton pectinatus* in a Mediterranean brackish lagoon, Lake Ichkeul, Tunisia. *Fundam. Appl. Limnol.* **2007**, *168*, 243–255. [CrossRef]

water

MDPI

Article

Classification and Assessment Methods for Mountain Channel Habitats in the Chishui River Basin, China

Fandong Yu [1,2], Fei Liu [1], Zhijun Xia [1,2], Pengcheng Lin [1], Chunsen Xu [1,2], Jianwei Wang [1,*], Miaomiao Hou [1,2] and Xinhua Zou [1,2]

1 Institute of Hydrobiology, Chinese Academy of Sciences, Wuhan 430072, China; yufd666@163.com (F.Y.); liufei@ihb.ac.cn (F.L.); Xiazj1995@163.com (Z.X.); linpc@ihb.ac.cn (P.L.); 18227588944@163.com (C.X.); houmiao006@163.com (M.H.); 18370085282@163.com (X.Z.)

2 University of Chinese Academy of Sciences, Beijing 100049, China

* Correspondence: wangjw@ihb.ac.cn; Tel.: +86-027-6878-0033

Abstract: Mountain channels have received relatively little study compared to lowland rivers due to their complicated fluvial geomorphology and inconvenient traffic. Classification schemes and habitat assessments in mountain channels should be strengthened to provide a scientific basis for river ecological restoration. Therefore, we tried to simplify the habitat assessment of mountain channels using a suitable habitat classification scheme based on high-resolution satellite imagery. We used China's Chishui River basin because it is a typical mountain river system. Five parameters (stream order, elevation, slope, sinuosity and river network density) and 120 sites were used for habitat classification. In addition, we recorded 20 metrics in four categories (water environmental status, river morphology, riparian zone and human disturbance). Our results identified a total of 40 representative sampling sections belonging to six habitat types that were useful for habitat assessment across the Chishui River basin. The basin was given a mean comprehensive habitat quality index (CHQI) score of 130.66 ± 24.14 and classified under the status "good." However, the headwaters, Tongmin River, Tongzi River and Xishui River were disturbed by various human activities. We conclude that the process of developing and simplifying our habitat assessment systems can be regarded as a reference for biomonitoring in other mountain river systems.

Keywords: mountain channels; habitat classification; habitat assessment; simplification; Chishui River basin

Citation: Yu, F.; Liu, F.; Xia, Z.; Lin, P.; Xu, C.; Wang, J.; Hou, M.; Zhou, X. Classification and Assessment Methods for Mountain Channel Habitats in the Chishui River Basin, China. *Water* **2022**, *14*, 515. https://doi.org/10.3390/w14040515

Academic Editors: Eva Papastergiadou and Kostas Stefanidis

Received: 25 January 2022
Accepted: 5 February 2022
Published: 9 February 2022

1. Introduction

Since the 19th century, ecologists and geographers have recognized fundamental differences between mountain channels and their lowland counterparts [1–3]. Compared to lowland rivers, mountain channels have more intense hydrologic changes, more variable gradient and morphology, poorer nutrition, and clearer spatial variation in the ecosystem that is prone to forcing by external influences [4]. In the past, mountain channels have received relatively little study compared to lowland rivers because the technology was not good enough to conduct in-depth studies in these areas [5]. Strengthened classification schemes and habitat assessments for mountain channels would help us better understand and predict their response to both human and natural disturbance [6,7]. Furthermore, classifying and assessing river habitats improves our understanding of riverine ecology.

Geomorphic units are the elementary spatial physical features of the river mosaic at the reach scale that are nested within the overall hydromorphological structure of a river and its catchment [8]. Principles of fluvial geomorphology have guided the development of riverine ecology over the past few decades [9]. One axiom associated with fluvial geomorphology is that what initially appears complex is even more so upon further investigation [6]. River habitats have diverse ecological characteristics due to their different aquatic organisms, physical environment, ecosystem pattern, etc., and to a certain extent they determine the state of the river ecosystem [10].

A suitable classification scheme would help simplify otherwise complex river systems. The effort to classify freshwater ecosystems is not new, and various classification approaches have been developed for lakes, streams and wetlands [11–13]. For example, Wolfgang et al. indicated five major classified systems of Amazonian white-water river floodplains [14]; Davenport et al. identified sites of urban rivers that have particular qualities or may require particular types of management [15]. Complementarily, biodiversity and ecosystem integrity are being degraded around the world [16,17]. Over the past decades, freshwater ecosystems have become increasingly threatened by various stressors, such as pollution, land use changes, dam construction and water extraction [18–20]. As a result, health assessments of freshwater ecosystems are becoming widespread [21]; one part of this evaluation is a habitat assessment. Since the 1980s, different models [22,23], protocols and frameworks [24,25] have been developed to assess river habitats, especially in Europe, North America and Australia [26–28]. These methods, to some extent, help measure the physical habitat characteristics of rivers and evaluate the corresponding characteristics. However, there are no systematic standards for habitat assessment. Especially for mountain channels, complicated fluvial geomorphology and inconvenient traffic limit the collection of data and planning needed to customize management activities for unique ecosystems. Our first priority for doing this is to improve environmental monitoring tools.

The upper Yangtze River supports a diverse aquatic fauna and is extremely rich in endemicity, with at least 286 fish species distributed throughout [29]. However, an increase in anthropogenic activities in the Yangtze River over the past decades has disrupted habitats and led to many species becoming extinct or highly endangered [30,31]. As a typical mountain river system, the Chishui River is the last free-flowing tributary of the upper Yangtze River and provides an ideal model to test river ecological principles, as no dams have been built on its main stream [32,33]. In addition, the Chishui River is an important "National Nature Reserve for Rare and Endemic Fishes of the Upper Yangtze River," a classification established by the Chinese Government in 2005 [33], and an ecology–conservation hotspot. In recent years, increasing research reports on the Chishui River have been proposed to capture the fish diversity patterns, community biology and conservation biology [32,33]. However, little research has been done on habitat classification and habitat assessment. Here, we attempt to simplify the evaluation steps, which we suggest have been insufficiently studied by ecologists and less used by ecosystem managers; we also make it easier to perform habitat assessments on mountain channels using a suitable habitat classification scheme based on high-resolution satellite imagery.

Our efforts to simplify the habitat assessment by categorizing river systems help us achieve, to some extent, the following objectives: (1) categorize river habitats of the Chishui River basin into reasonable ecoregions, (2) assess the habitat condition of rivers throughout the Chishui River basin, (3) identify the existing factors that hinder ecological health, and (4) provide a reference for those working on habitat assessments in other mountain river systems.

2. Materials and Methods

2.1. Study Region and Technical Procedures

The Chishui River basin (27°20′–28°50′ N; 104°45′–106°51′ E) has a drainage area of 20,440 km^2 and includes the Chishui River mainstream and its 11 tributaries (in order from upstream to downstream: Zhaxi River, Daoliu River, Tongche River, Baisha River, Erdao River, Wuma River, Tongzi River, Gulin River, Tongmin River, Datong River and Xishui River). The Chishui River originates from the Wumeng Mountains in Yunnan Province and flows through Yunnan, Guizhou and Sichuan Provinces for nearly 436.5 km before meeting the upper Yangtze River in Hejiang County, Sichuan Province, southwest China. These 11 s-order tributaries range from 35 to 150 km long. All of these streams are located in the eastern Yungui Plateau, Sichuan Basin or the transitional area between them. With a subtropical monsoon climate, the annual average rainfall of the Chishui River is about 1000 mm. The Chishui River contains a large amount of laterite soil, which can lead to extensive erosion; thus, it is the origin of the name "Chishui" (i.e., "red river" in Chinese). Karst landforms are mainly distributed in the upper and midstream of the river, and the river's downstream areas belong to the Sichuan Basin [32–34].

For the mountain channels, some scientific methods were needed before field sampling since it was impractical for us to reach all sampling sections. We used habitat classification results to select suitable sampling sections for subsequent habitat assessment. Remote sensing technology was initially used to categorize the river systems; based on these categories, we then selected representative sampling sections from the Chishui River basin that showed a strong capability to distinguish sites of human perturbation. Field habitat surveying and assessment work were then carried out, and the comprehensive habitat quality index (CHQI) was finally calculated. The detailed procedures are shown in Figure 1.

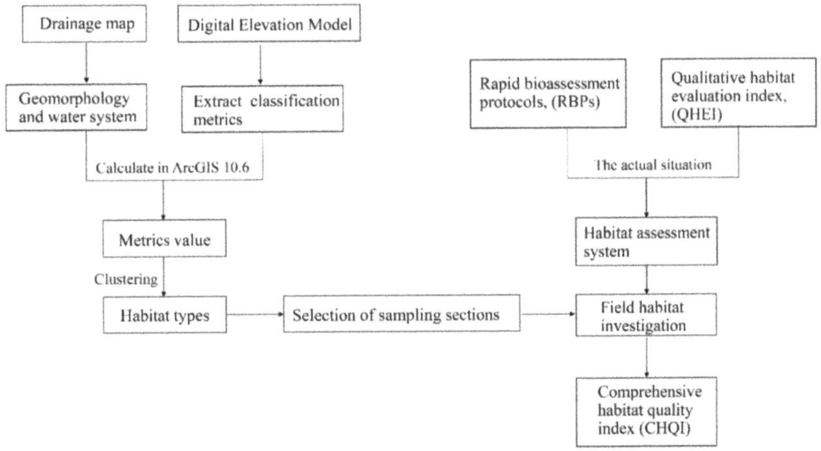

Figure 1. Technical procedure for habitat assessment in the Chishui River basin.

2.2. Habitat Classification

2.2.1. Extraction of Basic Data

As many data sites as possible in the Chishui River basin were initially selected based on Google Earth Pro software and a 1:250,000 high-definition drainage map. About three data sites were collected for each 10 km of riverbank for a total of 120 data points (Figure 2). The 30 m resolution Digital Elevation Model (DEM) was downloaded and cut via Global Mapper software (v. 21.0). With reference to the monograph [35], basic data were used to calculate parameters that, for the habitat classification, were extracted for each data site with ArcGIS software (v.10.6.1) based on 30 m-resolution DEM databases. The specific steps included filling the sink, analysis of flow direction, analysis of flow accumulation, reclassification of flow, river linking, vectorization (extracting water systems) and classification of the stream net. An algorithm for D8 flow direction, spatial analysis and the Strahler stream net classification method were used in the above steps. These basic data provided river information, subbasin boundaries and altitude for parameter calculations.

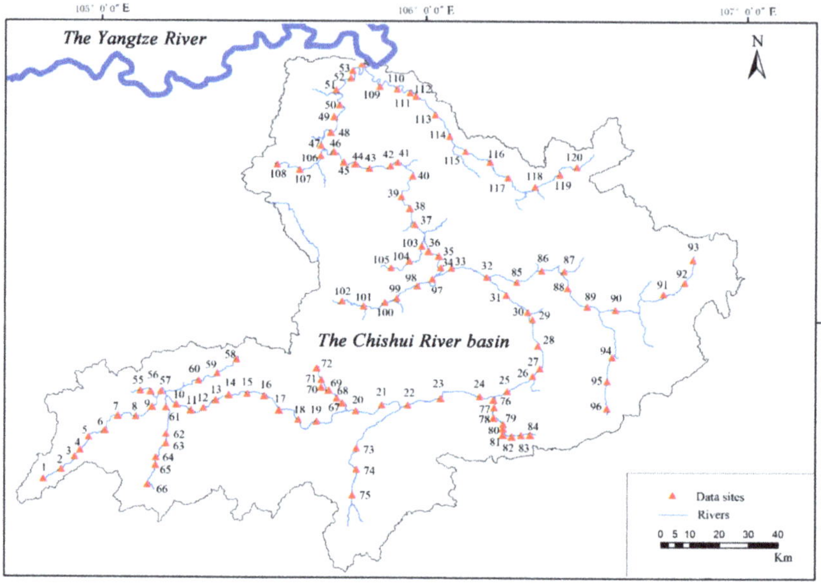

Figure 2. Locations of data sites along the Chishui River basin.

2.2.2. Parameters Used for the Habitat Classification

Classification parameters that reflect fluvial geomorphology, physical form and hydrological characteristics were selected based on the inherent attributes of classification parameters and the habitat characteristics of mountain channels. With reference to the previous studies [36,37], five parameters were used: stream order, elevation (m), slope (km/km), sinuosity (km/km) and river network density (km/km^2). Elevation reflects the topographic conditions, slope and sinuosity show the river's physical form, and river network density and stream order illustrate the river system's structure. Datasets with elevation and stream order were directly extracted from ArcGIS 10.6, and the other three parameters were defined as follows:

$$P = \frac{(Eu - Ed)}{Lv} \quad (1)$$

where P is the slope, Eu is the elevation of river inlet, Ed is the elevation of river outlet and Lv is the basin centerline (the straight length between the inlet and outlet of the river).

$$S = \frac{Lr}{Lv} \quad (2)$$

where S is the sinuosity, Lr is the river centerline (the actual length between the inlet and outlet of the river) and Lv is the basin centerline (the same as above).

$$D = \frac{L}{A} \quad (3)$$

where D is the river network density, L is the total length of the river network and A is the area of the basin.

2.2.3. River Habitat Classification

To determine the spatial patterns of habitats in the Chishui River basin, data on the above five parameters were used in the following analyses: (1) Analysis of similarity (ANOSIM) was carried out to determine the differences between different site-groups. (2)

Similarity of percentage analysis (SIMPER) was used to identify parameters that were principally responsible for similarities within site-groups [32]. (3) Cluster analysis (a group average hierarchical sorting strategy) and nonmetric multidimensional scaling (NMDS) ordination analysis were used to classify the spatial patterns of habitat in the Chishui River basin [38]. All these steps were performed with the PRIMER 5 software package [39].

Habitat types were classified using the clustering methods above and named based on the most significant characteristic parameter based on river hydromorphology and physical habitats. The rivers were divided into headwater, upstream, midstream, downstream, estuary and tributaries based on the location of the data sites; straight rivers and curved rivers based on sinuosity values; steep mountain rivers and flat rivers based on slope; and sparse river networks and dense river networks based on the river network density [37].

2.3. Habitat Evaluation

2.3.1. Metrics Used for the Habitat Assessment Criteria

We selected 20 metrics in four categories for habitat evaluation based on authoritative research [40,41]: water environmental status (pool form, transparency, water smell, flow regime, water color); river morphology (riverbed type, sedimentation characteristics, silt coverage, embeddedness, sinuosity, river harden and canalization); riparian zone (riparian stability, riparian plant width, riparian plant coverage, dominant vegetation); and human disturbance (sewage outlet, solid waste point, dams and channel engineering, cross-river bridge, residential and industrial area). All of these metrics were finally used to establish habitat assessment criteria as shown in Table 1. We then calculated the metric values for each sampling section.

2.3.2. Habitat Sampling

Using the results of habitat classification and depending on sampling operability, we collected data on habitat types and stream morphology, physical habitats and hydrological characteristics at different spaces from the mainstream and 11 tributaries for a total of 40 sampling sections along the Chishui River (mainstream = 18, tributaries = 22) (Figure 3). Field habitat surveys were conducted in March–May 2021.

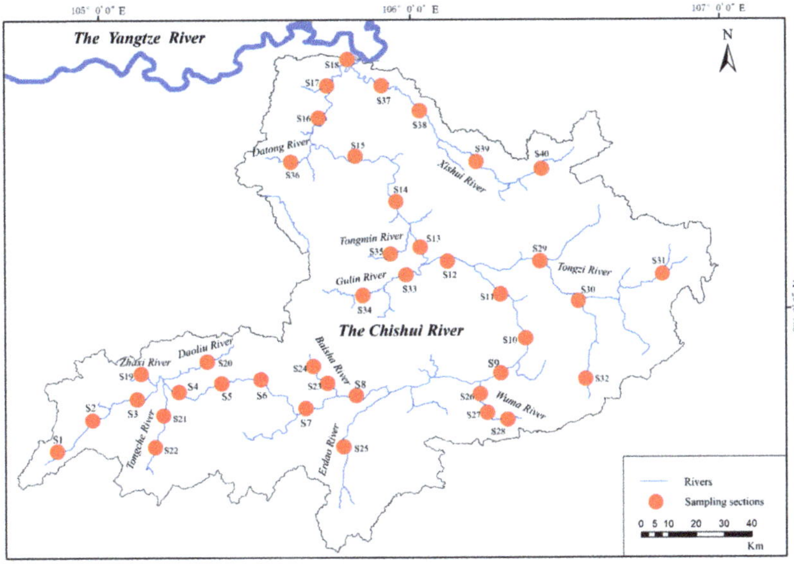

Figure 3. Locations of sampling sections along the Chishui River basin.

Table 1. Twenty metrics and the habitat assessment system.

Habitat Metrics	Environmental Grades			
	Excellent	Good	Fair	Poor
		1 Water environmental status		
Pool form	Balanced combination of deep and shallow pools	Deep pools are more than 50%	Shallow pools are more than 50%	Pools are less than 20% or there is no pools
Transparency	Clear, transparency > 1 m or	Relatively clear, transparency is 0.5–1 m	Relatively turbid, transparency is 0.3–0.5 m	Very turbid, transparency < 0.3 m
Water smell	No odor	Has a slight peculiar smell	Has obvious smell	Pungent smell
Flow regime	All 4 flow regimes appear (slow-deep, slow-shallow, fast-deep, fast-shallow), slow is <0.3 m/s, and deep is >0.5 m	Only 3 flow regimes appear (lack of fast-shallow flow regimes have lower scores than others)	Only 2 flow regimes appear (lack of fast-shallow or slow-shallow have lower scores)	Dominated by 1 flow regime (deep-slow have lower score)
Water color	Water color is transparent, no precipitation after standing	Water color is basically transparent, and there are some precipitation after standing	The water is darker and more algae	Water color is black or eutrophication
		2 River morphology		
Riverbed type	Deep pools and shallow pools	Flat riverbed	Ladder flow or reservoir	Dry riverbed
Sedimentation characteristics	70% of the sediment suitable for aquatic organisms to settle (root plants, or gravels with a diameter of 20–256 mm)	40–70% of the sediment suitable for aquatic organisms to settle (newly fallen trees are not suitable for settlement and have lower score)	The stable habitat is 20–40%, and the heterogeneity of bottom is low (such as the bedrock > 4000 mm)	The stable habitat is only less than 20%, the availability of substrate is low, and it is often severely disturbed or lacked.
Silt coverage	Silt substrate is less than 10%	Silt substrate is about 10–40%	Silt substrate is about 40–60%	Silt substrate is more than 60%
Embeddedness	0–25% fine-grained sediments (0.06–2.0 mm in diameter) are embedded around pebbles and boulders, and the sediments show spatial diversity	25–50% fine-grained sediment are embedded around pebbles and boulders	50–75% fine-grained sediment are embedded around pebbles and boulders	More than 75% of fine-grained sediments are embedded around pebbles and boulders
Sinuosity	Sinuosity is higher than 2.0, and the river course is extremely curved	Sinuosity is 1.5–2.0, and the river course is curved	Sinuosity is 1.2–1.5, the river course is slightly curved	Sinuosity is lower than 1.2, and the river course is straight
River harden and canalization	Without channelization and hardening, river remains in natural condition	Hardened river course is less than 40%	Hardened river course is 40–80%	Hardened river course is more than 80%
		3 Riparian zone		
Riparian stability	River bank is stable and the erosion area is less than 10%	River bank is relatively stable and the erosion area is 10–30%	River bank is unstable and the erosion area is less than 30–60%	River bank has collapsed and the erosion area is more than 60%
Riparian plant width	Riparian plant width is higher than 18 m	Riparian plant width is 12–18 m	Riparian plant width is 6–12 m	Riparian plant width is less than 6 m
Riparian plant coverage	90% of the coastal zone with 50m of vegetation cover and various species	70–90% of the coastal zone with 50m of vegetation cover and single species	50–70% of the coastal zone with 50m of vegetation cover and partly exposed	Less than 50% of the coastal zone with 50m of vegetation cover and most exposed
Dominant vegetation	There are more than 50% arbor forest	There are more than 50% shrubbery	There are more than 50% grassland	There are more than 50% farmland
		4 Human disturbance		
Sewage outlet	There is no any sewage outlet	There is 1 sewage outlet in the river section, which is slightly polluted	There are 2 sewage outlets in the river section with obvious pollution	There are more than 3 sewage outlets in the river section
Solid waste point	Keep neat and no garbage	There are scattered garbage fragments	There is 1 solid waste point, but not yet extended into the river	There are more than 2 solid waste points and have spread to the river
dams and channel engineering	There is no dam or channel engineering, no shipping activities	There is 1 dam or channel engineering, and occasional shipping activities	There are 2 dams or channel engineering and shipping activities are more frequent	There are 3 dams or channel engineering and some shipping terminals
Cross-river bridge	There is no bridge across the river	There is 1 bridge across the river	There are 2 bridges across the river, and at least one bridge has occasional traffic	There are 3 bridges across the river, and at least one bridge has frequent traffic with loud noise
Residential area and industry	There are only less than 10% residential or industrial areas in the 50 m of coastal zone	There are 10–30% residential or industrial areas in the 50 m of coastal zone	There are 30–50% residential or industrial areas in the 50 m of coastal zone	There are more than 50% residential or industrial areas in the 50 m of coastal zone
Score	10 9	8 7 6	5 4 3	2 1 0

For each sampling section, three survey units within the visible range (approximately 500–800 m) were randomly selected, and each unit was treated as a sample square (including the left and right banks) 100 m long. Due to the high heterogeneity of the river habitat, multiple sampling tools were adopted during surveys, included cameras, GPS devices, laser rangefinders, telescopes, water harvesters, bottom dip nets and mud harvesters.

2.3.3. Determination of Comprehensive Habitat Quality Index (CHQI)

To quantify the assessment process, we calculated the comprehensive habitat quality index (CHQI) to evaluate how degenerated the river habitat was using the formula below. To avoid any subjectivity in the discrete scoring, all metrics were scored on a continuous scale from 0 to 200 [42,43]. Each metric received a score of 0–10 based on the habitat assessment criteria (Table 1). The scores for each metric were summed to obtain CHQI, which were divided into five grades based on the relevant habitat score standard: (1) CHQI > 150: excellent, (2) 120 < CHQI ≤ 150: good, (3) 90 < CHQI ≤ 120: fair, (4) 60 < CHQI ≤ 90: poor and (5) CHQI ≤ 60: bad [42–44].

$$CHQI(\text{section}) = \frac{\sum_i^{n=20} Gi\ (\text{three units})}{3} \tag{4}$$

where *CHQI* is the value of the sampling section habitat and *Gi* denotes the value for each metric. *CHQI* can be determined as the average of the scores from all three units in each sampling section.

3. Results

3.1. Spatial Habitat Types

River fragments and subbasins were calculated by spatial analysis in ArcGIS software. A total of 109 river fragments 0.08–59.45 km long were obtained and finally combined into the mainstream and 11 tributaries of the Chishui River in this study. In addition, the Chishui River basin was divided into eight subbasins with areas of 896.59–4732.95 km^2 and circumferences of 140.18–427.62 km.

All the 120 data sites were divided into six sub-groups based on the spatial patterns of the river habitats: group 1 (G1): steep tributaries habitat (15 data sites); group 2 (G2): high-altitude headwater habitat (41 data sites); group 3 (G3): upstream dense river net habitat (12 data sites); group 4 (G4): midstream low-curved habitat (27 data sites); group 5 (G5): low-altitude estuary tributaries habitat (3 data sites); and group 6 (G6): downstream flat habitat (22 data sites). These are shown in Figure 4 and Table 2 based on cluster and ordination analyses. ANOSIM ($p < 0.05$) and the stress value of NMDS was 0.05 (less than 0.2), which further confirmed that the six sub-groups were significantly different [32]. The sub-groups were named based on the following characteristics: group 1 had the highest mean slope value (0.0186) and lowest mean stream order value; data sites in group 2 were characterized by high elevation (996.49 m); group 3 had a high mean river network density value (0.10) whereas group 4 had a low mean river network density value (1.24); groups 5 and 6 were characterized by low elevation and low slope value, and group 5 was located in an estuary of Chishui River (Table 2). The results of SIMPER analysis showed that data sites within each group have high similarity (from 96.37–98.77%). Group 2 and group 6 have the highest average dissimilarity (14.12%) (Table 3).

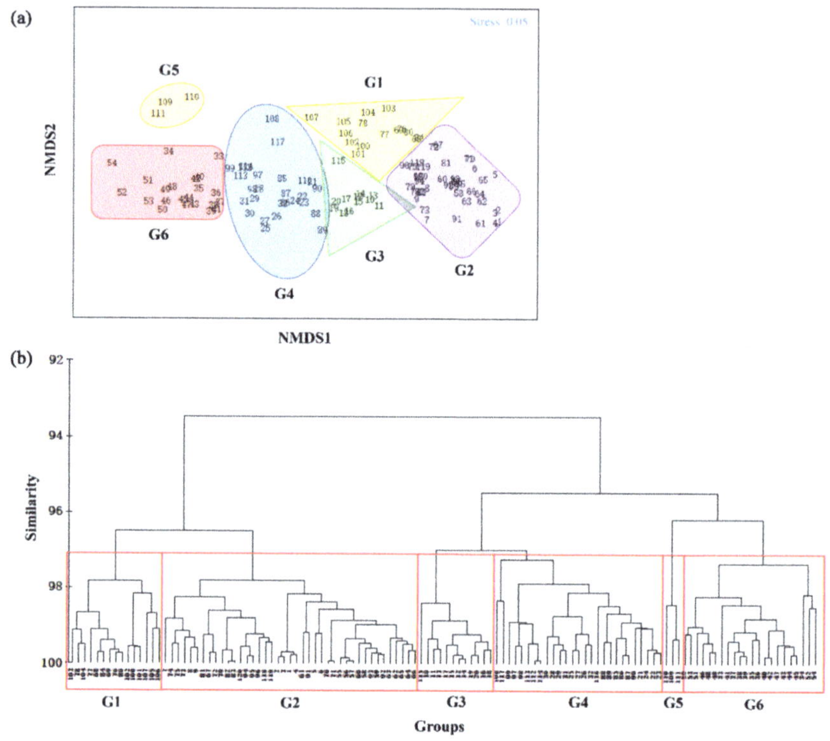

Figure 4. The NMDS ordination (**a**) and classification (**b**) plots of the river habitats in the Chishui River basin.

Table 2. Five parameters of habitat classification among the six sub-groups.

Habitat Types	Number of Data Sites	Proportion	Average Elevation (m)	Average Sinuosity (km/km)	Average Slope (km/km)	Average River Network Density (km/km^2)	Average Stream Order
Steep tributaries habitat	15	12.50%	484.60	1.41	0.0186	0.062	1
High-altitude headwater habitat	41	34.17%	996.49	1.28	0.0119	0.073	1.15
Upstream dense river net habitat	12	10.00%	709.67	1.24	0.0081	0.100	2.92
Midstream low-curved habitat	27	22.50%	397.26	1.24	0.0031	0.088	3.04
Low-altitude estuary tributaries habitat	3	2.50%	227.67	2.16	0.0020	0.164	2
Downstream flat habitat	22	18.33%	258.64	1.60	0.0021	0.089	4

Table 3. Similarity percentages of SIMPER analysis.

Groups	Average Similarity (%)	Groups	Average Dissimilarity (%)
Group 1	97.89	Groups 1 and 2	5.39
Group 2	97.13	Groups 1 and 3	7.16
Group 3	98.47	Groups 1 and 4	6.99
Group 4	96.37	Groups 1 and 5	9.69
Group 5	98.77	Groups 1 and 6	10.74
Group 6	97.78	Groups 2 and 3	6.09
		Groups 2 and 4	9.82
		Groups 2 and 5	13.48
		Groups 2 and 6	14.12
		Groups 3 and 4	4.92
		Groups 3 and 5	10.75
		Groups 3 and 6	8.3
		Groups 4 and 5	7.79
		Groups 4 and 6	5.14
		Groups 5 and 6	6.36

Our results also showed that the high-altitude headwater habitat (G2) has the highest proportion (34.17%), while the low-altitude estuary tributaries habitat (G5) had the lowest proportion (2.50%), so we created 14 and three sampling sections, respectively, in these habitat types during field surveys for habitat assessment. In addition, five, four, nine and seven sampling sections were adopted in the corresponding habitat types below: steep tributaries habitat (G1), upstream dense river net habitat (G3), midstream low-curved habitat (G4) and downstream flat habitat (G6). A total of 40 representative sampling sections belonging to six habitat types were finally set to simplify habitat assessment across the Chishui River basin (Figure 3, Table 2).

3.2. CHQI and Habitat Health

The final CHQI scores for the 40 sampling sections ranged from 75 to 120 with the mean $\pm$ SD of 130.66 $\pm$ 24.14. The mean score was between 120 and 150, meaning that the ecological health of the Chishui River basin habitats was classified as good: nine sampling sections were excellent, 18 were good, 10 were fair and three were poor (Figure 5). Among them, a unit of S1 was scored 87 due to poor water quality; units in both S31 and S35 ranged from 75 to 82; and S2–S8, S11 and S25 were classified as having an excellent habitat status with high CHQI scores.

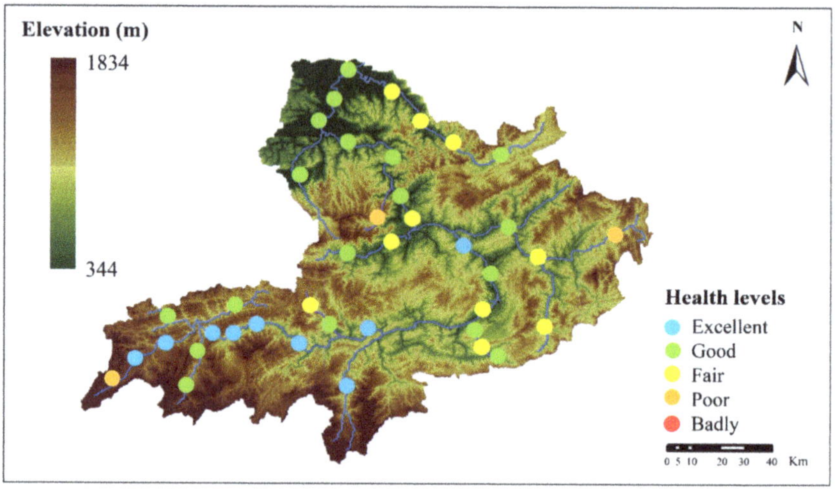

Figure 5. Habitat health status of the 40 sampling sections in the Chishui River basin.

According to the mean ± SD values of CHQI scores, the habitat health conditions for the Chishui River's mainstream and eleven tributaries were: mainstream: 142.74 ± 18.94, Zhaxi River: 136.00 ± 13.49, Daoliu River: 137.00 ± 10.19, Tongche River: 137.67 ± 3.99, Baisha River: 124.00 ± 18.52, Erdao River: 163.67 ± 3.86, Wuma River: 131.11 ± 13.24, Tongzi River: 105.00 ± 20.16, Gulin River: 111.50 ± 16.88, Tongmin River: 79.67 ± 1.69, Datong River: 139.67 ± 3.29 and Xishui River: 110.33 ± 18.85. All three units of S25 in the Erdao River were classified as excellent and no site was considered poor. S31 of Tongzi River and S35 of Tongmin River had six units that were classified as having poor status, whereas eight units of Xishui River were classified as having fair status, accounting for 66.67% of the total units (Figure 6).

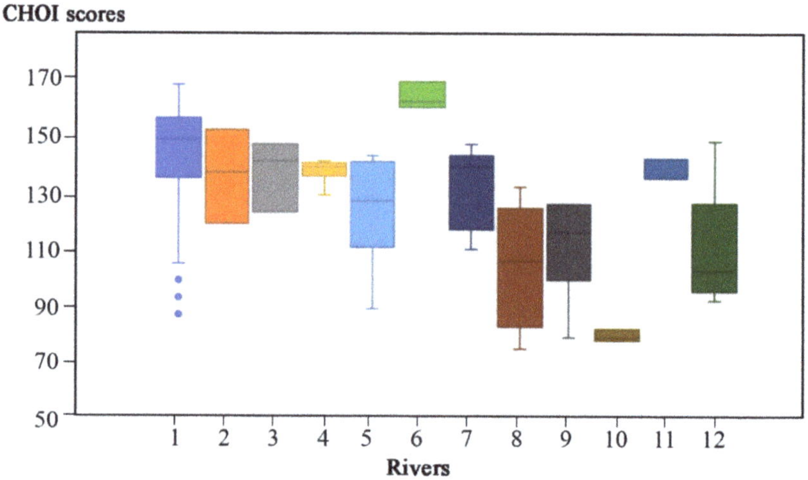

Figure 6. Box plot showing the CHQI score of each stream (1: mainstream, 2: Zhaxi River, 3: Daoliu River, 4: Tongche River, 5: Baisha River, 6: Erdao River, 7: Wuma River, 8: Tongzi River, 9: Gulin River, 10: Tongmin River, 11: Datong River, 12: Xishui River).

3.3. Traits of the Metrics and Habitat Characteristics

The five parameters used for the habitat classification differed among the headwater, upstream, midstream, downstream and tributaries of the Chishui River [32]. Our results from the 120 data sites showed that the average elevation of headwater (data sites 1–15, above Potou Town), upstream (data sites 16–30, Potou Town to Maotai Town), midstream (data sites 31–42, Maotai Town to Hushi Town), downstream (data sites 43–54, Fuxing Town to Hejiang Town) and tributaries (data sites 55–120) was 1101.6 ± 299.92 m, 509.73 ± 113.02 m, 295.87 ± 46.19 m, 226.56 ± 11.81 m and 652.92 ± 289.18 m, respectively. Among the 11 tributaries, Tongche River had the highest average elevation (1138.33 m), whereas Datong River had the lowest average elevation (302.33 m). The mean sinuosity was 1.36±0.38, ranging from 1.09 to 2.81; sinuosity was highest in Hejiang Town and lowest in Maotai Town. On the whole, tributaries and downstream had higher sinuosity, followed by the midstream, whereas upstream and the headwater had the lowest. The highest slope (0.034) appeared in Jianzhu Town (Baisha River), and the lowest (0.0001) appeared in Changsha Town (Xishui River). In general, the slopes were highest in the headwater, Baisha River, Erdao River and Wuma River, and lowest in the midstream, downstream, Xishui River, Datong River and Tongmin River. The average river network density was 0.083 ± 0.02 (km/km^2), between 0.04 and 0.16. Generally, the headwater and upstream had higher densities of river networks, followed by the midstream, downstream, and tributaries. The Chishui River basin consisted of four grades of stream order due to the Strahler algorithm and the natural growth of the river system (Figures 2 and 7).

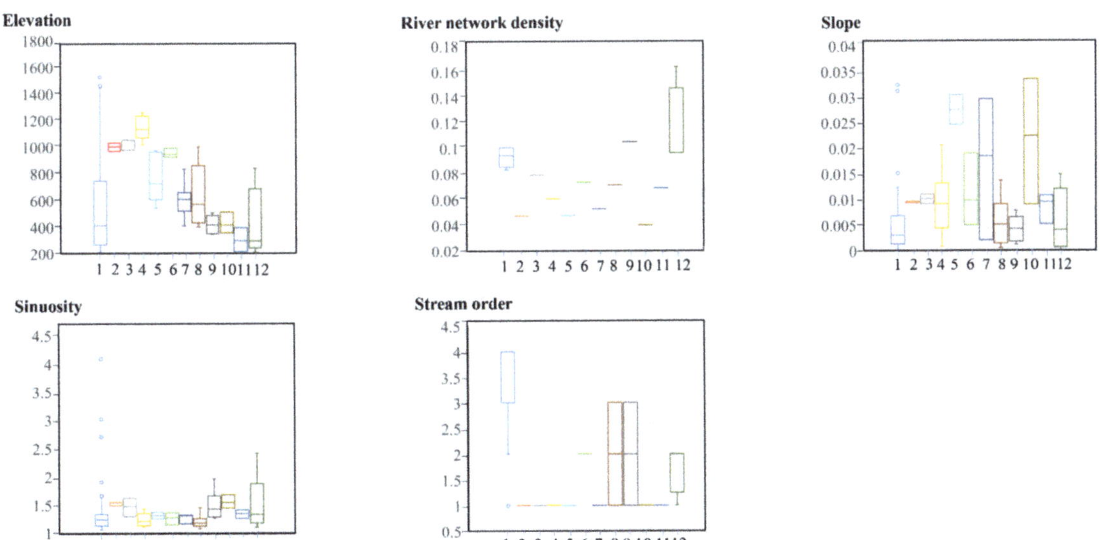

Figure 7. Box plots showing the five classification parameters for each stream (1: mainstream, 2: Zhaxi River, 3: Daoliu River, 4: Tongche River, 5: Baisha River, 6: Erdao River, 7: Wuma River, 8: Tongzi River, 9: Gulin River, 10: Tongmin River, 11: Datong River, 12: Xishui River).

Our results showed that the 20 metrics in four categories that we chose for the habitat assessment ranged in average scores between 3.70 and 7.86, with river harden and canalization being the greatest and sinuosity being the lowest, of which metrics (riverbed type, river harden and canalization, riparian stability, riparian plant coverage, dominant vegetation) had the higher scores, all greater than 7. However, the average scores of the metrics (flow regime, sinuosity, dams and channel engineering) were lower, all less than 6. In addition, results of these four metric categories showed that the average score of the riparian zone was 7.17, whereas the parallels of water environmental status, river morphology and human disturbance were 6.28, 6.46 and 6.36, respectively. In summary, their higher scores suggested that riparian stability and riparian plant width were the most stable metrics in the Chishui River basin. The results are shown in Figure 8.

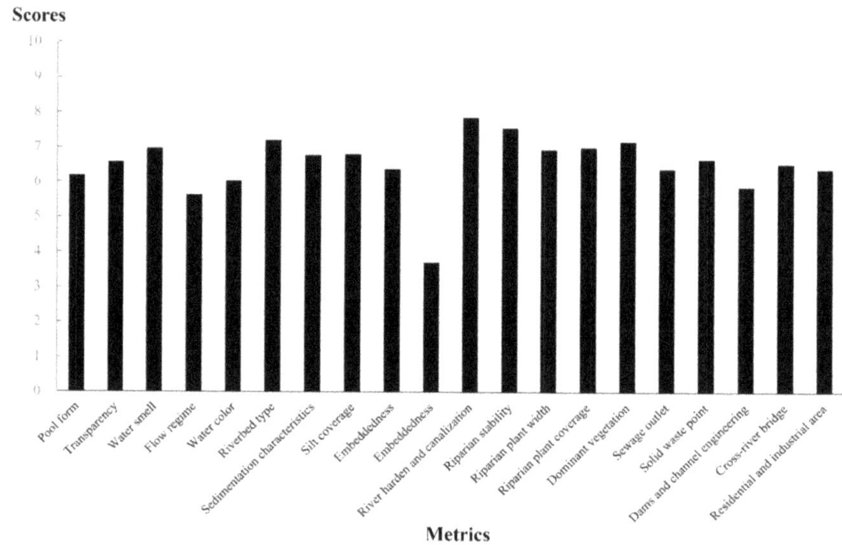

Figure 8. Scores of the 20 metrics that were used for the habitat assessment.

We also found spatial differences among the 20 metrics used for the habitat assessment in the headwater, upstream, midstream, downstream and tributaries. The scores of some metrics (pool form, transparency, water smell, flow regime, water color) in the Chishuiyuan Town section (S1) were significantly lower than those of other sampling sections. The scores of some metrics (sewage outlet, residential and industrial area) in Maotai Town (S9) were lower than those of other sampling sections. Among the tributaries, Erdao River had the best habitat health due to its high-scoring metrics, whereas Tongmin River, Tongzi River and Xishui River had the poorest scores due to disturbances from various human activities.

4. Discussion

4.1. Performance of Our Habitat Assessment System

A more refined and convenient habitat assessment system is required to research river hydromorphology and the physical habitats of mountain channels [5]. Principal characteristics of our habitat assessment system are that it (1) simplifies the habitat assessment by using a suitable habitat classification scheme, (2) classifies this basin into six types of habitats based on high-resolution satellite imagery, (3) synthesizes the cumulative metrics of a wide variety of environmental disturbances that match river hydromorphology and physical habitats and (4) provides universal research methods for other mountain river systems. However, there are some potential limitations of this research to be aware of, including (1) dynamic changes in the river habitat could not be reflected due to a lack of historic data and (2) there was no photographic coverage for a small portion of a few habitat areas, such as data sites 21–24.

Mountain channels are a unique ecosystem characterized by complex and varied habitats. A previous study performed qualitative habitat classification on the Chishui River mainstream based on the longitudinal variations in topography, altitude, climate and vegetation; it concluded that the mainstream is divided into four natural regions: headwater, upstream, midstream and downstream [45]. Our study complements Wang et al. by quantitatively identifying the differences in habitat characteristics in the Chishui River basin (the mainstream and 11 tributaries) [45]. First, six habitat sub-groups that our study defined provide a basis for managing and assessing environmental protection activities. We found considerable differences in the habitat characteristic parameters among the six habitat sub-groups. Of the six sub-groups studied, the high-altitude headwater habitat was the most widespread. Table 2 presents the most characteristic and distinctive attributes.

Most of our study area is located in the eastern Yungui Plateau and is characterized by high mountains, deep valleys and scarce vegetative cover. Upstream dense river net habitats are located on the sloping Yungui Plateau, with a predominance of riffles, rocky bottoms, scarce vegetative cover and developed water system. Midstream low-curved habitat mainly lies in the transitional area between the Yungui Plateau and the Sichuan Basin, with a relatively soft slope and moderate vegetative cover. Downstream flat habitat is located on the edge of the Sichuan Basin, with a predominance of sandy substrate, reduced water flow and lush vegetative cover [33,45]. Steep tributary habitats emerge from each tributary, while low-altitude estuary tributary habitats only appear downstream of Xishui River with the narrowest distribution. In many respects the habitat sub-groups were therefore quite distinct from each other. Additionally, the habitat sub-groups defined by this study could also be useful as the criteria for selecting sampling points scientifically, which would simplify subsequent habitat assessments. Based on experience from previous surveys of the Chishui River basin, some sampling sections were found to be difficult to reach and collect data from due to their complicated fluvial geomorphology and inconvenient traffic, e.g., areas that were between data sites 21 (Qingchi Town) and 24 (Maotai Town) (Figure 2) and areas that lay upstream of Tongzi River. All these areas are characterized by steep mountains and long canyons. Therefore, we used remote sensing technology to identify sampling sections that were representative of these areas. Data sites 20 and 25–32, having convenient traffic, were selected to represent data sites 21–24 since all data sites from 21 to 32 belonged to group 4: midstream low-curved habitat (Figures 2 and 4). We thus sampled the sections S8–S11 to reflect the habitat of these areas where we went to actually sample (Figure 3). Similarly, 120 data sites were finally combined into 40 sampling sections to create a scientifically simplified habitat assessment of mountain channels.

Using 20 metrics that reflect the quality of hydrology, channel, riparian and direct human activities, we created a habitat assessment system that is more comprehensive than previous ones [40,41]. The results obtained using these metrics are easy to translate into values that are meaningful to the general public [43]. It is worth mentioning that quantitative metrics based on different factors of interference—such as transparency, silt coverage and sewage outlet—responded noticeably to human disturbance. Unlike the quantitative metrics, several qualitative metrics also appeared in our studies, such as water smell, water color and riverbed type, because they are capture ecological and environmental differences among habitats that other metrics do not [46]. Therefore, we argue that these 20 metrics complement each other well. Additionally, we only sampled in March–May because, in reality, physical habitats are stable since they have a longer cycle time of change, whereas hydrological characteristics are highly dynamic and change seasonally as rainfall varies. Due to rainfall, mountain rivers are divided into wet periods and dry periods, and the Chishui River is a typical rain-source river [47]. Even though changes in natural water levels also alter habitats, we wanted to track how human disturbances, but not natural changes, impact these river habitats. A previous study suggested that, to reflect general river habitat and hydrological characteristics objectively, habitat sampling should be conducted during periods with stable hydrology, such as at the transition of spring and summer (March–May), because flow conditions and hydrological fluctuations were less dynamic then [46].

Similarly, there are a large number mountain channels like the Chishui River in the upper Yangtze River, such as the Han River, the Jinsha River, and the Dadu River [31,32,48]. Characterized by complex and varied river habitats, they are also hard to research since complicated fluvial geomorphology and inconvenient traffic limit the collection of data and the planning needed to customize management activities for unique ecosystems. Therefore, the steps of developing and simplifying our habitat assessment systems presented herein will be helpful for habitat assessments not only in this region, but also in other mountain channels with similar characteristics related to human disturbance.

4.2. Habitat Status of the Chishui River Basin

Unlike most rivers in China, as a recognized ecological river basin, the Chishui River basin was generally classified as having a good ecological status with a mean CHQI score of 130.66 based on our results. Previous research has shown that areas with relatively low human population density and lush forests are generally healthy, whereas sites with a poor status were densely populated with a high degree of clustering and had certain human activities around that seriously impacted them [49]. Similarly, in our study we found that sites with "excellent" and "good" status were all distributed in sparsely populated regions with a good vegetation coverage and little industrial or agricultural activity, such as in S2 (Guozhu Town), S4 (Shuitian Town), S14 (Hushi Town) and S25 (Malu Town). This may be because they were not impacted much by anthropogenic factors [48]. By comparison, sampling sections with "fair" and "poor" status differed primarily in the extent to which they were urbanized and had industrial and agricultural activity, such as in S1 (Chishuiyuan Town), S31 (Guancang Town), S35 (Tongmin Town) and S37 (Changsha Town).

The mainstream of the Chishui River basin had an average CHQI score of 142.74, which means that it is healthy; nevertheless, it is experiencing a variety of problems. Several sampling sections of the mainstream (S1, S9 and S12) had lower CHQI scores (102.33, 109.33 and 118.67, respectively). Although it had good vegetation cover, S1 (Chishiuiyuan Town) had little runoff and was seriously polluted by local domestic sewage. Based on our investigation, river channels with smelly and polluted water, low transparency, hardening and canalization are poor habitat environments. We found that excessive domestic sewage from Chishiuiyuan Town was leaking into the river channel and destroying its ecological balance. Now, the local government has taken measures to remedy this damage; for example, it prohibited domestic sewage from being discharged into the river channel and built sewage treatment plants in proximity to the stream channel. In addition, S9 (Maotai Town) and S12 (Taiping Town) were classified under the "fair" ecological status and shown to suffer from frequent industries. Without management measures, these areas will likely be classified as "poor" in the near future. Maotai Town is well-known around the world for its famous white spirit [33]. In recent years, pollution from wineries and excessive construction activities have enveloped Maotai Town and further damaged its river's health [33]. Commercial shipping and industrial and mining enterprises of Taiping Town also have an impact on habitat health. According to our survey, damage from frequent water transportation and various mining operations are having serious effects on the region's waterways.

Of the 11 tributaries of the Chishui River basin, streams with "excellent" and "good" condition accounted for 4.55% and 54.55%, respectively, whereas streams with a "fair" and "poor" status accounted for only 27.27% and 13.63%, respectively, and no streams were rated "bad." Erdao River and Datong River are the healthiest according to their CHQI scores, mainly because their surrounding areas have lots of forest and little anthropogenic impact from industry and agriculture [49]. Regrettably, Tongmin River, Tongzi River, Xishui River and Gulin River were classified as "fair" or "poor," with average CHQI scores of 79.67, 105.00, 110.33 and 111.55, respectively. As the largest tributary of the Chishui River, several upstream sampling sections of Tongzi River—such as S30 (Huoshigang Town) and S31 (Guancang Town)—were classified as "poor," with CHQI scores of 97.33 and 76.67, respectively. Tongzi River is characterized by high dams and large reservoirs, such as Yangjiayuan Dam and Yuanmanguan Dam, the construction of which has led to the fragmentation of habitats that aquatic organisms depend on and the increasing nonrheophilic and pollution-tolerant species [48] Moreover, S30 and S31 suffer from both industrial sewage (coal industries) and domestic sewage. S35 (Tongmin Town) of Tongmin River suffers severely from human interference, construction and industrial activities. Construction activities have recently increased in Tongmin River, which is further degrading its habitats. Similarly, Gulin River is suffering from other human disturbances, in spite of its current marginal good condition. We found that domestic sewage is gradually polluting the channel of Gulin River. In addition, S37 (Changsha Town) and S39 (Shibao Town) are affected by dams and at least 15 hydropower stations have been built on the Xishui River [32]. It was shown that these cascade dams not only block the migration routes of fishes and reduce the heterogeneity of the habitat, but also cause frequent and irregular fluctuations in water level, habitat size and food resources [50]. More seriously, Xishui River below the Gaodong Dam often dries up. Therefore, effective measures need to be implemented immediately to deal with these problems.

5. Conclusions

The development of classifications and assessments for mountain channel habitats is an ongoing issue, and convenient and effective methods are needed. To solve the limitations of complicated fluvial geomorphology and inconvenient traffic in mountain channels, a suitable habitat classification scheme based on high-resolution satellite imagery was used to simplify the habitat evaluation steps. A total of 40 representative sampling sections belonging to six habitat types were used for habitat assessment across the Chishui River basin. Among them, the high-altitude headwater habitat (G2) had the highest proportion (34.17%), whereas the low-altitude estuary tributaries habitat (G5) had the lowest proportion (2.50%). Data sites 20 and 25–32, having convenient traffic, were selected to represent data sites 21–24, which had complicated fluvial geomorphology, since all data from sites 21 to 32 belonged to group 4: midstream low-curved habitat. The basin was given a mean comprehensive habitat quality index (CHQI) score of 130.66 ± 24.14 and classified under the status "good." However, the headwaters, Tongmin River, Tongzi River and Xishui River were disturbed by various human activities. We believe that the process of developing and simplifying our habitat assessment systems presented herein will be helpful for ecosystem assessment, not only in this region but also in other mountain channels with similar characteristics related to human disturbance.

Author Contributions: J.W., F.Y. and F.L. conceived and designed the investigation; F.Y., F.L., Z.X., C.X., and P.L. performed field work; F.Y., Z.X. and J.W. analyzed the data; M.H. and X.Z. contributed materials and analysis tools; Writing—original draft preparation, F.Y. and J.W.; Writing—review and editing, J.W. and F.Y.; project administration, J.W. and F.L. All authors have read and agreed to the published version of the manuscript.

Funding: This study was supported by grants from the Biodiversity Survey and Assessment Project of the Ministry of Ecology and Environment, China (2019HJ2096001006), the Ministry of Agriculture and Rural Affairs of China (CJDC-2017), the China Three Gorges Corporation (0799574) and Sino BON-Inland Water Fish Diversity Observation Network.

Institutional Review Board Statement: Not applicable.

Informed Consent Statement: Not applicable.

Data Availability Statement: The datasets used and/or analyzed during the current study are available from the corresponding author upon reasonable request.

Acknowledgments: The authors thank constructive advice from Wenjing Li, Zhen Wang and Xiao Qu for improving the manuscript.

Conflicts of Interest: The authors declare that they have no conflict of interest.

References

1. Surell, A. *Étude sur les Torrents des Hautes-Alpes*; Carilian-Goeury: Paris, France, 1841.
2. Dana, J.D. On denudation in the Pacific. *Am. J. Sci.* **1850**, 2, 48–62.

3. Shaler, N.S. *The Origin and Nature of Soils*; U.S. Geological Survey 12th Annual Report. 1891; pp. 213–345. Available online: https://xueshu.baidu.com/usercenter/paper/show?paperid=893ebca413bf00a03b4c0fa65ca4a84e&site=xueshu_se&hitarticle=1 (accessed on 3 February 2022).
4. Flotemersch, J.E.; North, S.; Blocksom, K.A. Evaluation of an alternate method for sampling benthic macroinvertebrates in low-gradient streams sampled as part of the National Rivers and Streams Assessment. *Environ. Monit. Assess.* **2014**, *186*, 949–959. [CrossRef] [PubMed]
5. Montgomery, D.R.; Buffington, J.M. Channel-reach morphology in mountain drainage basins. *Geol. Soc. Am. Bull.* **1997**, *109*, 596–611. [CrossRef]
6. Rosgen, D.L. A Classification of Natural Rivers. *Catena* **1994**, *22*, 169–199. [CrossRef]
7. Hughes, R.M.; Herlihy, A.T.; Kaufmann, P.R. An Evaluation of Qualitative Indexes of Physical Habitat Applied to Agricultural Streams in Ten U.S. States1. *JAWRA J. Am. Water Resour. Assoc.* **2010**, *46*, 792–806. [CrossRef]
8. Belletti, B.; Rinaldi, M.; Bussettini, M.; Comiti, F.; Gurnell, A.; Mao, L.; Nardi, L.; Vezza, P. Characterising physical habitats and fluvial hydromorphology: A new system for the survey and classification of river geomorphic units. *Geomorphology* **2017**, *283*, 143–157. [CrossRef]
9. Lee, B.; Leroy, P.N.; Daniel, M.; Thomas, D.; Gordon, R.; George, P.; Michael, P. The Network Dynamics Hypothesis: How Channel Networks Structure Riverine Habitats. *BioScience* **2004**, *54*, 413–427.
10. Soranno, P.A.; Spence, C.K.; Webster, K.E.; Bremigan, M.T.; Wagner, T.; Stow, C.A. Using Landscape Limnology to Classify Freshwater Ecosystems for Multi-ecosystem Management and Conservation. *BioScience* **2010**, *60*, 440–454. [CrossRef]
11. Brinkhurst, R.O. *The Benthos of Lakes*; St. Martin's Press: New York, NY, USA, 1974.
12. Shuter, B.J.; Jones, M.L.; Korver, R.M.; Lester, N.P. A general, life history based model for regional management of fish stocks: The inland lake trout fisheries (Salvelinus namaycush) of Ontario. *Can. J. Fish. Aquat. Sci.* **1998**, *55*, 2161–2177. [CrossRef]
13. Euliss, N.H.; LaBaugh, J.W.; Fredrickson, L.H.; Mushet, D.M.; Laubhan, M.K.; Swanson, G.A.; Winter, T.C.; Rosenberry, D.O.; Nelson, R.D. The wetland continuum: A conceptual framework for interpreting biological studies. *Wetlands* **2004**, *24*, 448–458. [CrossRef]
14. Wolfgang, J.J.; Maria, T.; Fernandez, P.; Schoengart, J.; Wittmann, F. A classification of major natural habitats of Amazonian white-water river floodplains (va´rzeas). *Wetl. Ecol. Manag.* **2012**, *20*, 461–475.
15. Davenport, A.J.; Gurnella, A.M.; Armitage, P.D. Habitat survey and classification of urban rivers. *River Res. Appl. River Res. Appl.* **2004**, *20*, 687–704. [CrossRef]
16. Dudgeon, D.; Arthington, A.H.; Gessner, M.O.; Kawabata, Z.; Knowler, D.J.; Leveque, C.; Naiman, R.J.; Prieur Richard, A.H.; Soto, D.; Stiassny, M.L. Freshwater biodiversity: Importance, threats, status and conservation challenges. *Biol. Rev.* **2010**, *81*, 163–182. [CrossRef]
17. Corlett, R.T. Tropical rainforests and climate change. *Ref. Modul. Earth Syst. Environ. Sci.* **2016**, 1–5.
18. Ormerod, S.; Dobson, M.; Hildrew, A.; Townsend, C. Multiple stressors in freshwater ecosystems. *Freshw. Biol.* **2010**, *55*, 1–4. [CrossRef]
19. Gomes-Silva, G.; Cyubahiro, E.; Wronski, T.; Riesch, R.; Apio, A.; Plath, M. Water pollution affects fish community structure and alters evolutionary trajectories of invasive guppies (Poecilia reticulata). *Sci. Total Environ.* **2020**, *730*, 138912. [CrossRef] [PubMed]
20. Ngor, P.B.; McCann, K.S.; Grenouillet, G.; So, N.; McMeans, B.C.; Fraser, E.; Lek, S. Evidence of indiscriminate fishing effects in one of the world's largest inland fisheries. *Sci. Rep.* **2018**, *8*, 1–12. [CrossRef] [PubMed]
21. Ruaro, R.; Gubiani, É.A. A scientometric assessment of 30 years of the Index of Biotic Integrity in aquaticecosystems: Applications and main flaws. *Ecol. Indic.* **2013**, *29*, 105–110. [CrossRef]
22. Bovee, K.D. *A Guide to Stream Habitat Analysis Using the Instream Flow Incremental Methodology*; Instream flow information paper No.12. FWS/OBS-82-26; US Fish and Wildlife Service Biological Services Program: Bailey's Crossroads, VA, USA, 1982.
23. Hill, J.; Grossman, G.D. An energetic model of microhabitat use for rainbow trout and rosyside dace. *Ecology* **1993**, *74*, 685–698. [CrossRef]
24. Maddock, I. The importance of physical habitat assessment for evaluating river health. *Freshw. Biol.* **1999**, *41*, 373–391. [CrossRef]
25. Vaughan, I.P. Habitat indices for rivers: Derivation and applications. *Aquat. Conserv.* **2010**, *20*, 4–12. [CrossRef]
26. Barbour, M.T. Biological assessment strategies: Applications and limitations. In *Whole Effluent Toxicity Testing: An Evaluation of Methods and Prediction of Receiving System Impacts*; Grothe, D.R., Dickson, K.L., Reed, D.K., Eds.; SETAC Press: Pensacola, FL, USA, 1996; pp. 245–270.
27. Davies, N.M.; Norris, R.H.; Thoms, M.C. Prediction and assessment of local stream habitat features using large-scale catchment characteristics. *Freshw. Biol.* **2000**, *45*, 343–369. [CrossRef]
28. Oliveira, S.V.; Cortes, R.M.V. A biologically relevant habitat condition index for streams in northern Portugal. *Aquat. Conserv.* **2005**, *15*, 189–210. [CrossRef]
29. He, Y.F.; Wang, J.W.; Lek, S.; Cao, W.X.; Lek-Ang, S. Structure of endemic fish assemblage in the upper Yangtze River Basin. *River Res. Appl.* **2011**, *27*, 59–75. [CrossRef]
30. Lin, P.C.; Liu, F.; Li, M.Z.; Gao, X.; Liu, H.Z. Spatial pattern of fish assemblages along the river-reservoir gradient caused by the Three Gorge Reservoir (TGR). *Acta Hydrobiol. Sin.* **2018**, *42*, 1124–1134.
31. Zhang, X.; Gao, X.; Wang, J.W.; Cao, X.W. Extinction risk and conservation priority analysis for 64 endemic fishes in the Upper Yangtze River, China. *Environ. Biol. Fishes* **2015**, *98*, 261–272. [CrossRef]

32. Liu, F.; Wang, J.; Zhang, F.B.; Liu, H.Z.; Wang, J.W. Spatial organization of fish assemblages in the Chishui River, the last free-flowing tributary of the upper Yangtze River, China. *Ecol. Freshw. Fish* **2020**, *30*, 1–13.
33. Wu, J.; Wang, J.; He, Y.; Cao, W. Fish assemblage structure in the Chishui River, a protected tributary of the Yangtze River. *Knowl. Manag. Aquat. Ecosyst.* **2011**, *400*, 170–181. [CrossRef]
34. Jiang, X.M.; Xiong, J.; Xie, Z.C.; Chen, Y.F. Longitudinal patterns of macroinvertebrate functional feeding groups in a Chinese river system: A test for river continuum concept (RCC). *Quat. Int.* **2011**, *244*, 289–295. [CrossRef]
35. Tang, G.A.; Yang, X. *Experimental Tutorial on Spatial Analysis of Geographic Information System*; Science Press: Beijing, China, 2012.
36. Higgins, J.V.; Bryer, M.T.; Khoury, M.L.; Fitzhugh, T.W. A freshwater classification approach for biodiversity conservation planning. *Conserv. Biol.* **2005**, *19*, 432–445. [CrossRef]
37. Kong, W.J.; Zhang, Y.; Wang, Y.H. River habitats classification in Taizi River based on spatial data. *Res. Environ. Sci.* **2013**, *26*, 487–493.
38. Oksanen, J.; Blanchet, F.G.; Friendly, M.; Kindt, R.; Legendre, P.; McGlinn, D.; Solymos, P. Vegan: Community Ecology Package; R package version 2.5-7. 2020. Available online: https://xueshu.baidu.com/usercenter/paper/show?paperid=1h4c0en0bb6y0ar0052v0ae03k594560 (accessed on 3 February 2022).
39. Clarke, K.R.; Warwick, R.M. *Change in Marine Communities: An Approach to Statistical Analysis and Interpretation*; Plymouth Marine Laboratory: Plymouth, UK, 1994; p. 144.
40. Barbour, M.T.; Gerritsen, J.; Snyder, B.D.; Stribling, J.B. *Rapid Bioassessment Protocols for Use in Stream and Wadeable Rivers: Periphyton, Benthic Macroinvertebrates and Fish*, 2nd ed.; EPA 841-B-99-002; U.S. Environmental Protection Agency, Office of Water: Washington, DC, USA, 1999.
41. Taft, B.; Koncelik, J.P. *Methods for Assessing Habitat in Flowing Waters: Using the Qualitative Habitat Evaluation Index (QHEI)*; OHIO EPA Technical Bulletin EAS.; State of Ohio Environmental Protection Agency, Division of Surface Water: Washington, DC, USA, 2006.
42. Zheng, B.H.; Zhang, Y.; Li, Y.B. Study of indicators and methods for river habitat assessment of Liao River Basin. *Acta Sci. Circumstantiae* **2007**, *27*, 928–936.
43. Zhou, X.; Yang, R.H.; Yang, Z.; Zheng, Z.W.; Shi, F.; Chi, S.Y.; Zhu, A.M.; Shao, K.; Yuan, Y.J.; Wan, C.Y. Habitat health assessment of typical tributaries of the Yangtze River. *J. Hydroecology* **2021**, *42*, 11.
44. Kwang, G.A.; Seok, S.P.; Shin, J.Y. An evaluation of a river health using the index of biological integrity along with relations to chemical and habitat conditions. *Environ. Int.* **2002**, *28*, 411–420.
45. Wang, Z.S.; Jiang, L.G.; Huang, M.J.; Zhang, C.; Yu, X.B. Biodiversity status and its conservation strategy in the Chishui River basin. *Recourse Environ. Yangtze Basin* **2007**, *16*, 175–180.
46. Zhang, X. Setting conservation priority to endemic fishes and their habitats during the hydropower development in the Lower Jinsha River. Master's Thesis Dissertation, Chinese Academy of Sciences, Beijing, China, 2018.
47. Wang, J.; Huang, Z.L.; Li, H.Y.; Jiang, X.M.; Li, Z.F.; Meng, X.L.; Xie, Z.C. Construction of Macroinvertebrate Integrity-Based Health Assessment Framework for the Chishui River. *Environ. Monit. China* **2018**, *34*, 69–79.
48. Chen, K.; Jia, Y.; Xiong, X.; Sun, H.; Chen, Y. Integration of taxonomic distinctness indices into the assessment of headwater streams with a high altitude gradient and low species richness along the upper Han River, China. *Ecol. Indic.* **2020**, *112*, 106106. [CrossRef]
49. Li, Z.; Zeng, B. Health assessment of important tributaries of Three Georges Reservoir based on the benthic index of biotic integrity. *Sci. Rep.* **2020**, *10*, 18743. [CrossRef]
50. Liu, F.; Yang, G.H.; Liu, D.M.; Li, L.; Wang, X.; Zhang, Z.; Liu, H.Z. Current situation and conservation strategies of fish resources in the Xishui River. *Freshw. Fish.* **2019**, *49*, 36–43.

water

MDPI

Article

Functional Responses and Additive Multiple Predator Effects of Two Common Wetland Fish

Linton F. Munyai [1,*], Tatenda Dalu [2,3,*], Ryan J. Wasserman [3,4], Lutendo Mugwedi [1], Farai Dondofema [1], Gordon O'Brien [2] and Ross N. Cuthbert [3,5]

1 Aquatic Systems Research Group, Department of Geography and Environmental Sciences, University of Venda, Thohoyandou 0950, South Africa; lutendo.mugwedi@univen.ac.za (L.M.); farai.dondofema@univen.ac.za (F.D.)
2 School of Biology and Environmental Sciences, University of Mpumalanga, Nelspruit 1200, South Africa; gordon.obrien@ump.ac.za
3 South African Institute for Aquatic Biodiversity, Makhanda 6140, South Africa; r.wasserman@ru.ac.za (R.J.W.); rossnoelcuthbert@gmail.com (R.N.C.)
4 Department of Zoology and Entomology, Rhodes University, Makhanda 6140, South Africa
5 GEOMAR Helmholtz-Zentrum für Ozeanforschung Kiel, 24105 Kiel, Germany
* Correspondence: munyailinton@gmail.com (L.F.M.); dalutatenda@yahoo.co.uk (T.D.)

Abstract: Understanding trophic interactions is essential for the prediction and measurement of structure and function in aquatic environments. Communities in these ecosystems may be shaped by variables such as predator diversity, prey density and emergent multiple predator effects (MPEs), which are likely to influence trophic dynamics. In this study, we examined the effect of key predatory fish in floodplain wetlands, namely *Oreochromis mossambicus* and *Enteromius paludinosus*, towards Chironomidae prey, using a comparative functional response (FR) approach. We used single predator species as well as intra- and interspecific paired species to contrast FRs under multiple predator scenarios. Attack rate and handling time estimates from single predator FRs were used to predict multiple predators' feeding rates, which were compared to observe multiple predators' feeding rates to quantify potential MPEs. From single fish trials, each species displayed a significant Type II FR, characterized by high feeding rates at low prey densities. *Oreochromis mossambicus* had a steeper (initial slope, i.e., higher attack rate) and higher (asymptote of curve, i.e., shorter handling time and higher maximum feeding rate) FR, whereas *E. paludinosus* exhibited lower-magnitude FRs (i.e., lower attack rate, longer handling time and lower feeding rate). In multiple predator scenarios, feeding rates were well-predicted by those of single predators, both in conspecific and interspecific pairs, and thus we did not find evidence for antagonistic or synergistic MPEs. Predator–prey interactions in wetland systems can have significant consequences on the structure and dynamics of ecological communities. In turn, this could have destabilizing effects on resources in tropical wetlands. These results, although experimental, help us understand how trophic interaction among conspecific or interspecific fish species in Austral tropical wetlands might influence their aquatic prey species. This will help us to understand food web dynamics better.

Keywords: consumer-resource dynamics; feeding rates; *Oreochromis mossambicus*; *Enteromius paludinosus*; predator-prey dynamics; prey risk

Citation: Munyai, L.F.; Dalu, T.; Wasserman, R.J.; Mugwedi, L.; Dondofema, F.; O'Brien, G.; Cuthbert, R.N. Functional Responses and Additive Multiple Predator Effects of Two Common Wetland Fish. *Water* **2022**, *14*, 699. https://doi.org/10.3390/w14050699

Academic Editors: Eva Papastergiadou and Kostas Stefanidis

Received: 22 January 2022
Accepted: 21 February 2022
Published: 23 February 2022

1. Introduction

Predator–prey dynamics are central to our understanding of how species interact and are pervasive determinants of community structure [1,2]. However, these dynamics in tropical and subtropical wetlands systems have received little scientific interest [3,4]. In most tropical and subtropical regions, wetland environments are widespread and highly diverse, representing a myriad of systems ranging from permanent to ephemeral [5,6]. Floodplain wetlands are especially common in tropical and subtropical regions and are characterized by seasonal rainfall [7]. Unlike most endorheic systems, floodplain connection to permanent water typically facilitates small-bodied fish presence in their food webs during their hydroperiod [8]. In these shallow systems, small-bodied fish typically represent the top of the aquatic food web, often exploiting the rich productivity associated with wetlands during the wet season [9,10]. Multiple predatory species exploit these environments, but little is known on how predator–predator interactions may facilitate or disrupt predator–prey dynamics.

Trophic interactions and food webs are characterized by trait-(non-consumptive) and density-mediated (consumptive) processes, with the former particularly pervasive in aquatic environments due to the presence of water-borne cues [11,12]. Although consideration of both processes is crucial for a holistic understanding of trophic dynamics, most attention has been directed at density-mediated effects. One way in which trait-mediated effects can manifest is through so-called multiple predator effects (MPEs). Most communities are comprised of more than one predator, with predator–predator interactions potentially resulting in altered prey risk [13]. There are three forms in which predator–predator interactions classically manifest, and these include (*i*) additive, where predators interact independently with their prey irrespective of predator density, and thus multiple predator feeding rates are predictable based on individuals (i.e., a lack of MPE); (*ii*) antagonistic MPEs, where predator–predator interference reduces impact and thus alleviates prey risk; and (*iii*) synergistic MPEs, where interactions enhance predatory impacts and, therefore, increase prey risk [1,14].

Functional response quantification is a classical approach in determining consumer-resource dynamics in ecosystems. Functional responses describe consumption rates (e.g., by predators) as a function of resource density (e.g., prey) [15,16]. In doing so, FRs can inform whether the consumer will have the ability to regulate, stabilize or de-stabilize the resource (prey) populations, with implications for population viability in an ecosystem [17]. Several studies (e.g., [18–20]) have used FRs to understand predator–prey interactions. The significance of floodplain fish and their interaction with the natural aquatic ecosystem is less well understood than in other inland wetland habitats. Even though human modifications are well-known to have significantly altered fish community structures and reduced their diversity [21,22], floodplains continue to support diverse fish assemblages and provide an important habitat for many fish species [23]. Furthermore, available data on floodplains habitat and wetland fish assemblages and interactions are still quite limited and much information on the basic ecology of wetland fish associated with these habitats is less studied.

Functional responses have also been used in MPE frameworks, because predation risk from multiple predators may also be inherently influenced by prey density [24,25]. For example, studies by [1,14,25] reported that the MPE sign and strength differed with prey density and that the influence of prey density varied for different prey species. Specifically, at low prey densities, all prey are typically extirpated in non-prey replacement experimental designs, and thus there is little capacity to detect non-trophic interactions (e.g., interference between predator individuals). At intermediate prey densities, competition between predators for limited resources is high, resulting in antagonisms. Conversely, at high prey densities, prey are abundant and thus not extirpated, with predator–predator interactions, and thus MPEs, potentially less pertinent [26].

The use of such an approach for floodplain wetland fauna could thus link the density-dependent dynamics of predator and prey populations [27] and help to understand direct and indirect food-web interactions [28,29] among key individuals (such as macroinvertebrates, freshwater fish and plants) in these systems. Previous studies on predator–prey interactions have mostly focused on MPE relationships between macroinvertebrate predators and their prey [30,31], however, higher trophic levels have lacked extensive examination (but see Wasserman et al. [32] and Mofu et al. [33]). In floodplain wetlands, fish predation is a major pressure on invertebrate prey, particularly during seasons where large numbers of small bodies and young fish species are prevalent. Determining the strength of *per capita* interactions, and predator–predator dynamics, from these fish is thus fundamental for informing comprehensive floodplain wetland food web modelling, providing valuable information on top-down control dynamics in these systems.

This study thus aimed to assess the feeding interactions of the native Mozambique tilapia (*Oreochromis mossambicus*) and straight-fin barb (*Enteromius paludinosus*), towards a readily consumed prey (Chironomidae), under multi-predator scenarios using the comparative FR approach. We sought to examine the potential importance of conspecific and interspecific interaction dynamics for the nature and strength of MPEs towards prey, across a range of prey densities. *Oreochromis mossambicus* and *E. paludinosus* have both been found to be among the most common fish species present in tropical floodplain wetlands' systems [34,35]. We predicted that *O. mossambicus* would likely be more efficient at finding prey at low densities (i.e., higher attack rate) and would have a higher feeding rate throughout than *E. paludinosus*, given that the former is more carnivorous when they are juveniles, while the latter is omnivorous [36,37]. This prediction is supported by the evidence from a previous study which proved that *O. mossambicus* resembles a higher attack rate compared to *E. paludinosus* [38]. We also predicted that interspecific MPEs would be more pronounced than conspecific through interference, since interspecific combination resembles a high consumption rate towards their prey due to competition in the given ecosystem.

2. Materials and Methods

2.1. Animal Collection

In December 2020, experiments were performed in the Pollution Laboratory Atrium at the University of Venda, South Africa. Juvenile *Oreochromis mossambicus* and *Enteromius paludinosus* were collected from local wetland systems around Thohoyandou (i.e., Tshifulanani (−23.041668; 30.400553) and Duthuni (−22.965715; 30.395720)). Fish were captured using a 30-metre seine net (mesh 0.5 cm) and transported in 25 L plastic containers filled with source water to the laboratory. The two species were kept separately in 4 × 25 L open buckets with 10 fishes being placed per bucket in borehole/wetland (50:50) water. All fish were acclimatized and starved for 48 h prior to experiments at 26 ± 1.5 °C. This temperature was chosen based on the recorded temperature of water where fish were collected. Experiments were conducted in individual 10 L polyethylene buckets (navy-blue; 20 cm diameter at the base, 24 cm high) containing 6 L of borehole/wetland (50:50) water. All fish used in the experiment were size-matched according to total length (TL) (*O. mossambicus* (mean ± SD) = 6.4 ± 0.3 cm TL; *E. paludinosus* = 6.9 ± 0.3 cm TL), ensuring no substantial differences in total length between species that might affect feeding rates. Four hours prior to the experiment, random fish of each species were collected from the 25 L buckets and transferred individually and in conspecific/heterospecific combinations into experimental arenas (navy-blue plastic buckets; 20 cm diameter at the base, 24 cm high, containing 9 L borehole/wetland (50:50)) for further acclimatization to experimental arenas in a randomized array.

2.2. Experimental Design

The experimental treatments were (1) *O. mossambicus*, (2) *E. paludinosus*, (3) *O. mossambicus* + *O. mossambicus*, (4) *E. paludinosus* + *E. paludinosus*, (5) *O. mossambicus* + *E. paludinosus* (Figure 1). During the FR experiment, dead whole prey (Chironomidae (mean ± SD) = 10 ± 1.4 mm; Aquav freeze-dried bloodworms-AQUAV (Xiamen Mincheng Imp and Exp Co., Ltd, Beijing, China), were used for all predator treatments. Previous studies have also used dead prey to quantify trophic interaction strengths (e.g., Boets et al. [39]. For each treatment, six prey densities were used, i.e., 2, 4, 8, 16, 32 and 64, consisting of four to six randomized replicates per each fish group (i.e., 5 fish treatments × 6 prey densities × 4–6 replicates). Prey were transferred to the buckets with fish inside and after four h of feeding (12:00 to 16:00), fish were removed from the experimental arenas and the total number of prey remaining enumerated. Conspecific and interspecific pairs were treated as two individuals in one arena (bucket) consuming supplied prey, i.e., as a single predatory unit (Figure 1). In all replicates conducted, fish were only used once. No predators were added for controls ($n = 5$ per prey density, where n is the sample size that quantify prey consumption). After the experiments, all the fish used for experiments were euthanized humanely following recommendations by Weyl et al. [40] and discarded as a biohazard, as stipulated in our approved animal ethics application.

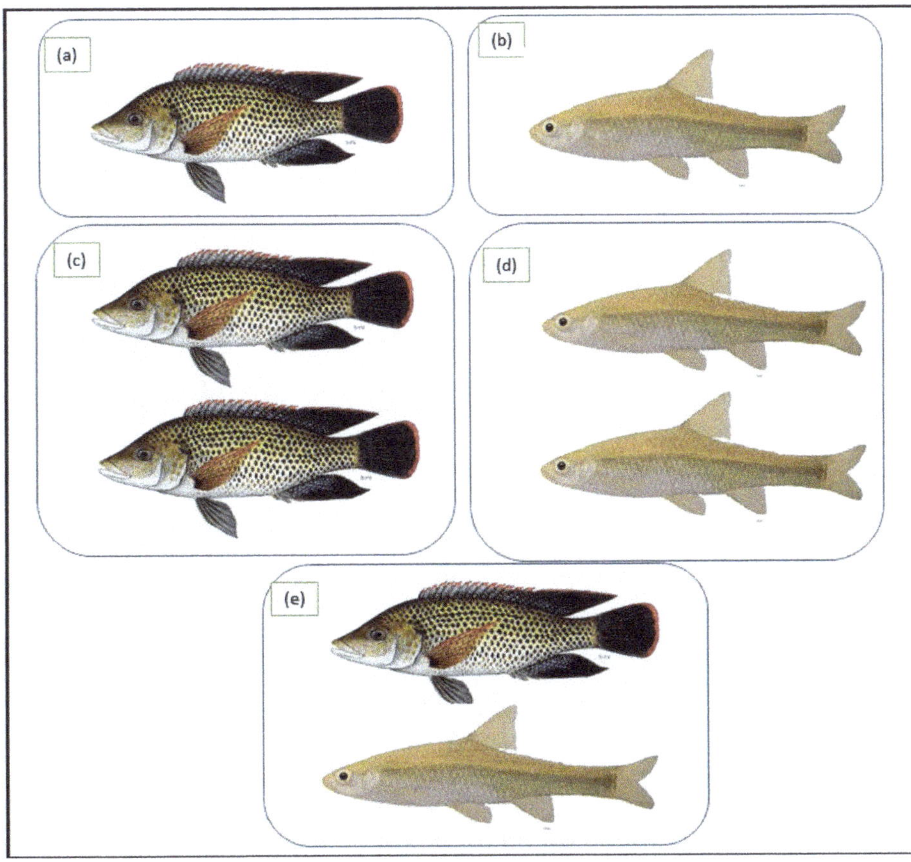

Figure 1. Experimental predator treatments comprising individual predator, conspecific and interspecific pairs of (**a**) *Oreochromis mossambicus;* (**b**) *Enteromius paludinosus;* (**c**) *Oreochromis mossambicus* + *Oreochromis mossambicus;* (**d**) *Enteromius paludinosus* + *Enteromius paludinosus;* (**e**) *Oreochromis mossambicus* + *Enteromius paludinosus.*

2.3. Data Analysis

Differences in proportional feeding rates were examined using a generalized linear model assuming a quasi-binomial error distribution, given residual deviances exceeded degrees of freedom. Predator treatment (five levels) was included as a predictor variable, alongside prey supply (continuous). The interaction term between these factors was not included. Analysis of deviance was used to compute *F*-tests for the resulting model. Tukey comparisons were used *post-hoc* for pairwise comparisons of predator treatments [41].

Binomial generalized linear models were additionally used to categorize FR types for both predator treatments at the single predator density [42,43]. A Type II FR was indicated through the presence of a significantly negative linear coefficient in response to increasing prey density, while a Type III functional response would be indicated by a significant positive first-order term and significant negative second-order term. Given that the prey were not replaced following consumption over the course of the experiment, Rogers' random predator equation was used to model FRs [44]:

$$N_e = N_0(1 - \exp(a(N_e h - T))) \tag{1}$$

where N_e is the number of prey eaten, N_0 is the initial density of prey, a is the attack constant, h is the handling time and T is the total experimental period. The Lambert W function was used to fit the model to the data [43,45]. The random predator equation is robust to prey depletion in parameter estimation [46].

We then used the attack rate and handling time estimates from single predator FRs (1) to predict multiple predators feeding rates, which were next compared to observed multiple predators feeding rates. This was completed separately for multiple predator groups (i.e., Om + Om, Ep + Ep and Om + Ep) using the corresponding single predator FR parameters. Estimations of IS_T (i.e., predicted interactions) were calculated following McCoy et al. [24] and Sentis and Boukal [25]:

$$\frac{dN}{dt} = -\sum_{i=1}^{n} f_i(N) P_i \tag{2}$$

where N is the prey population density, Pi $(i = 1, 2, \ldots, n)$ are the population densities of predators i, and $f_i(N)$ is the functional response of predator i (i.e., Equation (1)). This model assumes no emergent MPEs and its predictions can be compared to multiple predators feeding trials to assess the sign and strength of MPEs. To generate predictions of expected prey survival in the multi-predator experiments, initial values of N and P are set at the experimental initial prey and predator densities corresponding to the experimental treatment. For each predator treatment and prey density, Equation (2) was integrated over the full experimental time to obtain the expected numbers of surviving prey. To estimate the variance around the predictions, we used a global sensitivity analysis that uses the 95% confidence intervals of each FR parameter estimate and their variance-covariance matrix (covariance is assumed to be zero when unknown) to generate 100 random parameter sets using a Latin hypercube sampling algorithm [47]. For each parameter set (n = 100), Equation (2) was then integrated over time and expected prey survival was calculated using the 'sensRange' function in the R package 'FME' [47]. We thus compared the confidence intervals between predicted and observed FRs to discern differences (i.e., multiple predator effects) across prey densities. All statistical analyses were performed in R v4.0.2 [48].

3. Results

Consumption rates differed significantly among predator treatments (F_{4128} = 8.937, $p < 0.001$). Single *O. mossambicus* consumed generally a greater proportion of available prey than single *E. paludinosus*, but this was not significantly different (Tukey test, p = 0.093). However, conspecific pairs of *O. mossambicus* consumed significantly more than conspecific pairs of *E. paludinosus* (Tukey test, $p < 0.001$). In turn, interspecific pairs of *O. mossambicus* and *E. paludinosus* consumed significantly more prey than conspecific pairs of *E. paludinosus* (p = 0.026), but not conspecific pairs of *O. mossambicus* (Tukey test, p = 0.895) (Figure 2). The rate of prey consumption related significantly negatively with prey supply overall (F_{1128} = 146.239, $p < 0.001$) (Figure 2).

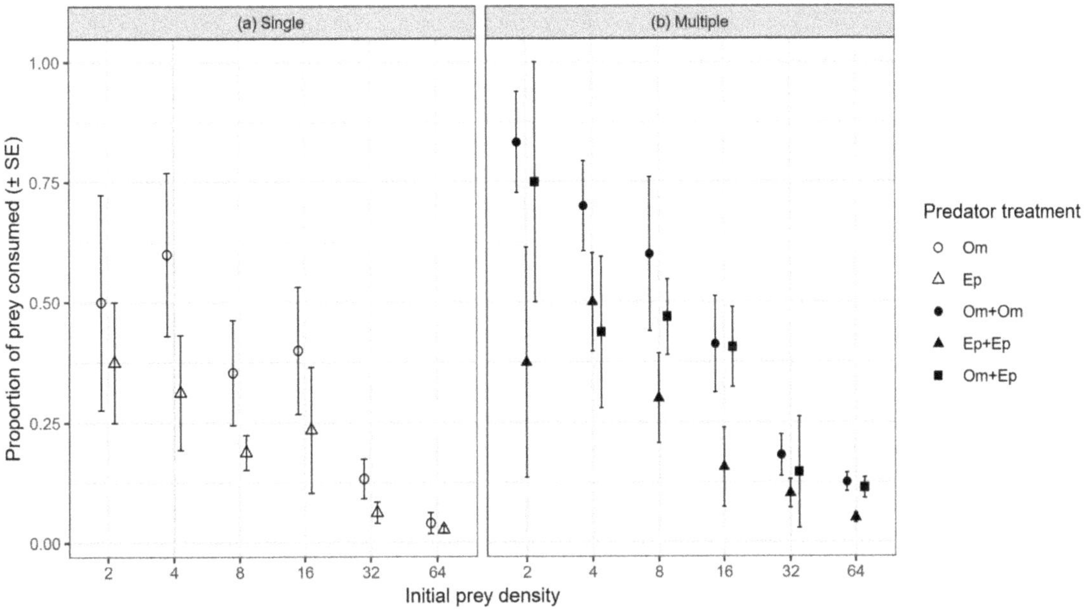

Figure 2. Mean (±SE) proportions of prey consumed by single (**a**) and multiple (**b**) predator groups according to predator treatment and prey density. Om = *Oreochromis mossambicus*, Ep = *Enteromius paludinosus*.

Type II FRs were evidenced by both fish species given significantly negative linear coefficients considering feeding rates as a function of prey density (*O. mossambicus*: estimate = −0.054, $p < 0.001$; *E. paludinosus*: estimate = −0.048, $p < 0001$). Considering single predators, attack rates of *O. mossambicus* ($a = 1.490$, $p = 0.001$, SE = 0.456) were higher than *E. paludinosus* ($a = 0.794$; $p = 0.058$, SE = 0.418), reflecting a steeper FRs slope at low densities (Figure 3a). Handling times of *O. mossambicus* ($h = 0.207$, $p < 0.001$, SE = 0.029) were also shorter than *E. paludinosus* ($h = 0.377$, $p < 0.001$, SE = 0.082) and, therefore, *O. mossambicus* had a substantially greater maximum feeding rate over the experimental duration ($1/h = 4.842$) than did *E. paludinosus* ($1/h = 2.653$) (Figure 3a). As such, the asymptote of the functional response for *O. mossambicus* was substantially higher. However, confidence intervals of FRs overlapped across all prey densities in the single predator groups (Figure 3a).

Considering multiple predator FRs, predicted feeding rates of conspecific *O. mossambicus* exceeded those observed, however, confidence intervals overlapped at all prey densities (Figure 3b). Likewise, for *E. paludinosus*, predictions generally exceeded observed FRs, but confidence intervals again overlapped across prey densities (Figure 3c). For interspecific predator pairs, predicted FRs tracked more closely with the experimental observations (Figure 3d). Accordingly, we did not find significant statistical evidence for MPEs in any of the multiple predator groupings, with predation rates generally combining additively between fishes. However, interspecific groupings tended to exhibit the highest predictability (i.e., fewest non-trophic interactions between predators that influence feeding rates).

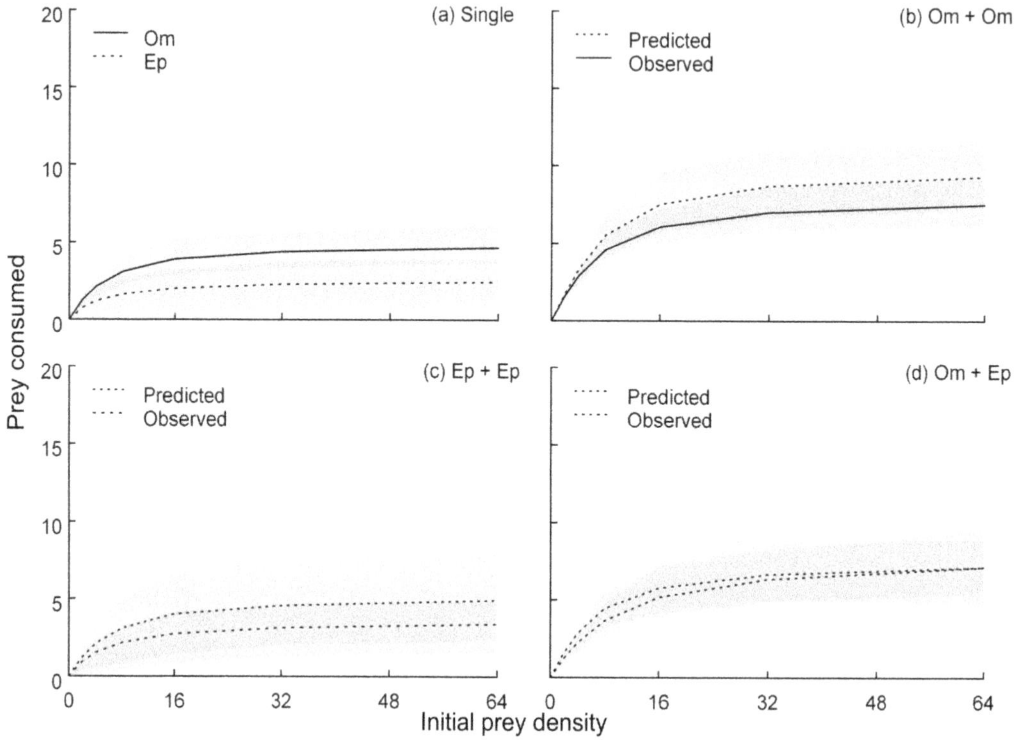

Figure 3. Functional responses of *Oreochromis mossambicus* (Om) and *Enteromius paludinosus* (Ep). As single predators (**a**) and in pairs (**b**–**d**). Observations correspond to experimental data from paired predator groups, whilst predictions in each panel were modelled from functional responses in single predator groups (**a**) using a population dynamic approach. Shaded areas are 95% confidence intervals.

4. Discussion

In this study, we combined experiments and modelling to assess potential MPEs between two common and often sympatric species in floodplain wetlands. In predator-prey interactions with single predators, the predictions that *E. paludinosus* would have higher FR and likely to be more efficient at finding prey at low densities than *O. mossambicus* were not supported. The two fish species both displayed potentially destabilizing consumption rates individually (Type II FRs) [49], with *O. mossambicus* generally exhibiting a higher maximum feeding rates and attack rates than *E. paludinosus*. Our results revealed that, for conspecific interactions, feeding rates combined additively across prey density treatments since expected proportions of prey were consumed relative to those upscaled from single predators. Furthermore, feeding rates originating from interspecific groups of *O. mossambicus* and *E. paludinosus* combined additively, thus again resulting in an absence of MPEs. Given the important role of predation in the structuring and functioning of aquatic systems [50,51], understanding the implications of predator–prey dynamics is critical for robust interaction strength quantifications between trophic levels; here we found that key wetland fishes display additive feeding interactions.

The study suggests that both *O. mossambicus* and *E. paludinosus* could destabilize prey populations, given the high consumption rates at low densities under laboratory conditions. However, the FR results also showed that in single fish trials, *O. mossambicus* did not have significantly higher FRs (shorter handling times and higher maximum feeding rates) than

E. paludinosus. A growing literature on predator–prey interaction has shown that predation can potentially yield both positive and negative MPEs [33,52] and a number of ways to model MPE has been employed [1,32]. The population dynamic model employed in this study has been shown to be more robust to non-replacement experimental data than other approaches, such as the multiplicative risk model [25]. When we consider Chironomidae predation by *O. mossambicus* and *E. paludinosus*, our study showed additive effects for conspecific pairs, and again additive interactions were observed for intraspecific pairs of these species.

In terms of feeding rates, it was observed that the proportion of prey consumed fell significantly as prey densities increased across treatments, with feeding rates at low densities highest (i.e., Type II FR), suggesting that handling time-limited consumption rates at higher prey densities in all predator treatments. The overall proportion of prey consumed by single predators was generally lower compared to that of multiple predators. *Enteromius paludinosus* also generally exhibited low resource use compared to *O. mossambicus*. This reduced consumption by *E. paludinosus* is due to less efficient foraging tactics that lessens attack rates (i.e., behavioral differences related to search efficiency) and also to a longer handling time (i.e., physiological differences) for prey [53]. *Enteromius paludinosus* exhibited reduced FRs in comparison with *O. mossambicus*, and this difference was pronounced when comparing intra- and interspecific pairs, whereby conspecific paired *O. mossambicus* and interspecific fishes were significantly higher than paired *E. paludinosus*. This study further suggests that combinations of different fish species of the same sizes exhibit limited interference when foraging towards their prey, thus showing particularly minimal predator–predator interactions. However, Mofu et al. [33] showed that net prey consumption under multiple predator scenarios can be misleading if individual consumption is not catered for. That study employed a similar design to the present one, but showed, using post-experimental gut content analysis, that one species elevated its feeding while the other reduced its feeding, resulting in a net consumption that revealed no MPEs. This has implications for predator competition dynamics, and such a distinction cannot be made by measuring FRs alone (i.e., with prey consumption pooled between predators). This was also previously attested with a study carried out by Labropoulou and Eleftheriou, [54], whereby two pairs of closely related demersal fish were used to determine the foraging efficiency towards single and multiple prey.

Although studies that model multispecies predator–prey interactions remain rare, the presence of additivity greatly simplifies the construction of predictive models [55,56]. This is because simple additivity decreases the need for detailed prior knowledge about all the interactions (both direct and indirect) that can occur between species in predator-prey systems. The findings of this study emphasize the importance of assessing predator–predator interactions, as these offer insights into resource-use differentiation in multiple predator scenarios. The study further supports the direct FR approach by integrating con-interspecific analysis into the FRs procedure when assessing multiple species interactions, instead of focusing solely on a single or interspecific predators without additional interactions. Given the evidence that a change in FRs of individual species may result from MPEs [24], this study further demonstrates how single and multiple predators can cause additive responses in the aquatic system and how intensely they interact with prey within these systems. The present study further indicates that predator-prey interactions are species-specific, as has been repeatedly evidenced in previous studies among taxa and contexts [57,58]; but there has remained a paucity of knowledge of interaction strength in floodplain wetlands in southern Africa [5].

Despite our results, it is important to highlight potential limitations in our approach. Whereas the present study used thawed frozen chironomid larvae during the experiment, we acknowledge that the use of live prey could have yielded different results given potential prey responses to different predator or prey densities. However, this design allowed us to exclude prey behavior and instead exclusively focus on predator–predator effects on the functional response between these wetland fish species. Furthermore, our choice of small-sized experimental arenas without habitat complexity could have resulted in confinement effects which may have intensified (or dampened) trophic (i.e., predator–prey) and non-trophic (i.e., predator–predator) interactions [59,60]. Therefore, while caution should be exerted when extrapolating these results to the floodplain wetland context, laboratory experiments can provide useful comparative insights between species, under controlled conditions, to discern interaction strengths in a controlled manner without wider mechanistic interpretations. A lack of non-trophic interactions found in the present study may thus change given greater volumes in empirical ecosystems; but we found no strong evidence for predator–predator interactions here.

In ecological studies, it is fundamental to understand the biotic processes which impact floodplain wetlands ecosystem structure and functioning. Recent studies have recognized the need to examine individual consumer variability within wetland populations [61,62]. Despite the existing data on trophic interaction of single predator–prey dynamics, which is helpful in determining the strength of MPEs, further exploration studies of interspecific and conspecific interaction relationships are critical, especially regarding fish predation, because fish drive top-down cascades and shape the structure and functioning of invertebrate communities, including the presence of Chironomidae in floodplain wetlands. As such, although often transient features of a community, fish cannot be excluded in the examination of factors that contribute to the functional ecology of floodplain wetlands.

5. Conclusions

We have observed that changes in prey can alter predator–prey interactions of fish, and this might lead to changes in predator species population dynamics [33], with implications for the broader aquatic ecosystem community structure and functioning [63]. Additionally, integration of habitat complexities based on the species' natural habitat would likely improve the understanding of processes involved in such interactions [64]. Future studies should be conducted focusing on the assessment of MPEs using FR approaches, not only considering predator–prey interactions of native species but also non-native species which increasingly threaten wetland ecosystems [65]. Furthermore, the present study recommends an intensive study which determines the distribution and abundance of *O. mossambicus* and *E. paludinosus* that are found in floodplain wetlands. This will strengthen the FR approach and the impacts which are brought by non-native species to floodplain wetlands systems.

Author Contributions: L.F.M.: Investigation, Data curation, Formal analysis, Writing—original draft; T.D.: Conceptualization, Investigation, Methodology, Resources, Data curation, Supervision, Writing—original draft; R.N.C.: Conceptualization, Methodology, Data curation, Formal analysis, Writing—original draft; L.M.: Investigation, Supervision, Writing—review and editing; R.J.W.: Conceptualization, Methodology, Writing, Supervision, Resources—review and editing; F.D.: Investigation, Writing—review and editing; G.O.: Resources, Writing—review, editing and original draft. All authors have read and agreed to the published version of the manuscript.

Funding: We greatly acknowledge the financial support of the University of Venda Niche Grant (SES/20/ERM/03) and NRF Thuthuka Grant (117700). L.F.M. and R.N.C. acknowledge funding from the National Research Foundation Postgraduate Bursary (129098) and the Alexander von Humboldt Foundation, respectively.

Institutional Review Board Statement: The study was conducted in accordance with the Declaration of Helsinki, and approved by the Ethics Committee of the University of Venda Research and Animal Ethics Committee (Ethical clearance number: SES/20/ERM/14/1611).

Data Availability Statement: The datasets generated and/or analyzed during the current study are not publicly available as they are part of larger study that is currently on-going but are available from the corresponding author on reasonable request.

Acknowledgments: We thank Thendo Mutshekwa for assisting with the experiments.

Conflicts of Interest: All co-authors have seen and agree with the contents of the manuscript and there is no financial interest to report. Thus, all authors have declared that no competing interests exist.

References

1. Sih, A.; Englund, G.; Wooster, D. Emergent Impacts of Multiple Predators on Prey. *Trends Ecol. Evol.* **1998**, *13*, 350–355. [CrossRef]
2. Barrios-O'Neill, D.; Dick, J.T.A.; Ricciardi, A.; MacIsaac, H.J.; Emmerson, M.C. Deep Impact: In Situ Functional Responses Reveal Context-Dependent Interactions between Vertically Migrating Invasive and Native Mesopredators and Shared Prey. *Freshw. Biol.* **2014**, *59*, 2194–2203. [CrossRef]
3. Kumar, R. Impacts of Predation by the Copepod, Mesocyclops Pehpeiensis, on Life Table Demographics and Population Dynamics of Four Cladoceran Species: A Comparative Laboratory Study. *Zool. Stud.* **2009**, *48*, 738–752.
4. Cuthbert, R.N.; Dalu, T.; Wasserman, R.J.; Sentis, A.; Weyl, O.L.F.; Froneman, P.W.; Callaghan, A.; Dick, J.T.A. Prey and Predator Density-Dependent Interactions under Different Water Volumes. *Ecol. Evol.* **2021**, *11*, 6504–6512. [CrossRef]
5. Dalu, T.; Wasserman, R.J. *Fundamentals of Tropical Freshwater Wetlands: From Ecology to Conservation Management*; Elsevier: Amsterdam, The Netherlands, 2022; ISBN 978-0-12-822362-8. [CrossRef]
6. Daniel, J.; Rooney, R.C. Wetland Hydroperiod Predicts Community Structure, but Not the Magnitude of Cross-Community Congruence. *Sci. Rep.* **2021**, *11*, 429. [CrossRef]
7. Sheaves, M.; Johnston, R.; Abrantes, K.; Sheaves, M.; Johnston, R.; Abrantes, K. Fish Fauna of Dry Tropical and Subtropical Estuarine Floodplain Wetlands. *Mar. Freshw. Res.* **2007**, *58*, 931–943. [CrossRef]
8. Carvalho, D.A.; Williner, V.; Giri, F.; Vaccari, C.; Collins, P.A.; Carvalho, D.A.; Williner, V.; Giri, F.; Vaccari, C.; Collins, P.A. Quantitative Food Webs and Invertebrate Assemblages of a Large River: A Spatiotemporal Approach in Floodplain Shallow Lakes. *Mar. Freshw. Res.* **2016**, *68*, 293–307. [CrossRef]
9. Middelburg, J.J. Stable Isotopes Dissect Aquatic Food Webs from the Top to the Bottom. *Biogeosciences* **2014**, *11*, 2357–2371. [CrossRef]
10. Rettig, J.E.; Smith, G.R. Relative Strength of Top-down Effects of an Invasive Fish and Bottom-up Effects of Nutrient Addition in a Simple Aquatic Food Web. *Environ. Sci. Pollut. Res.* **2021**, *28*, 5845–5853. [CrossRef]
11. Křivan, V.; Schmitz, O.J. Trait and Density Mediated Indirect Interactions in Simple Food Webs. *Oikos* **2004**, *107*, 239–250. [CrossRef]
12. Alexander, M.E.; Dick, J.T.A.; O'Connor, N.E. Trait-Mediated Indirect Interactions in a Marine Intertidal System as Quantified by Functional Responses. *Oikos* **2013**, *122*, 1521–1531. [CrossRef]
13. Buxton, M.; Cuthbert, R.N.; Dalu, T.; Nyamukondiwa, C.; Wasserman, R.J. Predator Density Modifies Mosquito Regulation in Increasingly Complex Environments. *Pest Manag. Sci.* **2020**, *76*, 2079–2086. [CrossRef] [PubMed]
14. Soluk, D.A. Multiple Predator Effects: Predicting Combined Functional Response of Stream Fish and Invertebrate Predators. *Ecology* **1993**, *74*, 219–225. [CrossRef]
15. Holling, C.S. Some Characteristics of Simple Types of Predation and Parasitism 1. *Can. Entomol.* **1959**, *91*, 385–398. [CrossRef]
16. Holling, C.S. The Components of Predation as Revealed by a Study of Small-Mammal Predation of the European Pine Sawfly 1. *Can. Entomol.* **1959**, *91*, 293–320. [CrossRef]
17. Dick, J.T.A.; Alexander, M.E.; Jeschke, J.M.; Ricciardi, A.; MacIsaac, H.J.; Robinson, T.B.; Kumschick, S.; Weyl, O.L.F.; Dunn, A.M.; Hatcher, M.J.; et al. Advancing Impact Prediction and Hypothesis Testing in Invasion Ecology Using a Comparative Functional Response Approach. *Biol. Invasions* **2014**, *16*, 735–753. [CrossRef]
18. Holling, C.S. The Functional Response of Invertebrate Predators to Prey Density. *Mem. Entomol. Soc. Can.* **1966**, *98*, 5–86. [CrossRef]
19. Abrams, P. The Functional Responses of Adaptive Consumers of Two Resources. *Theor. Popul. Biol.* **1987**, *32*, 262–288. [CrossRef]
20. DeLong, J.P. *Predator Ecology: Evolutionary Ecology of the Functional Response*; Oxford University Press: Oxford, UK, 2021; ISBN 9780192895509. Available online https://books.google.co.za/books?id=ytg9EAAAQBAJ (accessed on 14 December 2021).
21. Hermoso, V.; Clavero, M.; Blanco-Garrido, F.; Prenda, J. Invasive Species and Habitat Degradation in Iberian Streams: An Analysis of Their Role in Freshwater Fish Diversity Loss. *Ecol. Appl.* **2011**, *21*, 175–188. [CrossRef]
22. Leitão, R.P.; Zuanon, J.; Mouillot, D.; Leal, C.G.; Hughes, R.M.; Kaufmann, P.R.; Villéger, S.; Pompeu, P.S.; Kasper, D.; de Paula, F.R.; et al. Disentangling the Pathways of Land Use Impacts on the Functional Structure of Fish Assemblages in Amazon Streams. *Ecography* **2018**, *41*, 219–232. [CrossRef]
23. Hurd, L.E.; Sousa, R.G.C.; Siqueira-Souza, F.K.; Cooper, G.J.; Kahn, J.R.; Freitas, C.E.C. Amazon Floodplain Fish Communities: Habitat Connectivity and Conservation in a Rapidly Deteriorating Environment. *Biol. Conserv.* **2016**, *195*, 118–127. [CrossRef]
24. McCoy, M.W.; Stier, A.C.; Osenberg, C.W. Emergent Effects of Multiple Predators on Prey Survival: The Importance of Depletion and the Functional Response. *Ecol. Lett.* **2012**, *15*, 1449–1456. [CrossRef] [PubMed]
25. Sentis, A.; Boukal, D.S. On the Use of Functional Responses to Quantify Emergent Multiple Predator Effects. *Sci. Rep.* **2018**, *8*, 11787. [CrossRef] [PubMed]
26. Sentis, A.; Gémard, C.; Jaugeon, B.; Boukal, D.S. Predator Diversity and Environmental Change Modify the Strengths of Trophic and Nontrophic Interactions. *Glob. Change Biol.* **2017**, *23*, 2629–2640. [CrossRef]

27. Alsakaji, H.J.; Kundu, S.; Rihan, F.A. Delay Differential Model of One-Predator Two-Prey System with Monod-Haldane and Holling Type II Functional Responses. *Appl. Math. Comput.* **2021**, *397*, 125919. [CrossRef]
28. Calbet, A.; Landry, M.R. Mesozooplankton Influences on the Microbial Food Web: Direct and Indirect Trophic Interactions in the Oligotrophic Open Ocean. *Limnol. Oceanogr.* **1999**, *44*, 1370–1380. [CrossRef]
29. Schmitz, O.J.; Suttle, K.B. Effects of Top Predator Species on Direct and Indirect Interactions in a Food Web. *Ecology* **2001**, *82*, 2072–2081. [CrossRef]
30. Brooks, A.C.; Gaskell, P.N.; Maltby, L.L. Sublethal Effects and Predator-Prey Interactions: Implications for Ecological Risk Assessment. *Environ. Toxicol. Chem.* **2009**, *28*, 2449–2457. [CrossRef]
31. Klecka, J.; Boukal, D.S. Foraging and Vulnerability Traits Modify Predator–Prey Body Mass Allometry: Freshwater Macroinvertebrates as a Case Study. *J. Anim. Ecol.* **2013**, *82*, 1031–1041. [CrossRef]
32. Wasserman, R.J.; Alexander, M.E.; Dalu, T.; Ellender, B.R.; Kaiser, H.; Weyl, O.L.F. Using Functional Responses to Quantify Interaction Effects among Predators. *Funct. Ecol.* **2016**, *30*, 1988–1998. [CrossRef]
33. Mofu, L.; South, J.; Wasserman, R.J.; Dalu, T.; Woodford, D.J.; Dick, J.T.A.; Weyl, O.L.F. Inter-Specific Differences in Invader and Native Fish Functional Responses Illustrate Neutral Effects on Prey but Superior Invader Competitive Ability. *Freshw. Biol.* **2019**, *64*, 1655–1663. [CrossRef]
34. Barson, M.; Nhiwatiwa, T. Influence of Drought and Flooding on the Colonisation of Floodplain Pans by Riverine Fish in the Zimbabwean Lowveld. *Afr. J. Aquat. Sci.* **2010**, *35*, 205–208. [CrossRef]
35. Nhiwatiwa, T.; Maseko, Z.; Dalu, T. Fish Communities in Small Subtropical Reservoirs Subject to Extensive Drawdowns, with Focus on the Biology of Enteromius Paludinosus (Peters, 1852) and Clarias Gariepinus (Burchell, 1822). *Ecol. Res.* **2017**, *32*, 971–982. [CrossRef]
36. De Moor, I.J.; Bruton, M.N. *Atlas of Alien and Translocated Indigenous Aquatic Animals in Southern Africa*; National Scientific Programmes Unit: CSIR, 1988. Available online: http://hdl.handle.net/10204/2416 (accessed on 22 December 2021).
37. Mattson, N.S. Trophic Interactions between Two Tropical Omnivorous Fishes, Oreochromis Shiranus and Barbus Paludinosus: Feeding Selectivities and Food Web Responses. *Hydrobiologia* **1998**, *380*, 195. [CrossRef]
38. Rapatsa, M.M.; Moyo, N.A.G. The Functional Response of Two Aquatic Predatory Insects and Its Implications for the Biological Control of Coptodon Rendalli Fry. *Ecohydrol. Hydrobiol.* **2021**, *21*, 106–113. [CrossRef]
39. Boets, P.; Laverty, C.; Fukuda, S.; Verreycken, H.; Green, K.; Britton, R.J.; Caffrey, J.; Goethals, P.L.M.; Pegg, J.; Médoc, V.; et al. Intra- and Intercontinental Variation in the Functional Responses of a High Impact Alien Invasive Fish. *Biol. Invasions* **2019**, *21*, 1751–1762. [CrossRef]
40. Weyl, O.L.F.; Pattrick, P.; Ellender, B.R.; Miya, T.; Woodford, D.J.; Bennett, R.H.; Wasserman, R.J.; Äkinen, T.M. Ethical Considerations for Field Research on Fishes. *Koedoe Afr. Prot. Area Conserv. Sci.* **2016**, *58*, a1353. [CrossRef]
41. Length, R. Emmeans: Estimated Marginal Means, Aka Least-Squares Means. R Package Version 1.4.8. 2020. Available online: https://cran.r-project.org/web/packages/emmeans/emmeans.pdf (accessed on 27 December 2021).
42. Juliano, S. Non-Linear Curve Fitting: Predation and Functional Response Curve. *Des. Anal. Ecol. Exp.* **2001**, 178–196. [CrossRef]
43. Pritchard, D.W.; Paterson, R.A.; Bovy, H.C.; Barrios-O'Neill, D. FRAIR: An R Package for Fitting and Comparing Consumer. *Funct. Responses* **2017**, *8*, 1528–1534.
44. Rogers, D. Random Search and Insect Population Models. *J. Anim. Ecol.* **1972**, *41*, 369–383. [CrossRef]
45. Bolker, B.M. *Ecological Models and Data in R*; Princeton University Press: Vienna, Austria, 2008; ISBN 9781400840908.
46. Cuthbert, R.N.; Wasserman, R.J.; Dalu, T.; Kaiser, H.; Weyl, O.L.F.; Dick, J.T.A.; Sentis, A.; McCoy, M.W.; Alexander, M.E. Influence of Intra- and Interspecific Variation in Predator–Prey Body Size Ratios on Trophic Interaction Strengths. *Ecol. Evol.* **2020**, *10*, 5946–5962. [CrossRef] [PubMed]
47. Soetaert, K.; Petzoldt, T. Inverse Modelling, Sensitivity and Monte Carlo Analysis in R Using Package FME. *J. Stat. Softw.* **2010**, *33*, 1–28. [CrossRef]
48. R Core Team. R: A Language and Environment for Statistical Computing. R Foundation for Statistical Computing: Vienna, Austria, 2020. Available online: https://www.R-project.org/ (accessed on 27 December 2021).
49. Cuthbert, R.N.; Dalu, T.; Wasserman, R.J.; Callaghan, A.; Weyl, O.L.F.; Dick, J.T.A. Using Functional Responses to Quantify Notonectid Predatory Impacts across Increasingly Complex Environments. *Acta Oecologica* **2019**, *95*, 116–119. [CrossRef]
50. Heck, K.L.; Crowder, L.B. Habitat Structure and Predator—Prey Interactions in Vegetated Aquatic Systems. In *Habitat Structure: The Physical Arrangement of Objects in Space*; Bell, S.S., McCoy, E.D., Mushinsky, H.R., Eds.; Population and Community Biology Series; Springer: Dordrecht, The Netherlands, 1991; pp. 281–299. ISBN 9789401130769.
51. Wasserman, R.J.; Noyon, M.; Avery, T.S.; Froneman, P.W. Trophic Level Stability-Inducing Effects of Predaceous Early Juvenile Fish in an Estuarine Mesocosm Study. *PLoS ONE* **2013**, *8*, e61019. [CrossRef]
52. Palacios, M.M.; Malerba, M.E.; McCormick, M.I. Multiple Predator Effects on Juvenile Prey Survival. *Oecologia* **2018**, *188*, 417–427. [CrossRef]
53. Metcalfe, N.B.; Huntingford, F.A.; Thorpe, J.E. The Influence of Predation Risk on the Feeding Motivation and Foraging Strategy of Juvenile Atlantic Salmon. *Anim. Behav.* **1987**, *35*, 901–911. [CrossRef]
54. Labropoulou, M.; Eleftheriou, A. The Foraging Ecology of Two Pairs of Congeneric Demersal Fish Species: Importance of Morphological Characteristics in Prey Selection. *J. Fish Biol.* **1997**, *50*, 324–340. [CrossRef]

55. Wilbur, H.M.; Fauth, J.E. Experimental Aquatic Food Webs: Interactions between Two Predators and Two Prey. *Am. Nat.* **1990**, *135*, 176–204. [CrossRef]
56. Vucetich, J.A.; Hebblewhite, M.; Smith, D.W.; Peterson, R.O. Predicting Prey Population Dynamics from Kill Rate, Predation Rate and Predator–Prey Ratios in Three Wolf-Ungulate Systems. *J. Anim. Ecol.* **2011**, *80*, 1236–1245. [CrossRef]
57. Uiterwaal, S.F.; DeLong, J.P. Multiple Factors, Including Arena Size, Shape the Functional Responses of Ladybird Beetles. *J. Appl. Ecol.* **2018**, *55*, 2429–2438. [CrossRef]
58. Cuthbert, R.N.; Dalu, T.; Wasserman, R.J.; Weyl, O.L.F.; Froneman, P.W.; Callaghan, A.; Dick, J.T.A. Additive Multiple Predator Effects of Two Specialist Paradiaptomid Copepods towards Larval Mosquitoes. *Limnologica* **2019**, *79*, 125727. [CrossRef]
59. Barrios-O'Neill, D.; Dick, J.T.A.; Emmerson, M.C.; Ricciardi, A.; MacIsaac, H.J. Predator-Free Space, Functional Responses and Biological Invasions. *Funct. Ecol.* **2015**, *29*, 377–384. [CrossRef]
60. Uiterwaal, S.F.; Lagerstrom, I.T.; Lyon, S.R.; DeLong, J.P. Data Paper: FoRAGE (Functional Responses from Around the Globe in All Ecosystems) Database: A Compilation of Functional Responses for Consumers and Parasitoids. *BioRxiv* **2018**. [CrossRef]
61. Thorp, C.J.; Alexander, M.E.; Vonesh, J.R.; Measey, J. Size-Dependent Functional Response of Xenopus Laevis Feeding on Mosquito Larvae. *PeerJ* **2018**, *6*, e5813. [CrossRef]
62. Cuthbert, R.N.; Dalu, T.; Wasserman, R.J.; Monaco, C.J.; Callaghan, A.; Weyl, O.L.F.; Dick, J.T.A. Assessing Multiple Predator, Diurnal and Search Area Effects on Predatory Impacts by Ephemeral Wetland Specialist Copepods. *Aquat. Ecol.* **2020**, *54*, 181–191. [CrossRef]
63. Escalera-Vázquez, L.H.; Zambrano, L. The Effect of Seasonal Variation in Abiotic Factors on Fish Community Structure in Temporary and Permanent Pools in a Tropical Wetland. *Freshw. Biol.* **2010**, *55*, 2557–2569. [CrossRef]
64. Barrios-O'Neill, D.; Kelly, R.; Dick, J.T.A.; Ricciardi, A.; MacIsaac, H.J.; Emmerson, M.C. On the Context-Dependent Scaling of Consumer Feeding Rates. *Ecol. Lett.* **2016**, *19*, 668–678. [CrossRef]
65. Yam, R.S.; Huang, K.P.; Hsieh, H.L.; Lin, H.J.; Huang, S.C. An ecosystem-service approach to evaluate the role of non-native species in urbanized wetlands. *Int. J. Environ. Res. Public Health* **2015**, *12*, 3926–3943. [CrossRef]

water

MDPI

Article

Land Use Change to Reduce Freshwater Nitrogen and Phosphorus will Be Effective Even with Projected Climate Change

Andrew J. Wade [1,*], Richard A. Skeffington [1], Raoul-Marie Couture [2], Martin Erlandsson Lampa [3], Simon Groot [4], Sarah J. Halliday [5], Valesca Harezlak [4], Josef Hejzlar [6], Leah A. Jackson-Blake [7,8], Ahti Lepistö [9], Eva Papastergiadou [10], Joan Lluís Riera [11], Katri Rankinen [9], Maria Shahgedanova [1], Dennis Trolle [12], Paul G. Whitehead [13], Demetris Psaltopoulos [14] and Dimitris Skuras [15]

1 Department of Geography and Environmental Science, University of Reading, Reading RG6 6DW, UK; r.a.skeffington@gmail.com (R.A.S.); m.shahgedanova@reading.ac.uk (M.S.)
2 Département de Chimie, Université Laval, Québec, QC G1A 9A9, Canada; raoul.couture@chm.ulaval.ca
3 Länsstyrelsen Västmanlands län, 72211 Västerås, Sweden; martin.erlandsson.lampa@lansstyrelsen.se
4 Deltares, P.O. Box 177, 2600 MH Delft, The Netherlands; sgroot1953@gmail.com (S.G.); valesca.harezlak@deltares.nl (V.H.)
5 Department of Geography and Environmental Science, University of Dundee, Dundee DD1 4HN, UK; s.j.halliday@dundee.ac.uk
6 Biology Centre of the Czech Academy of Sciences, Institute of Hydrobiology, 37005 České Budějovice, Czech Republic; hejzlar@hbu.cas.cz
7 Norwegian Institute for Water Research NIVA, 0579 Oslo, Norway; leah.jackson-blake@niva.no
8 The James Hutton Institute, Aberdeen AB15 8QH, UK
9 Finnish Environment Institute SYKE, FI-00790 Helsinki, Finland; ahti.lepisto@syke.fi (A.L.); katri.rankinen@syke.fi (K.R.)
10 Department of Biology, University of Patras, GR26500 Patras, Greece; evapap@upatras.gr
11 Department of Evolutionary Biology, Ecology and Environmental Sciences, University of Barcelona, 08028 Barcelona, Spain; jlriera@ub.edu
12 Department of Ecoscience, Aarhus University, 8660 Silkeborg, Denmark; trolle@ecos.au.dk
13 School of Geography and the Environment, University of Oxford, Oxford OX1 3QY, UK; paul.whitehead@ouce.ox.ac.uk
14 School of Economics, Aristotle University of Thessaloniki, GR54124 Thessaloniki, Greece; dempsa@econ.auth.gr
15 School of Economics, University of Patras, GR26504 Patras, Greece; skuras@econ.upatras.gr
* Correspondence: a.j.wade@reading.ac.uk; Tel.: +44-(0)118-378-7315

Citation: Wade, A.J.; Skeffington, R.A.; Couture, R.-M.; Erlandsson Lampa, M.; Groot, S.; Halliday, S.J.; Harezlak, V.; Hejzlar, J.; Jackson-Blake, L.A.; Lepistö, A.; et al. Land Use Change to Reduce Freshwater Nitrogen and Phosphorus will Be Effective Even with Projected Climate Change. *Water* **2022**, *14*, 829. https://doi.org/10.3390/w14050829

Academic Editor: Per-Erik Mellander

Received: 11 January 2022
Accepted: 26 February 2022
Published: 6 March 2022

Abstract: Recent studies have demonstrated that projected climate change will likely enhance nitrogen (N) and phosphorus (P) loss from farms and farmland, with the potential to worsen freshwater eutrophication. Here, we investigate the relative importance of the climate and land use drivers of nutrient loss in nine study catchments in Europe and a neighboring country (Turkey), ranging in area from 50 to 12,000 km^2. The aim was to quantify whether planned large-scale, land use change aimed at N and P loss reduction would be effective given projected climate change. To this end, catchment-scale biophysical models were applied within a common framework to quantify the integrated effects of projected changes in climate, land use (including wastewater inputs), N deposition, and water use on river and lake water quantity and quality for the mid-21st century. The proposed land use changes were derived from catchment stakeholder workshops, and the assessment quantified changes in mean annual N and P concentrations and loads. At most of the sites, the projected effects of climate change alone on nutrient concentrations and loads were small, whilst land use changes had a larger effect and were of sufficient magnitude that, overall, a move to more environmentally focused farming achieved a reduction in N and P concentrations and loads despite projected climate change. However, at Beyşehir lake in Turkey, increased temperatures and lower precipitation reduced water flows considerably, making climate change, rather than more intensive nutrient usage, the greatest threat to the freshwater ecosystem. Individual site responses did however vary and were dependent on the balance of diffuse and point source inputs. Simulated lake chlorophyll-a changes were not generally proportional to changes in nutrient loading. Further work is required to accurately simulate the flow and water quality extremes and determine how reductions in freshwater N and P translate into an aquatic ecosystem response.

Keywords: water quality; eutrophication; Europe; Turkey; river; lake; nitrogen; phosphorus; chlorophyll

1. Introduction

Freshwater eutrophication due to nitrogen (N) and phosphorus (P) over-enrichment is a major problem worldwide [1]. The eutrophication effects include biodiversity loss, changes in aquatic plant assemblages, and increased primary production, leading to oxygen depletion in rivers, lakes, and wetlands through the microbial decomposition of dead plant matter. The increased nutrient flux from rivers to the sea can also cause estuarine and coastal eutrophication [2].

N and P over-enrichment is associated with farm and farmland runoff and wastewater effluent discharges, and N deposition can be important in upland areas [3]. Therefore, land use change will affect eutrophication trajectories through changes in nutrient inputs, source areas, and transport pathways [4]. Climate change threatens to worsen eutrophication through increased precipitation intensity, increasing nutrient loss, and lower summer precipitation, causing lower flows which, in turn, reduce effluent dilution [5–7].

National and international policy has been implemented to reduce eutrophication, with emphasis on measures for farm and farmland nutrient source control and transport limitation and wastewater treatment. To inform policy development, multiple assessments of the integrated impact of climate and land use change on eutrophication are needed to determine the effectiveness of measures for nutrient source reduction and catchment retention, given projected changes in precipitation and air temperatures, with such assessments accounting for hydrological change [5,6,8]. The evidence from model-based assessments of eutrophication trajectories for different climate and land use scenarios continues to grow and includes studies focused on individual catchments [9,10], with the recent consideration of detailed sediment and phosphorus transport processes in small (<50 km^2) catchments and regional and global-scale assessments of total nitrogen (TN) and total phosphorus (TP) fluxes using empirical and process-based models [11–14]. Collectively, these studies demonstrate that major agricultural changes are needed to reduce freshwater nutrient inputs, though uncertainty remains about the precise magnitude of the changes needed, especially in large catchments with a mix of nutrient sources and delivery pathways [4,11,15]. Furthermore, it is uncertain if plans to reduce nutrient loss will remain effective under projected climate change. Thus, the aim of this study was to quantify how the catchment-based changes, envisaged by stakeholders in nine study areas across Europe and Turkey (referred to hereafter as the 'study region'; Figure 1) in response to four socio-economic storylines, would alter river and lake N and P concentrations and lake chlorophyll-a (chl-a) concentrations, given projected climate change. The purpose is to provide an assessment of whether land use changes, proposed by stakeholders in catchments representative of key land use types in Europe and Turkey, will be effective to reduce stream and lake N and P concentrations and lake chl-a concentrations or whether climate-induced changes in water availability or flow pathways change will confound the effects from the proposed land cover change. To achieve the study aim, four objectives were defined:

1. To apply, and assess the performance of, process-based, dynamic catchment water quality models at nine sites for the simulation of daily river flow, nitrate (NO_3^--N), and total phosphorus (TP), as well as the soluble reactive phosphorus (SRP) concentrations in rivers and lakes and the lake chl-a concentrations.
2. To quantify how climate change alone will affect river N and P concentrations and loads, using the current land cover (business as usual) and the down-scaled outputs of three Global Circulation Model-Regional Climate Model combinations, driven by

the A1B scenario, to provide projections of future climate (2031–2060) representative of an ensemble mean and extremes.

3. To quantify how four scenarios of climate, land use, N-deposition and water use change would affect river N and P concentrations and loads.
4. To quantify the effect of the four scenarios on lake chl-a concentrations for five of the study areas, which include major lakes.

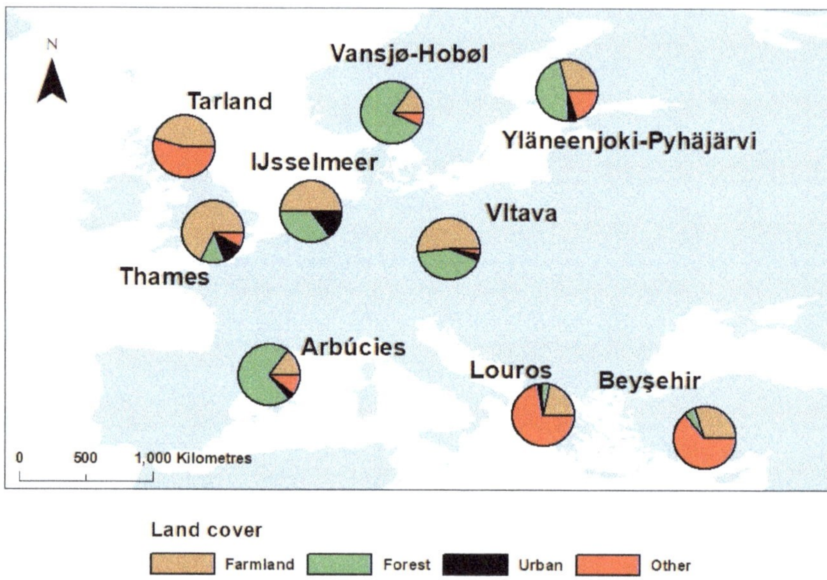

Figure 1. Catchment study areas and predominant land use types. Arbúcies, Louros, Tarland, and Thames are river catchment sites. The other sites are catchments which include a major lake. For land cover details see Table 1.

The 2031–2060 period was chosen as the future period of interest, thought to be sufficiently far into the future for discernable climate and land use change effects, yet not so far into the future that agricultural changes were unrelated to current technologies.

Table 1. Catchment climate and land use summary of the nine river catchments.

Study Catchment	Lat., Long.	Koppen-Geiger	Area	Altitude		Precipitation	Discharge	Land Use			
				Min	Max	Mean	Mean	Farmland	Forest	Urban	Other
	Decimal Degrees		km^2	m	m	$mm\ y^{-1}$	$m^3\ s^{-1}$	%	%	%	%
Yläneenjoki (FIN)	60.99, 22.30	Cold, without dry season, warm or cold summer (Dfb, Dfc)	233	50	100	630	2.1	29	47	4	20
Hobøl (NOR)	59.45, 10.67	Temperate, without dry season, warm summer (Cfb)	301	0	200	800	4.5	15	78	0	7
Tarland (GBR)	57.15, −3.30	Temperate, without dry season, cold summer (Cfc)	74	150	610	901	0.73	53	19	1	27 [1]
Thames (GBR)	51.40, −1.32	Temperate, without dry season, warm summer (Cfb)	9931	3	330	717	80	68	13	11	8
Vltava (CZE)	50.33, 14.47	Temperate, without dry season, warm summer (Cfb)	12,116	354	1378	700	90	52	42	3	3
Arbúcies (ESP)	41.83, 2.47	Temperate, dry summer, hot or warm summer (Csa, Csb)	112	65	1700	830	0.57	15	71	3	11
Louros (GRC)	39.16, 20.75	Temperate, dry summer, hot summer (Csa)	977	0	2000	1367	24	22	4	2	72 [2]
Beyşehir (TUR)	37.67, 31.62	Temperate, dry summer, warm summer (Csb)	4600	1050	3000	490	19	30	6	0	64 [3]
Rhine flowing to Ijsselmeer via the River Ijssel (NLD)	52.82, 5.25	Temperate, without dry season, warm summer (Cfb)	185,000 (Rhine)	0	4059	968 ***	340 **	50	35 *	15	*

* the forest land class includes other fallow land and water. ** Estimated flow to IJsselmeer from Rhine via the River IJssel. [16] *** Modelled estimate [16]. The 'Other' land use category includes: (1) Tarland, uncultivated rough grazing and moorland heath; (2) Louros, shrubland and estuary marshland; (3) Beyşehir, rangeland.

Through consideration of medium (50 km^2) to large (c. 10,000 km^2) catchments, the study investigates how different nutrient sources and transport pathways dominate in larger catchments and whether different land cover changes might be effective to reduce nutrient loss in different geographic settings [14]. All model applications were performed using a common modelling framework to provide consistency in the model performance assessment and scenario application [17]. The models used account for N and P retention in the river system through storage in groundwater and accumulation in the soil and bed sediment (i.e., legacy P). With respect to the ecological consequences of eutrophication, silica and the ratio of silica to P have been observed to control diatom growth and the balance between diatoms and blue-green algae in phytoplankton assemblages [18]. However, silica is not considered in this study as the focus is on understanding the links between climate and land use change and the N and P response as a first step, and the silica data are not yet as extensive as those for N and P.

2. Study Areas and Observational Datasets

Nine study sites were used to represent the climate and land-management types found across the study region, and whilst this is not an exhaustive representation of the all the different land use types present, it represents a trade-off between the resources available to perform the work and the study of the key catchment types found from north to south (Figure 1; Tables 1–3). Eight catchments were chosen because they had been studied previously and had both significant datasets and previous model applications (i.e., Yläneenjoki, Hobøl, Tarland, Thames, Vltava, Arbúcies, and Beyşehir), or had been the focus of socio-economic studies of catchment interventions to reduce freshwater nutrient concentrations (i.e., Louros). Of the nine sites, the stream water N and P concentrations and loads were modelled in eight catchments (i.e., all except IJsselmeer), and of these eight, four of the catchments included lakes (Table 3). The lake N, P, and chl-a concentrations were modelled for these. IJsselmeer was included as one of Europe's largest lakes with extensive algal and zooplankton monitoring and because a lake model was already set-up [19]. However, simulating the effect of land use change in the River Rhine, which flows into IJsselmeer via the River Ijssel, was beyond the project resources and the input N and P concentrations under the scenarios of land cover change were used from a previous project [16,19]. Across the nine sites, the $NO3^{-}$-N concentrations were higher in the north than the south, and these correlate weakly and non-significantly with the percentage of arable agriculture (Spearman's, $r = 0.37$, $n = 6$, $p = 0.25$); stream water TP concentrations also correlated weakly with the percentage of arable agriculture ($r = 0.37$, $n = 5$, $p = 0.23$).

2.1. *Yläneenjoki-Pyhäjärvi*

Lake Pyhäjärvi (154 km^2) is a large, shallow, mesotrophic lake in SW Finland with a mean depth of 5.5 m and a deepest point of 26 m (Figure 1; Table 1). The lake is used for water supply and recreation, and increased eutrophication of the lake has been a major concern since the late 1980s as cyanobacteria blooms have become more frequent. Two major rivers, the Yläneenjoki and the Pyhäjoki, discharge into the lake. The Yläneenjoki river basin is considerably larger (233 km^2) than the Pyhäjoki (78 km^2) and is considered in this study (Table 1). The land use in the Yläneenjoki is mainly managed forest, arable (spring cereals and root crops), and grassland agriculture.

Table 2. Summary of catchment water quality, including the mean SRP, TP, and NO_3^--N concentration for the period used for catchment model calibration and testing.

Catchment	SRP	TP	NO_3^-	Period	Land Cover/Use Description	Data References
	mg P L^{-1}	mg P L^{-1}	mg N L^{-1}			
Yläneenjoki (FIN)	0.02	0.08	2.4	2003–2008	Arable, mainly forest, mire, some settlements.	[20–22]
Hobøl (NOR)	-	0.04	-	1992–1995	Arable and grassland, mainly forest, some settlements, low relief.	[23–25]
Tarland (GBR)	0.01	0.05	2.9	1999–2010	Arable, heather heath, a little woodland, some settlements, low hills.	[26,27]
Thames (GBR)	0.19	-	-	2001–2008	Mainly arable, some woodland and grassland, large population, low hills and floodplain.	[10]
Vltava (CZE)	0.07	0.15	1.6	1991–2010	Arable, forest and grassland, aquaculture, settlements, mountains.	[28]
Arbúcies (ESP)	0.04	0.09	1.1	2001–2011	Small arable area, forested, small settlement, mountainous.	[29]
Louros (GRC)	0.05	0.06	0.7	2005–2010	Arable on floodplains, no significant settlements, mainly shrubland on karstic uplands.	[30]
Beyşehir (TUR)	0.10	-	0.4	2010–2012	Arable irrigated mostly, settlements, mainly rangeland, high altitude, mountainous.	[31–33]

Table 3. Summary of lake characteristics, including the mean SRP, TP, and NO_3^--N concentration for the period used for lake model calibration and testing.

Catchment	Study Lake	Area	Depth	Retention Time	SRP	TP	NO_3^-	Chl-a	Period	Data References
		km^2	m	Years	mg P L^{-1}	mg P L^{-1}	mg N L^{-1}	μg L^{-1}		
Yläneenjoki (FIN)	Pyhäjärvi	154	5.5	3.2	0.001	0.018	0.45 *	7	1980–2009	[20–22]
Hobøl (NOR)	Vansjø	36	3.8	0.21	0.014	0.035	-	13	2005–2012	[23–25]
Vltava (CZE)	Orlík reservoir	27	27.0	0.25	0.030	0.048	1.45	10	1991–1995	[28]
Beyşehir (TUR)	Beyşehir	650	5.0	5.1	-	-	-	3	2010–2012	[31–33]
IJsselmeer (NLD)	IJsselmeer	1140	4.5	0.30	0.030	0.116	1.81	26	2000–2013	[19]

* Total nitrogen.

Regular monitoring of the water quality of the river Yläneenjoki started in the 1970s. The nutrient load into Lake Pyhäjärvi via the river has been estimated from the (generally) fortnightly water sampling results and daily water flow records at the Vanhakartano measuring site. The water quality was monitored on a monthly basis at three additional points in the main channel in the 1990s and in 13 tributaries flowing into the Yläneenjoki [22].

2.2. Vansjø-Hobøl

The Vansjø-Hobøl catchment (690 km^2) in southeastern Norway is in one of Norway's most agricultural regions. This, together with P-rich clay soils, has led to a long history of eutrophication and problematic cyanobacteria blooms in lake Vansjø, the main lake in the catchment. The Hobøl River is the main tributary to Lake Vansjø (sub-catchment 301 km^2). Lake Vansjø is used for drinking water and hydropower generation and is an important recreational area. The lake has a surface area of 36 km^2 and consists of several sub-basins, the two largest being Storefjorden (eastern basin) and Vanemfjorden (western basin; Tables 1 and 2). The Storefjorden basin flows into the Vanemfjorden basin through a shallow channel and, ultimately, the Vansjø-Hobøl catchment discharges into the Oslo Fjord.

The observed daily meteorology data for Lake Vansjø were obtained from the Norwegian Meteorological Institute from stations located between Vanemfjorden and Storefjorden. The daily flow data were obtained from the Norwegian Water Resources and Energy Directorate, NVE) gauging station at Høgfoss.

The river TP and suspended sediment data are monitored downstream of Høgfoss, at Kure [34]. For the Vanemfjorden and Storefjorden lake basins, water chemistry data and temperature profiles are monitored by Bioforsk and NIVA and available via https://vannmiljo.miljodirektoratet.no/ (last accessed on 10 January 2022). Land use mapping for the Vansjø-Hobøl catchment was provided by the Norwegian Forest and Landscape Research Institute and complemented by a report on the fertilization regimes of agricultural fields [35]. The historical nutrient outputs from the sewage treatment plants were obtained from the online database KOSTRA (http://www.ssb.no/offentlig-sektor/kostra, accessed on 10 January 2022). P loadings from scattered dwellings were taken from an online GIS information system maintained by Bioforsk (http://www.bioforsk.no/webgis, last accessed on 10 January 2022). The land use of the Vansjø-Hobøl catchment is dominated by forestry (78%) and agriculture (15%), predominately cereal production (89%), with a smaller production of grass (9.8%) and vegetables (0.7%). Together, agricultural practices contribute an estimated 48% of the total P input to the river basin, followed by natural runoff (39%), and wastewater treatment plants (WWTPs; 13%) [23].

2.3. Tarland Burn

The Tarland Burn tributary drains the most westerly area of intensive agriculture in the River Dee catchment, northeast Scotland. The catchment has an area of approximately 74 km^2 and the stream itself is around 17 km long. In 2008, the Tarland Burn was classified as being of 'moderate' ecological status, primarily due to morphological alterations, namely channel straightening and resultant loss or degradation of habitat. Water quality is also of concern, primarily due to diffuse inputs of nutrients and sediments from agriculture. Agriculture in the tributary comprises a mosaic of arable and grassland, including beef cattle, sheep, barley, and small areas of other crops. The village of Tarland has a wastewater treatment works (600-person input), and septic tanks are common. The Tarland has been the focus of the Tarland Catchment Initiative since 2000. The initiative aims to provide a scientific assessment of the efficacy of the various measures used to improve the aquatic and riparian habitat, as well as build relationships with landowners and the local community.

For model calibration and testing, routine Scottish Environmental Protection Agency (SEPA) monitoring data from Aboyne, near the catchment outflow, were supplemented by data gathered by the James Hutton Institute (JHI; formerly Macaulay Institute) as part of the Tarland Catchment Initiative. The discharge data were available at Aboyne from

2003 to 2013 (SEPA) and from Coull, in the centre of the catchment, from 2000 to 2013 (JHI data). At Coull, daily and sub-daily (during storm events) water chemistry samples were gathered for the period February 2004–June 2005. Fortnightly/monthly/bi-monthly frequency sampling took place during the rest of the period of 2004–2010, both at Coull and at other points within the catchment, including at SEPA's routine monitoring site at Aboyne. Hourly data were converted to daily means for hydrology and water quality model calibration and testing [26].

2.4. River Thames

The River Thames is the principal river system in southern England and provides the main water supply for London and drains an area of approximately 10,000 km^2. The river source is at Cricklade in the Cotswold Hills and the freshwater boundary downstream at Teddington, below which the Thames discharges into the North Sea. The bedrock geology varies from high-permeability chalk to low-permeability clays. The catchment is predominantly rural in the upper reaches and becomes more urban further downstream. It is heavily farmed, with approximately 36% of the catchment used for intensive agriculture, and densely populated.

The water quality is characterised by high pH and high base cation concentrations where chalk aquifers are present. The mean annual flow (1999–2008) ranges from about 1.5 m^3 s^{-1} at Cricklade in the headwaters to 66 m^3 s^{-1} at Teddington. Seasonally, high flows normally occur in the winter and early spring (January to April) and low flows in the summer and late autumn (July–November), with a significant groundwater component. Mean rainfall for the catchment is approximately 700 mm year^{-1} (1961–1990 record) at Teddington. Daily stream flow is measured at eight sites, and monthly stream water NO_3^--N and SRP concentrations are measured at 17 sites on the main river channel. Data from 2001–2008 were used for the model calibration and testing [10,36]. The river is eutrophic with significant algal blooms.

2.5. IJsselmeer

Lake IJsselmeer is the largest shallow lake in western Europe and was constructed by the cutting off of the South Sea from the open ocean in 1932. Lake IJsselmeer provides fresh water supply, recreation and fisheries and is classified as a NATURA 2000 site. The lake has a surface area of approximately 1140 km^2 and a mean depth of 4.5 m, but with some gullies of approximately 10 m. The river IJssel, a distributary of the River Rhine with an average discharge of 300 m^3 s^{-1}, is the main water input and the lake discharges to the Wadden Sea through two sluices in the Afsluitdijk. The hydrodynamics of the lake are wind-dominated. The retention period of the lake is approximately 3 to 4 months, and the water-level is relatively stable at 20 and 40 cm below sea level during summer and winter, respectively [19].

Data from the National Surface Water Monitoring Program were used, including data on the daily river discharges and the nutrient concentrations (biweekly) to Lake Ijsselmeer and from Ketmwt, Steilbk, and Vrouwzd from 2000 to 2013. Data from the water boards also include direct discharges from wastewater water treatment plants and pumping stations to the lake. Daily meteorological data were available daily from 1950 (precipitation and temperature) and 1975 (radiation) to 2013.

2.6. Vltava-Orlík

The upper River Vltava catchment (12,116 km^2) extends from the border mountain range of the Bohemian Forest between the Czech Republic, Austria, and Germany to the Orlík Reservoir, situated approximately 70 km south (and upstream) of Prague [28]. The landscape is characterized by numerous artificial reservoirs. Most are shallow and were created in the Middle Ages/early Modern Ages for fish production. Four large and deep hydropower, water supply, and flood protection reservoirs were constructed by damming the Vltava River and its tributaries during the second half of the 20th century (Lipno,

Orlík, Římov, and Hněvkovice). The bedrock of the catchment is mostly gneiss, mica-schist, and granite. The main land cover classes are farmland, forest (>80% of Norway spruce plantation), and urban area (Table 1), with farmland dominance (60%) below an elevation of 600 m, and forest dominance (88%) at elevations greater than 900 m. The water quality of the Vltava River has suffered from high P loading, mainly from point sources (WWTPs), and from diffuse sources from agriculture and fishpond fisheries production. Exceedance of the phosphorus concentration standards is the largest single reason for Czech water bodies not reaching good ecological status as defined in the EU Water Framework Directive (WFD).

Land cover was determined from the current database of the Czech Republic (ZABAGED, www.cuzk.cz, last accessed on 10 January 2022) and LANDSAT 7 ETM+ satellite images (for 2007–2009). Land management (fertilization rate, livestock numbers, and crop yields) and demographic data (population and wastewater discharges) for the period 1950–2010 were obtained from the Czech Statistical Office, with mean values at the district level (approximately 1500 km^2).

Daily hydrological data at the Orlík, Lipno, and Římov reservoirs (period 1961–2010) and at 30 additional hydrological stations (10 to 25 years of data) were obtained from the Czech Hydrometeorological Institute and the Vltava River Basin Authority. Daily climate data were obtained from the Czech Hydrometeorological Institute for the period of 1961 to 2010 for eight stations spread across the catchment. Monthly water quality data were obtained from the Institute of Hydrobiology BC AV CR (BCAS) and from the Czech Republic (CR) State Surface Water Quality Monitoring System. The BCAS datasets include long-term monitoring data at the Římov and Lipno reservoirs. The CR datasets include 25 stations distributed across the catchment with monthly sampling for 1975–2010. In addition, the data from 194 stations for 2000 to 2009 are available from the Vltava River Basin Authority. Atmospheric deposition data were derived from BCAS monitoring for the upper mountainous part and the lower part.

2.7. Arbúcies

The Arbúcies river, a tributary of the La Tordera river, is situated by the coast of Catalunya in northeast Spain, approximately 55 km northeast of Barcelona in the Montseny mountain range (Figure 1). The Arbúcies (catchment area of 112 km^2) is approximately 28 km long. Its headwaters in the west, at Font de Regàs, emerge from numerous springs, from which water is bottled and commercially sold. At the small town of Arbúcies (6700 inhabitants), around halfway down the river, a small sewage treatment plant discharges into the stream and overflows of untreated sewage occur during heavy rainstorms when treatment capacity is exceeded. The western part of the catchment is occupied by the Montseny National Park, a biosphere reserve which has been protected since 1978. The large altitudinal difference creates a mosaic landscape, with Mediterranean tree species prevailing at lower altitudes and a sub-alpine climate with typically Central European species and heather (*Calluna vulgaris*) heathlands at higher altitudes. The bedrock in the area is mainly granitic. The catchment is mostly semi-natural, with 80% of it forested (predominantly evergreen broadleaf). Agricultural areas are mainly near the river in the valley plains, and a small (5 km^2) extended plain at the bottom of the catchment, near the confluence with La Tordera, comprises more than half of the total agricultural area in the catchment [29]. Infiltration in the stream bed can occur during the dry summers, causing the stream channel to dry out.

There is one gauging station approximately 3 km upstream of the confluence with La Tordera where flow was measured daily from October 1994 to March 2011. There are three Catalan Water Agency (ACA) sites with chemistry observations in the catchment, including NO_3^--N, NH_4^+-N, and PO_4^{3-}-P concentrations. Between 1995 and 2006, samples were taken at a site at Hostalric, near the confluence with La Tordera. The sampling frequency was irregular, approximately monthly during the period of 1995–2002 and seasonally from 2003 to 2006. In 2007, the sample site was moved to the gauging station, and an additional, new site located 3 km upstream of the Arbúcies was started. These sites were sampled seasonally from 2007 to 2011. Each sample was analysed for NO_3^--N, NH_4^+-N, and phosphate (PO_4^{3-}-P) and also for TP for the period of 2000–2006.

2.8. Louros

The Louros river (977 km^2) is a Mediterranean river situated in the Epirus region in the western part of the Greek mainland (Figure 1). The river emerges from a spring-fed lake (Terovo, 300 masl) near the mountain of Tomaros, which has an elevation of almost 2000 masl. The river is fed by numerous springs on its course to the sea, the largest one being close to the village of Agios Georgios, approximately halfway to the outlet. Downstream of Agios Georgios is a small hydroelectric dam. The river then flows through an agricultural plain before it discharges in the Amvrakikos Gulf, a Ramsar and Natura 2000 site. The approximate total length of the river is 73 km, and the mean annual discharge is approximately 24 m^3 s^{-1}. The catchment is largely rural and relatively sparsely populated; the largest settlements are the towns of Fillipiada (8400 inhabitants) and Louros (5200 inhabitants). A large part of the catchment is mountainous with steep slopes and sparse vegetation, but a substantial part of the lowland is arable land. The Louros provides water for irrigation of about 120 km^2 of cultivated land. In the region, more than 100 large and small agriculture industries and fish farms are operating. Water is also taken for drinking water supply and industrial uses. It receives treated domestic effluent and effluent from light industrial activities, including meat processing, abattoirs, and pig farms, and a small quantity of olive mill wastewater, mostly during the autumn and early winter months [30].

The only available discharge data is from a dam outlet from a hydroelectrical power plant. From the dam, the water outflow is regulated by a sluice, from which the water flows to the plant turbines. Dam overflow typically happens once or a few times each year, and the recorded discharge is the sum of the flow through the turbines and the dam overflow. Discharge here is thus partly regulated by the capacity at which the power plant is operating. However, the dam is shallow and has limited storage capacity (an approximate retention time of one day) and therefore provides a reasonable approximation of river flow.

For the water chemistry observations, different data sources were available from agencies and the scientific publications and projects of the University of Patras, Department of Biology (UPAT-BIO). The data are from a number of sites, generally with sampling at monthly to seasonal frequency and for a maximum of 3 years for any given site [30,37].

2.9. Beyşehir

Lake Beyşehir is the largest freshwater lake in Turkey. The lake is located in the southwest of the country (Figure 1) and is primarily fed by waters from the Sultan and Anamas mountains. The relatively high-altitude (1050–3000 masl) catchment (area 4600 km^2, consists mostly of range land (48%) and agriculture (30%), the latter mostly wheat, barley, chickpeas, or sugar beet. The trophic status of the lake is within an oligotrophic to mesotrophic range, with low phytoplankton biomass and nutrient concentrations [31]. The lake is used for irrigation and aquaculture and is a national park.

Few discharge or water chemistry data were available for the lake, and so, these were collected as part of this study. Daily discharge was measured in five tributaries from May 2010 to April 2012, supplemented with monthly data from 1995–2001. For both main tributaries and in-lake stations, conductivity, total dissolved solids estimate (TDS), pH, dissolved oxygen, temperature, and salinity were measured in the field using a YSI 556 MPS multi-probe field meter (YSI Incorporated, Yellow Springs, OH, USA). Water samples from 14 tributary channels (13 inflows and one outflow) and from two stations within the lake were collected to measure monthly TN, nitrite-nitrate (NO_2^--NO_3^- as N), (NH_4^+-N), TP, SRP, alkalinity, and total suspended solids. From the lake stations, water samples were collected using a Ruttner sampler covering the entire column, and these water samples were also used for chl-a and plankton analysis. In addition, Secchi disc depth and maximum water depth was measured at each sampling period. To determine TP, the acid hydrolysis method was used, and to determine SRP, filtered water was processed using the molybdate reaction method [38]. N analysis was performed using the Scalar Autoanalyzer Method (San++ Automated Wet Chemistry Analyzer, Skalar Analytical, B.V., Breda, The Netherlands). The chl-a was measured spectrophotometrically after using the ethanol extraction method [39].

3. Materials and Methods

3.1. Modelling Workflows

Most of the catchment biophysical simulations were performed with the INCA-N [40] and INCA-P [27] models (Table 4). SWAT [41] was used at Beyşehir and Stream-N [42] was used in addition to INCA-P for the Tarland Burn because these model applications were well developed with good representation of the catchment N and P sources and transport pathways. Similarly well-established lake models were used and drew on existing work. Three lake models were used to simulate the nutrient and chlorophyll dynamics in Lake Beyşehir, (DYRESM-CAEDYM [43,44], PROTECH [45] and PCLake [46]), two were used for Lake Pyhäjärvi (Lake Load Response model [47], MyLake [48]), one for Lake Vansjø (MyLake) in the Hobøl catchment, and one for the Orlík reservoir (CE-QUAL-W2 [49]) in the Vltava catchment (Table 3). Delft3D was applied to simulate the effects of climate and nutrient changes in the IJsselmeer.

Table 4. Summary of catchment and lake models applied at each study site and references to the individual model applications.

Catchment	Catchment Models Applied	Lake	Lake Models Applied	Model Application Reference
Yläneenjoki (FIN)	INCA-N, INCA-P	Pyhäjärvi	Lake Load Response, MyLake	[20–22]
Hobøl (NOR)	INCA-N, INCA-P	Vansjø	MyLake	[23–25]
Tarland (GBR)	Stream-N, INCA-P	-	-	[26,27]
Thames (GBR)	INCA-P	-	-	[10]
Vltava (CZE)	INCA-N, INCA-P	Orlík reservoir	CE-QUAL-W2	[28]
Arbúcies (ESP)	INCA-N, INCA-P	-	-	[29]
Louros (GRC)	INCA-N, INCA-P	-	-	[30]
Beyşehir (TUR)	SWAT	Beyşehir	DYRESM-CAEDYM, PROTECH, PCLake	[31–33]
IJsselmeer (NLD)	None	IJsselmeer	DELFT-3D, HABITAT	[19]

Once calibrated and tested (Section 3.2), the models were run for a baseline period of 1981–2010 and then for a scenario period of 2031–2060. The difference in the mean flow and concentrations simulated for the scenario and baseline periods were used to derive percentage changes. Model calibration was conducted using MCMC-DREAM at the Hobøl and Tarland and manually elsewhere [50].

3.2. Model Performance—Calibration and Testing

All the models were calibrated and tested using a split sample, apart from the DELFT-3D model for Ijsselmeer, which was calibrated for one year (2006). Model performance was assessed using target diagrams of Normalised Root Mean Square Difference between the simulation and observation (RMSD'*(σ)) and the normalised bias, (Bias*) (Figures 2–4) [51]. Normalisation is by the standard deviation of the observations. Target diagrams allow for a quick visual comparison, on a common scale, of model performance for a range of determinands. The normalised bias is positive when the modelled output overestimates the observations and negative for a general underestimation. A perfect model fit would result in a zero bias. The RMSD'*(σ) is positive if the variance of the modelled output is less than the variance of the observed data and zero when the two variances equate. To help visual interpretation, two circles are drawn on the plots about the point (0, 0) with radii of 0.7 and 1.0. Simulations resulting in points inside the smaller circle were described as 'excellent' fits, whilst those that lay between the larger and smaller circle were considered 'good'. These diagrams were used alongside a visual comparison of the modelled and observed patterns in flow, stream water NO_3^--N, TP, and SRP concentrations and lake chl-a concentrations, estimates of model performance using the co-efficient of determination (r^2), and the use of RMSD′*(σ) and the normalised bias calculated from modelled loads [10,22,28,31]. TP captures both particulate and solute phosphorus dynamics, and SRP is a measure of the phosphorus assumed to be biologically available.

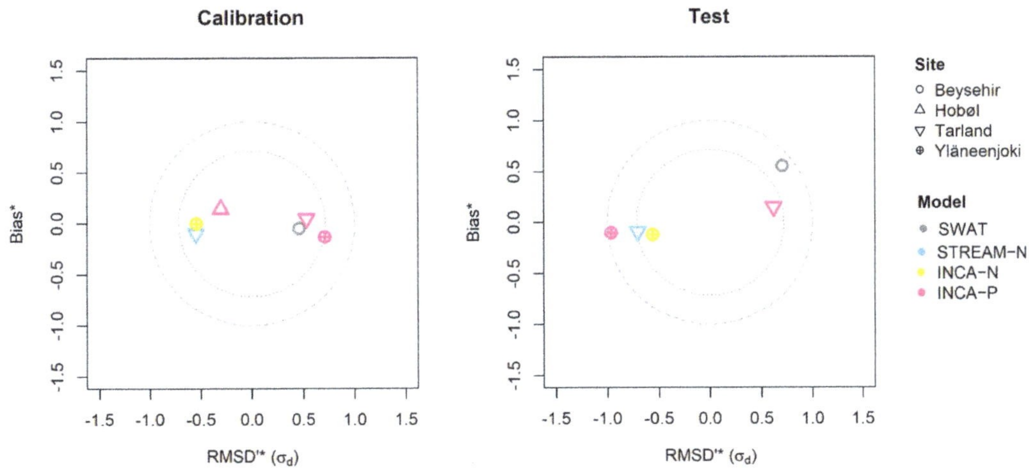

(**a**) Flow

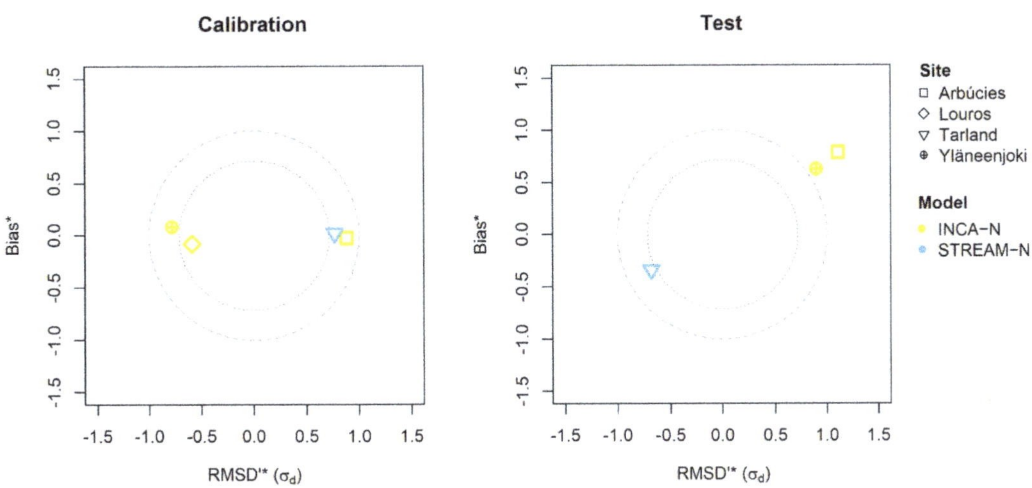

(**b**) Nitrate

Figure 2. Model calibration and test results expressed in terms of the normalised root mean square difference, RMSD′*(σ) and the normalised bias (Bias*) for (**a**) flow and (**b**) stream water nitrate concentrations. The inner and outer circles have radii of 0.7 and 1. At the Arbúcies and Louros catchments, both INCA-N and INCA-P were applied with the same hydrological model structure and parameters.

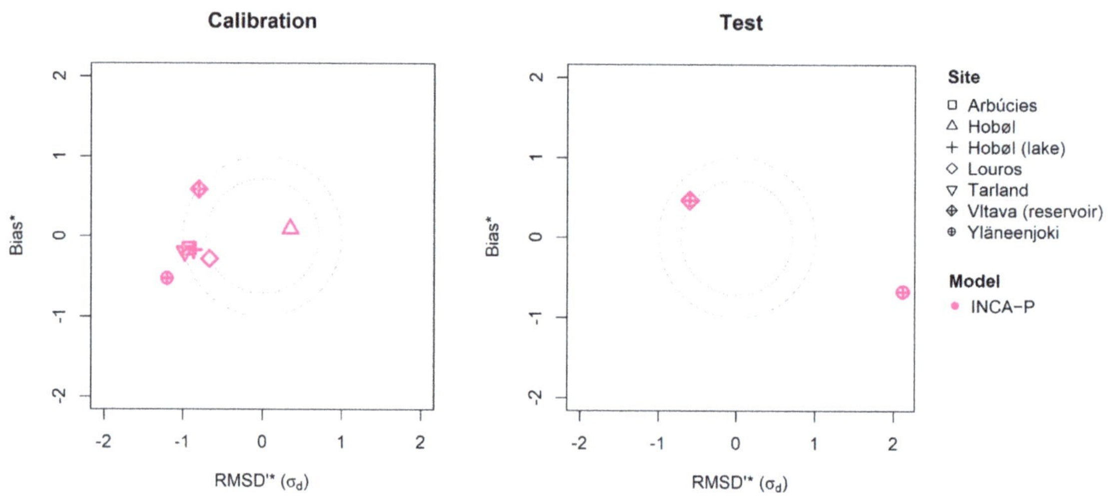

(**a**) Total phosphorus

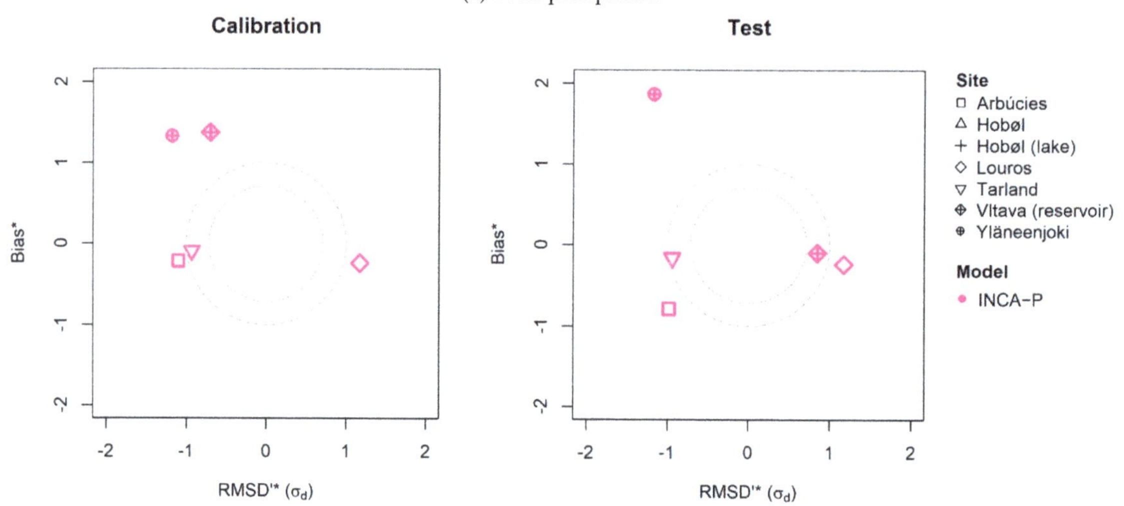

(**b**) Soluble reactive phosphorus

Figure 3. Model calibration and test results expressed in terms of the normalised root mean square difference, RMSD′*(σ) and the normalised bias (Bias*) for stream water (**a**) total and (**b**) soluble reactive phosphorus concentrations. The inner and outer circles have radii of 0.7 and 1.

3.3. Climate Change Projections

Three climate model-regional climate model combinations were used: ECHAM5-KNMI (abbreviated to KNMI in this text), HadRM3P-HadCM3Q0 (HadRM3), and SMHIRCA-BCM (SMHI) [52]. These combinations were selected because they show the best agreement with observations (KNMI) or tend to emphasize a move to hot and dry (HadRM3) or wetter (SMHI) conditions. The modelled projections from KMNI are close to the ENSEMBLES average (http://ensembles-eu.metoffice.com/, last accessed on 10 January 2022). HadRM3 represents one extreme in the ensemble, producing warmer, drier summers. SMHI represents the other extreme, being relatively cold and wet. The A1B scenario was chosen, and this represents rapid economic growth and balanced fossil and non-fossil fuel use to satisfy energy requirements [53]. Data were extracted for a current climate baseline period (1981–2010) and a future period (2031–2060). The climate model output is generally biased, and to take this into account and generate site-specific climate change projections, we used the delta change method using monthly mean change factors. This simple bias correction method assumes that the relative difference between the model baseline and the model future is realistic, despite any bias, and it only corrects for a bias in the mean. Actual evapotranspiration was estimated using the Penman–Monteith method and simulated air temperature.

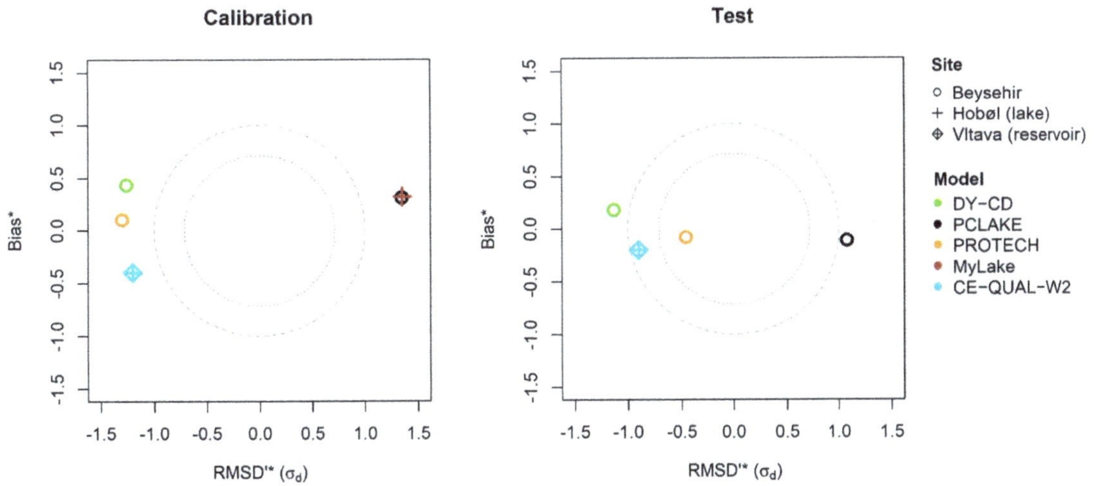

Figure 4. Model calibration and test results expressed in terms of the normalised root mean square difference, RMSD′*(σ) and the normalised bias (Bias*) for lake chlorophyll concentrations. The inner and outer circles have radii of 0.7 and 1.

3.4. Land Use, Atmospheric Deposition, and Water Use Projections

Land use change scenarios were derived for each catchment based on a local interpretation of the four IPCC SRES Storylines A1, A2, B1, and B2 and local land use capability for, and management of, agriculture and forestry. An interpretation of the land cover change was performed by local catchment stakeholders through workshops held in six of the catchments (Table S1). The storylines represent moves to food security in a heavily competitive world (LU1), or domestic market scenarios (LU2), or more collaborative international markets to promote global sustainability (LU3), and local stewardship based on more green agricultural policies to help reduce nutrient losses (LU4). These map to the IPCC SRES Storylines A1, A2, B1, and B2, respectively. For the 'world' (LU1) and 'domestic' (LU2) market scenarios, it is assumed that crops are grown wherever land capability for agriculture supports this. The N deposition change under the four storylines was derived from EMEP (European Monitoring and Evaluation Programme, http://www.emep.int/, last accessed 10 January 2022) using a combination of measured and modelled emissions and meteorological data, the projected land use, and a chemical transport model developed at the Meteorological Synthesizing Centre-West (MSC-W). Water use, in each storyline, was estimated from projected farm practice and population growth estimates. A summary of the climate and land use scenarios, the latter of which incorporates the N deposition and water use projections considered at each study area, is presented in Table S2 and varies in the number of scenarios considered due to the resources available in each study area, though at all sites the climate scenario which was most like present-day (KNMI at most sites, but HadRM3 in Scotland) was considered along with at least a 'best' (LU3 or LU4) and a 'worst' (LU1 or LU2) land use scenario.

4. Results

4.1. Model Performance and Uncertainty Evaluation

The evaluation of the model performance demonstrated that all the catchment models produced a generally excellent representation of the observed flows, measured at the most downstream gauging station in each demonstration catchment (Figure 2a; RMSD'*(σ): −1.0 to 0.7; Bias*: −0.1 to 0.1). This was true for both the model calibration and the test periods. There was little bias in model calibrations for flow, though the modelled flows tended to have a smaller variance than the observed flows as the extreme low and high flows were not simulated well. Differences in the performance between INCA-N and INCA-P arose due to the different calibration and test periods used in the applications of the two models to match the available stream water NO_3^--N, TP, and SRP data [20–22].

The simulations of the stream water NO_3^--N, TP, and SRP concentrations were satisfactory in general, though in all cases they were better at describing the seasonal (monthly) variations in concentrations than the daily extremes [10,18,22,23,26,28,30,31]. There was an apparent deterioration in the model performance during the testing (Figures 2 and 3), and this suggests that conditions were present in the test period which were generally not seen in the calibration period across all sites. Building models that adequately represent the flow and concentration extremes for rivers and lakes remains challenging due to issues around adequate characterization of the model inputs, the representation of catchment water and pollutant storage and transport, and the complex and multiple lake ecosystem interactions.

The simulations of the observed stream water nitrate concentrations (RMSD'*(σ): −1.0 to 1.0; Bias*: −0.4 to 0.8; Figure 2b) were better than those of the observed stream water TP (RMSD'*(σ): −1.2 to 0.4; Bias*: −0.5 to 0.1; Figure 3a) and SRP concentrations (RMSD'*(σ): −1.2 to 1.2; Bias*: −0.8 to 1.9; Figure 3b). There were too few stream water NH_4^+-N concentration data for a meaningful comparison of the model performance between sites. The highest TP and SRP concentrations, which typically coincide with the highest flows, were underestimated. The observed seasonal patterns in the PO_4^{3-}-P concentrations in the catchments draining to the Orlík reservoir were well represented by the lake models but again not in terms of the extremes [28]. The chl-a concentrations at Lake Beyşehir, Lake Vansjø, and the Orlík reservoir appear to be simulated poorly (RMSD'*(σ): −1.2 to 1.2; Bias*: −0.4 to 0.5; Figure 4); however, a closer inspection of the simulated chl-a concentration time-series for these shows that the seasonal patterns in chl-a concentration were modelled well, though the models do not reproduce the exact timing and magnitude of peaks in the chl-a concentrations, leading to poor values of the RMSD'*(σ) and normalised bias [23,28,31]. As when simulating the stream water NO_3^--N, TP, and SRP concentrations, due to the complex, time-varying nature of the controlling factors, it was difficult to capture the precise timing and magnitude of algal blooms. Thus, dynamic lake P and chlorophyll models were used only to explore the general trends in mean annual concentrations and seasonal patterns in concentrations and algal growth [54]. The coefficients of determination, where calculated, show the same results as for the RMSD'*(σ) and normalised bias; namely, the general seasonal patterns in the flow, stream water NO_3^--N, TP, and SRP concentrations and the lake chl-a concentrations are well represented, but the extremes are not [10,22,28,31].

4.2. Projected Change in Hydrology

All three GCM-RCM models predict a rise in temperature over the region, with HadRM3 consistently and significantly higher than the others, with a mean rise of 2.2 °C between the baseline (1981–2010) and scenario (2031–2060) periods, followed by KNMI (1.4 °C) and SMHI (1.0 °C) (Figure 5). This is the order generally observed for these models and for the CMIP5 models for the simulations of European climate change [55]. There is a general north–south gradient, with the greatest temperature rises in the south, though the differences between the models are generally greater than the differences between the sites (Figures 5 and S1). Finland is an exception, with a higher temperature rise (+1.3 to +3.0 °C) projected than in Norway or the UK.

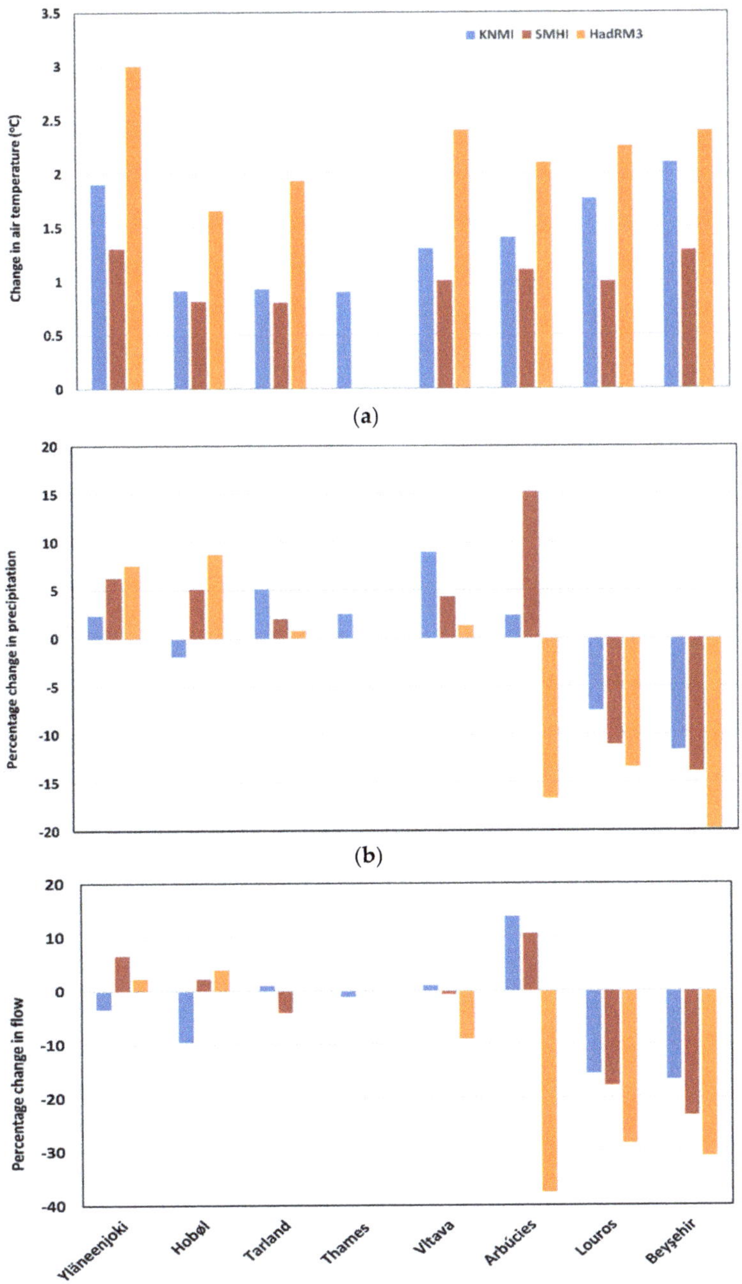

(a)

(b)

(c)

Figure 5. (**a**) Absolute change in air temperature and percentage change in (**b**) precipitation and (**c**) river flow, for the period 2031–2060 relative to the baseline period 1981–2010. Sites are arranged north to south along the x-axis. Climate models were: ECHAM5-KNMI (blue); SMHIRCA-BCM (red); and HadRM3P-HadCM3Q0 (orange). Blank columns indicate that the GCM-RCM combination was not used for a study site.

The northern sites of the Hobøl, Yläneenjoki, and Tarland generally have a small increase in precipitation, and the increase in air temperature, and therefore evapotranspiration, almost balances the precipitation increase, leading to a small percentage increase in discharge or even a slight reduction in some model-site combinations (Figures 5 and S2). HadRM3 produces the largest decreases in simulated discharge in the south and KNMI the smallest. For precipitation, there is a distinct north–south divide, with small increases in the north and mid-latitude sites and large decreases in the south at the Louros and Beyşehir (Figure 5). At the Arbúcies, there is considerable variability between the models with HadRM3 predicting a 17% decline in precipitation and SMHI a 15% increase. This reflects the issues around the robust simulation of precipitation in mountains and confirms the need to use an ensemble of GCM-RCMs or selected combinations that represent the median, wet, and dry extremes. All models concur, however, in predicting substantial decreases in precipitation in the Louros and Beyşehir.

The greatest declines in mean annual river flow occurred at the southern catchments, of the order of −16 to 38% at the Arbúcies, the Louros, and Beyşehir (Figure 5c). These sites had a large percentage decrease in precipitation (Arbúcies: −16% HadRM3 only; Louros: −7 to −13%; and Beyşehir: −12% to −20%) and higher temperatures (approximately +2 °C), which exaggerated the flow change due to increased summer evapotranspiration. Elsewhere, the changes in the flows were smaller and the mean annual flow tended to increase marginally (0–15%). To understand the flow changes, seasonal effects need to be considered. At the Arbúcies, for instance, KNMI predicts a 2% increase in precipitation, but a 14% increase in discharge is simulated. The greater increase in discharge compared to precipitation occurs because there is less summer precipitation but an increase in winter. Lower winter evapotranspiration means winter precipitation is more effective in generating runoff and replenishing soil moisture, hence the discharge increase. Similar, though less spectacular, effects are probably occurring at all the sites. At the Tarland Burn, for instance, KNMI, HadRM3, and SMHI all predict decreased summer and increased winter rainfall, resulting in an overall mean annual flow change of less than 6%. The HadRM3 model predicts a large decline in annual precipitation in the western Mediterranean, with a resultant 38% decline in discharge simulated at the Arbúcies in contrast to the simulations based on the KNMI and SMHI outputs, in which a 14 and 10% discharge increase was projected. This highlights greater uncertainty in the projected flow and water quality response at this mountainous site. In all cases, land use change had minimal impact on the simulated mean flows relative to climate (Figure 6).

4.3. *Projected Change in River Nutrient Concentrations and Loads: Climate Change Only*

This section reports the modelled effects of climate change alone ('climate' in Figures 7–9). In these analyses, changes in the mean concentration of 5% or less were assumed marginal and not indicative of any change. NO_3^--N concentrations were projected to generally decline due to climate change alone, but the changes are small, being more than −10% only at Tarland Burn and Beyşehir (Figure 7). The NO_3^--N loads show both marginal increases and decreases under climate change alone, depending on the site and climate model (Figure S3). Whilst the moderate increases in NO_3^--N concentration and load are likely due to increased precipitation during winter; it is not easy to pinpoint the causes of the lower NO_3^--N concentrations, though reduced leaching due to reduced water flux is one factor where hydrologically effective rainfall declines, and increased denitrification and plant uptake could be contributing factors.

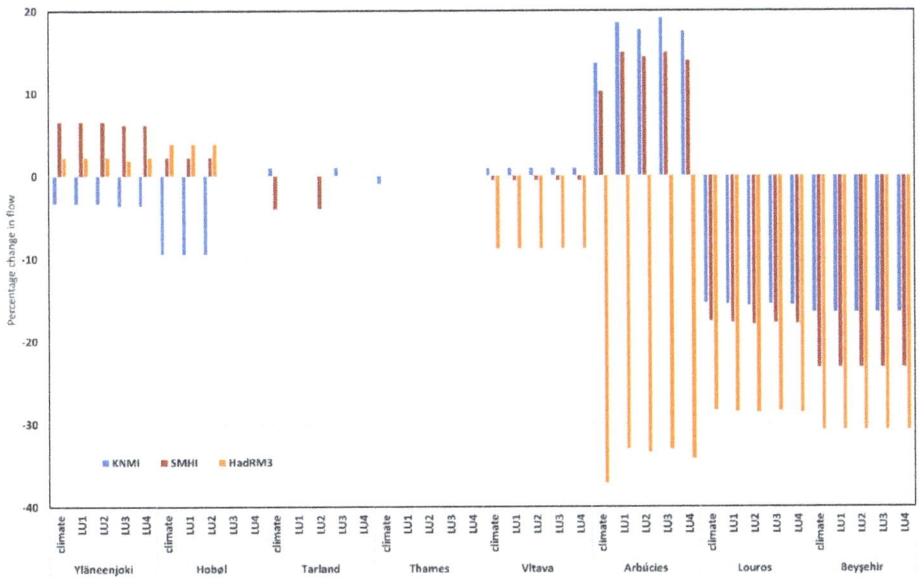

Figure 6. The percentage change in stream flow at each of the study areas between the baseline (1981–2010) and future (2031–2060) periods due to climate change (climate) and combined climate and land cover change (LU1–4). Blank columns indicate that the GCM-RCM combination was not used for a study site.

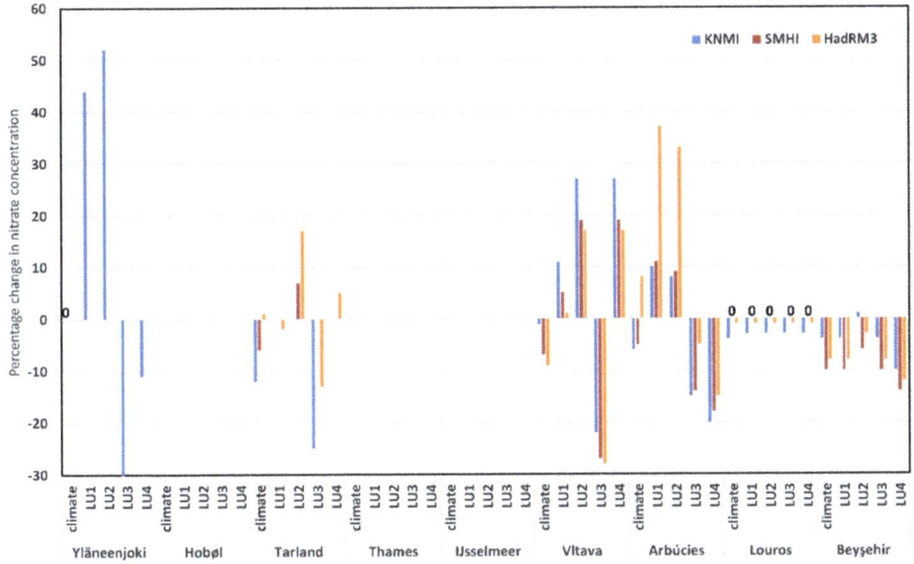

Figure 7. The percentage change in stream water nitrate concentration at each of the study areas between the baseline (1981–2010) and future (2031–2060) periods due to climate change (climate) and climate and land cover change (LU1–4). Blank columns indicate that the GCM-RCM combination was not used for a study site. A '0' denotes a percentage change of zero.

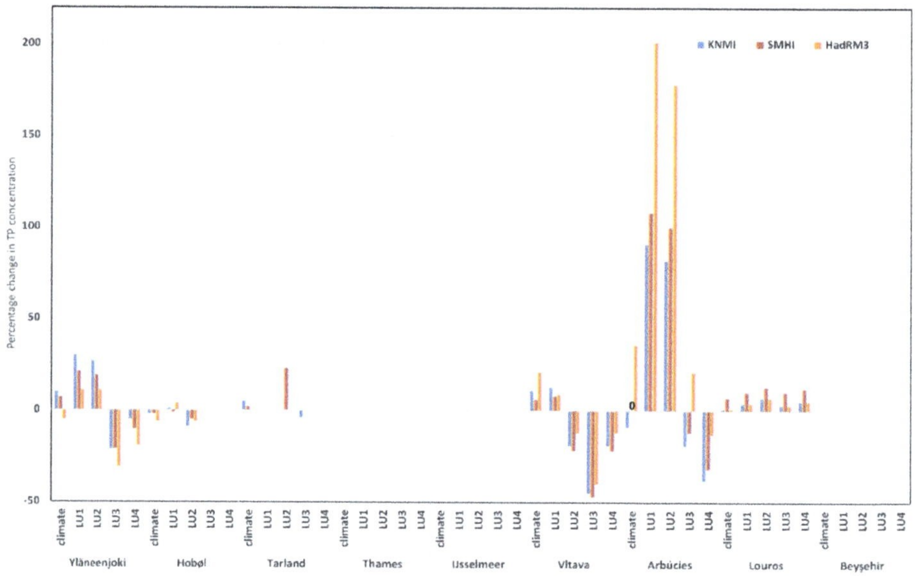

(a)

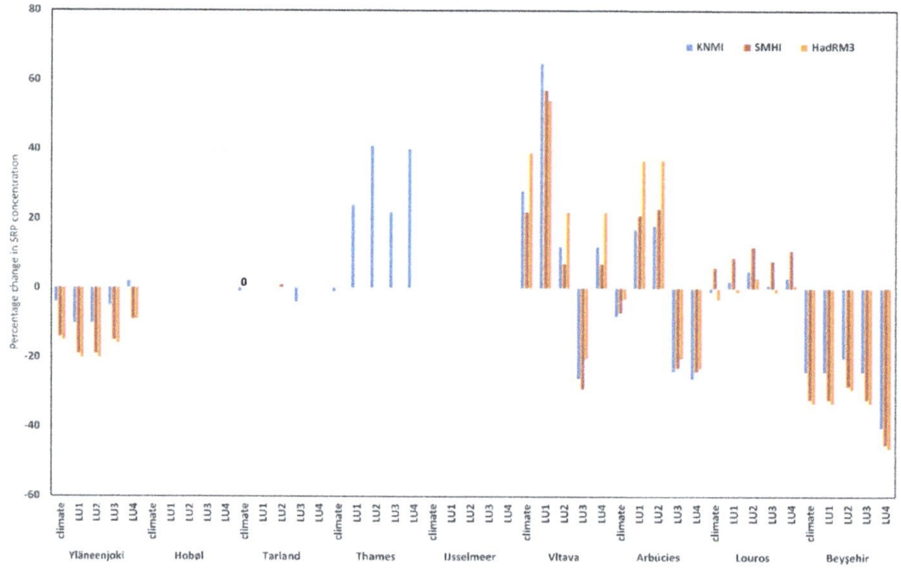

(b)

Figure 8. The percentage change in stream water (**a**) total phosphorus and (**b**) soluble reactive phosphorus concentration at each of the study areas between the baseline (1981–2010) and future (2031–2060) periods due to climate change (climate) and climate and land cover change (LU1–4). Blank columns indicate that the GCM-RCM combination was not used for a study site. A '0' denotes a percentage change of zero.

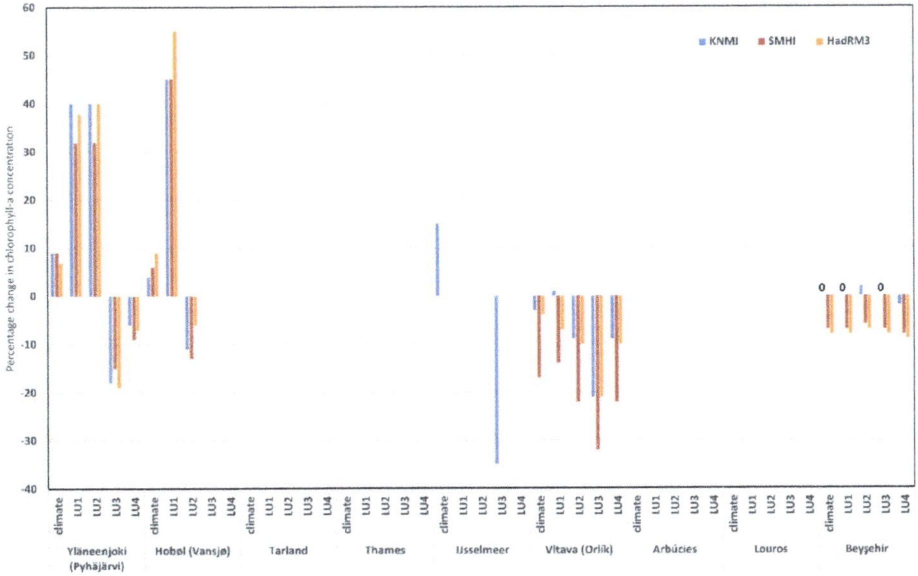

Figure 9. The percentage change in lake chlorophyll-a concentration at each of the study areas between the baseline (1981–2010) and future (2031–2060) periods due to climate change (climate) and climate and land cover change (LU1–4). The Tarland, Thames, Arbúcies, and Louros study areas do not contain lakes. A '0' denotes a percentage change of zero.

The changes in TP concentration due to climate alone were mostly small increases (1–10%) except for the HadRM3 simulations at the Arbúcies (40%) and the Vltava (20%). The latter had the highest TP concentrations to start with due to the mixture of TP inputs from agricultural, wastewater, and fish farming (Figure 8a). The increases in the Vltava and Arbúcies occur primarily in the summer months and are associated with low flow situations, indicating that they are due to less dilution of the wastewater inputs. At the Vansjø-Hobøl in Norway, in contrast, the stream water TP concentration and load declined with climate change, but in the lake, the TP concentration increased (not shown). This is a curious result that contrasts with other sites where stream water TP concentration increases when higher precipitation is projected under climate change. At the Vansjø-Hobøl, the reduced fertiliser rates, vegetated buffer strips, constructed wetlands, and improved wastewater treatment that are already in place have led to a declining trend in observed stream water TP concentrations from 1990–2010, and this was simulated by INCA-P [24]. Under climate change alone, the simulated stream water TP concentrations and loads continue to decrease due to these measures. The increased lake TP concentration under climate change alone suggests accumulation of TP in the lake or a possible internal source from the bed sediment, though this is not well resolved in the MyLake model [24].

SRP showed much larger effects than TP for climate change alone, both positive and negative, and the effects depended much more strongly on the climate model (Figure 8a,b). The highest SRP concentrations were found in the Thames, where the large population generates substantial SRP in wastewater. Second highest, perhaps surprisingly, was Beyşehir because the low specific runoff appears to make the lake water more concentrated. Concentration changes due to climate alone ranged from 39% increases at the Vltava to 33% decreases at Beyşehir, both with the HadRM3 Model. The increase at the Vltava is consistent with less dilution of the wastewater inputs, and the decreases at Beyşehir show that the P is not derived from a fixed volume source such as wastewater but is being leached from the land. Changes at the Tarland Burn and the Thames in contrast are very small. At the Yläneenjoki in Finland, SRP declines even though TP increases. This is likely to be because with greater discharge the soil erosion increases, leading to an increase in PP. Soil erosion is a recognised problem in this catchment. In catchments where agriculture is the major nutrient source, with climate change alone, the loads tend to change in proportion to the change in water flux, whereas concentrations can increase (e.g., TP and SRP at the Vltava due to increased winter transport by storms and reduced summer dilution), remain constant (nitrate at the Tarland), or even decline, as at Beyşehir where the reduction in SRP is approximately 40 to 50% and 20 to 40% for nitrate (Figures S3–S5).

4.4. Projected Change in River Nutrient Concentrations and Loads: Integrated Scenarios

In contrast to river flow, the additional effects of land use changes on river nutrient concentrations are larger than those of climate alone. In general, and given projected climate change, the land use changes representing the "sustainable production" storyline (LU3) reduce nutrient concentrations and loads, and those from the "intensification" storylines (LU1 and LU2) increase them (Figures 6–8). The response is more nuanced for "local stewardship" (LU4), depending on the local interpretation of the scenario, and in some cases, 'greener' agricultural practice lowers the nutrient concentrations, whilst in others the nutrient concentrations increase in response to increased fertiliser use to improve local yields (e.g., the Vltava, Figures 6–8). For all four scenarios, there are also considerable differences in response between sites, reflecting the local mixture of nutrient sources (e.g., agriculture versus wastewater). In most cases, the nutrient concentrations react in the same way independently of the climate model used, but in some instances (notably at the Louros for SRP), the differences between the predictions using different climate models are greater than those between the land scenarios (Figures 6–8).

The modelled changes in NO_3^--N concentrations are greater when there is land use change as well as climate change, except at Beyşehir and the Louros (where all the changes are small; Figure 7). At these sites, the NO_3^--N concentrations decline in virtually all scenarios, whereas further north there is a mixture of responses depending on the scenario, with in general, the "Environmental" scenarios (LU3, LU4) generating reductions. At the Vltava, the LU4 scenario, however, generated an increase in N as this scenario has a higher proportion of arable land (Table S1). The percentage changes in NO_3^--N concentrations are generally smaller than those of the TP or SRP concentrations, except at the Yläneenjoki, which may be because the land use scenarios in that catchment involve increases in agricultural area for LU1 and LU2, an increase in vegetation cover during winter (LU3), and the conversion of approximately 15–20% of arable land to forest (LU4; Figures 6–8; Table S1).

For TP, the modelled changes in concentration and load are also greater when there is land use as well as climate change (Figures 8a and S4). Clearly the effects of the land use changes will depend on the magnitude of the change modelled, but both the climate and the land use changes are best estimates of the likely scenarios; so, it seems valid to compare them. The "Economic" storylines (LU1 and LU2) usually lead to TP increases, whereas the "Environmental" storylines (LU3 and LU4) lead to declines. However, at the Vltava, the LU2 scenario leads to a decline as well, and at the Louros all scenarios lead to P increases where all scenarios involve an increase in agricultural production to some degree. At the Vansjø-Hobøl, the wetter projected futures (HadRM3, SMHI) result in an increase in TP concentration as the projected land use changes are unable to prevent TP transfers from land to stream during storms. At the Arbúcies catchment, the TP increases in the "Economic" scenarios are very large (Figure 8a). This is due largely to assumptions about population growth (larger in these scenarios) and the effectiveness of the sewage treatment works (poorer in the "Economic" scenarios due to lax regulation and smaller investment); here, the effects of agriculture are minor by comparison.

On the Thames, both the land use changes (LU1 and LU2) involve an increase in intensive agriculture and hence SRP output (Figure 8b). LU3 and LU4 represent the same land use change but with the construction of a reservoir, which changes the seasonal flow pattern but has only a small effect on annual P concentrations. Climate change alone has almost no effect on SRP concentrations in the Thames (Figure 8b). At Beyşehir, the SRP concentrations decline in all scenarios, in tandem with the reduction in water discharge. This includes the climate-change-only scenario and indicates less leaching from a catchment source. The "Environmental" scenarios have greater reductions. At the Vltava, for LU1, the change in SRP followed a similar pattern to TP, but the percentage changes were generally greater, again reflecting the importance of the sewage treatment works in this catchment. At the Finnish site of the Yläneenjoki, the pattern of change in SRP was strikingly different from that of TP, with the economic scenarios generating a larger reduction than the environmental scenarios. It is not clear which features of the catchment or scenario modelling are causing this. For SRP, the pattern whereby land use change has a greater effect than climate change is not as general as with TP. At Beyşehir, the modelled changes due to climate change are large; so, the agricultural changes have less effect relatively.

4.5. Projected Change in Lake Nutrient and Chlorophyll Concentrations

Mostly, the changes in nutrient concentrations in lakes due to climate change are only quite small and are smaller than the relative concentration changes in the rivers that feed the lakes [56]. The differences between the predictions of the GCMs are small too. This may be because lakes have mechanisms that buffer concentrations, such as longer residence times and P release from sediments and denitrification, and these reduce the differences between scenarios. Changes in discharge also tend to reduce the differences between scenarios, making lake loadings less variable than the concentrations. The exception is Beyşehir, where reductions in both river discharge and river concentration cause a large change in load (Figures S3 and S5) [32]. The changes in river discharge in the Yläneenjoki, Hobøl, and Vltava are small (1–10%), which makes the input loads of P and N less variable than the concentrations (Figures S3–S5).

Generally, those land use changes designed to reduce lake TP concentrations do so, sometimes considerably, and the pursuit of the "Economic" land use scenarios increases concentrations [56]. The differences between the GCMs are reduced compared to those in the driving variables, with little consistent pattern. At Pyhäjärvi, the SMHI model leads to smaller changes than the other models, except with LU4. The lake SRP concentrations show a higher percentage of change than for TP at Vansjø-Høbol and the Vltava. In the Vltava catchment (the Orlík Reservoir), the TP increases mostly occur in summer and are associated with low flow situations and lower dilutions of wastewater inputs. In Vansjø-Høbol, the nutrient retention measures employed are responsible for the increases and decreases of SRP [23,56]. In the Orlík reservoir (Vltava), the phosphorus concentrations were highest in the LU1 scenarios, exceeding the 1991–1995 state by approximately 60%, which is more than the corresponding increase in the inflow concentrations. In contrast, the LU3 scenarios showed a smaller decrease than that in the TP and SRP concentrations in the inflow. This suggests that the P retention in the reservoir is controlled by a non-linear function of P loading that is apparently also influenced by the changes in the seasonal hydrological pattern and/or climate variables.

Changes in the chl-a concentration due to land use are generally considerably greater than those due to climate change alone (Figure 9); however, the projected chlorophyll changes are less certain than the other variables. In the northern catchments, Lake Pyhäjärvi in Finland and lake Vansjø in Norway, climate change leads to a small (5–10%) increase in chl-a, intensification land use scenarios produce a further large (30–55%) increase, and sustainable production land use scenarios lead to a reduction (between 1 and 20%; Figure 9). In each case, the changes closely mirror those of TP. Temperature-induced increases in algal growth cause an increase in chl-a at the IJsselmeer due to climate change alone, whereas a move to sustainable production leads to a 35% reduction in primary productivity due to decreased external nutrient loading. The IJsselmeer simulations show that higher temperatures increase stress on *Dreissena polymorpha* (Zebra mussel) and *Osmerus eperlanus* (European smelt), reducing food availability for benthivorous and piscivorous birds. At the central European site of Orlík Reservoir (Vltava), a 6–17% chlorophyll decrease due to climate change is attributed to increased P retention in the reservoir due to lower summer flows and higher delivery due to increased winter flows, and land use changes generally caused a 6–32% decrease in chl-a, even when TP increased, again most likely due to increased P retention in the reservoir. Finally, at Beyşehir, the changes in chl-a are largely negative (c. −10%) due to reduced nutrient loading and water flux, and the chl-a changes are substantially smaller than the changes in the water (−16 to −31%) and nutrient (1 to −45%) inputs. At both the Vltava and Beyşehir, there are considerable differences in chl-a concentrations in response to different climate model projections, with the response to KNMI forcing showing a mix of small (0 to 3%) increases and decreases and more consistent and larger decreases (4–17%) in response to SMHI and HadRM3 forcing. Again, this highlights a different lake response, dependent on the driving variables of temperature, precipitation, and discharge, and that HadRM3 has the largest effect on these.

5. Discussion

5.1. The Effect of Climate and Land Cover Change on Stream Water N and P Concentrations

The modelled outcomes suggest that large-scale land use change under "sustainable production" scenarios will reduce nutrient concentrations and loads in general, and there is no evidence that "sustainable production" changes would be less effective under a future climate. In general, the modelled NO_3^--N concentration changes are greater when there is land use change as well as climate change, except at Beyşehir and the Louros (where all the changes are small). At these two sites, the NO_3^--N concentrations decline in virtually all scenarios, whereas further north there is a mixture of responses depending on the scenario with, in general, the "sustainable production" scenarios (LU3 and LU4) generating reductions. At the Vltava, the LU4 scenario generated an increase in NO_3^--N because this scenario has a higher proportion of arable land to provide food locally. For TP, the modelled changes in concentration due to land use change are generally greater than with climate change alone (Figure 8a). For SRP, land use change also generally causes a greater change than climate alone, but at Beyşehir the change due to climate is substantial (less precipitation and higher temperatures); so, the agricultural changes have relatively little effect except for the move to global sustainability (LU4) that shows the greatest SRP percentage decline for concentration. The percentage changes in the nitrate concentrations are generally somewhat smaller than those of the TP or SRP concentrations, except at the Yläneenjoki, where the localised land use scenarios involve large changes in agricultural practice, including greater use of winter cover crops, and a high level of sewage treatment of phosphorus is already in place. The result that 'local stewardship' can in some cases increase stream and lake nutrient concentrations is important. This outcome highlights that growing crops in areas with a greater capability for agricultural production, though with higher energy requirements for transportation of crops to market, but with lower nutrient input requirements, might be a more sustainable strategy overall than a focus on local production where more nutrient inputs are needed to improve yields.

The changes in TP concentration due to climate change alone are mostly small, whereas changes in SRP concentration are larger and more variable, both between sites and between climate models. The differences seem mostly due to differences in the types of P source present at each site. Where there is a substantial wastewater input, as at the Vltava site, reducing water volumes increases the instream nutrient concentrations as wastewater inputs are a reasonably constant volume and the reduced dilution of these causes higher TP and SRP concentrations, especially during summer. Where agriculture is the major source of nutrients, loads can decline in proportion to the declining water flux, and hence, concentrations do not change or even decline, as at Beyşehir, where the reduction in SRP is substantial. The NO_3^--N concentration and load shows little change due to climate change alone, the modelled changes being mostly small declines. Given that the mix of point and diffuse sources is likely to change as catchment size increases, and nutrient retention in lakes, reservoirs, and the groundwater will be important in the largest basins, then a more nuanced understanding of how sources and stores change with catchment size is needed, and current modelled projections that consider diffuse sources only are likely to be too generalized. There is the danger that we are drawing the wrong conclusions because the mix of sources and retention is not yet adequately characterized in our assessments of future nutrient source and transport. It is unclear if larger catchments are as sensitive to nutrient loss during precipitation events as has been observed in small scale studies of agricultural catchments. This further emphasizes the need for more studies of nutrient loss across spatial scales. Given the low explained variance of the nutrient simulations, large uncertainty remains regarding nutrient availability and transport under high flows in large catchments, and more field data are needed from nested catchment studies to determine if greater precipitation intensity will increase nutrient loading along a continuum of small (<10 km^2) to large (>10,000 km^2) catchment areas [57,58].

5.2. The Strength of Observational Evidence to Support the Model-Based Assessments

A 140-year dataset for the River Thames, UK, provides evidence for a strong link between stream water nitrate concentrations and land use, rather than climate, where increased arable area and fertiliser application rates correlate to a four-fold increase in stream water NO_3^--N concentrations [59]. At the global scale, the density of water quality observations is low outside developed countries, and therefore, it is difficult to determine current trends in stream and lake water N, P, and chl-a concentrations. Those data that do exist also show a strong link between stream water N and P concentrations and land use and land cover change, with increases in total phosphorus loadings to lakes evident in east Asia and South America where fertiliser application is high, and some reductions in central Asia are likely due to lower fertiliser application following the separation of the former Soviet Union into multiple, independent states [60]. In the USA, there is little evidence of a clear trend in stream water N and P concentrations [61]. Across Europe, recent decreases in stream and lake water N and P have been observed, and these have largely been ascribed to more efficient, or lower, fertiliser and manure use and a greater removal of nutrients during wastewater treatment [62,63]. In China, reductions in stream water P have also been observed in response to improved wastewater treatment [64]. Thus, at the larger (>50 km^2) spatial scales, there is observational evidence that supports the model-based assertions made here and elsewhere [12], namely that land use is the dominant control on stream water nutrient concentrations, rather than climate, and a stream water nutrient response can be expected in response to land use change if the change is sufficient to outweigh any P legacy effect [65,66].

5.3. Comparison with Other Projections of Future Change in Nutrient Concentrations

Comparison of the modelled trend gradient and absolute change in the stream water nutrients, between this study and others, is difficult and highlights a current lack of consistency in approach because such assessments are at an exploratory stage. Current studies tend to focus on either nutrient concentration or load, report results either for rivers or lakes, but not both, focus on catchments of a specific scale only, ignore the mixture of diffuse and point sources, and consider different future periods for scenario analysis. Some comparison was possible with the study of Mack et al., 2019 [11], and there is agreement on the small degree of expected change in catchments in Finland and Norway in response to climate and land use change. Comparison of general trends (i.e., an increase or decrease) is possible and observational and model-based studies demonstrate that stream water concentrations do respond to land cover change and that there is societal choice on the balance between food production, profitability, urbanisation, and the environment [4,11,14,67]. However, questions remain about the degree of land use change necessary and about how long before a response is observed. Reductions in P inputs of between 20 to 80% and in N inputs of between 9 to 57% have been suggested in general to mitigate increased losses due to projected increases in precipitation intensity [14]. In the Louros catchment, discussions with stakeholders suggest that a reduction of 50 to 75% in P input is feasible [30]. Decreases in SRP have already been observed in the Vansjø-Høbol and Ylaneejoki catchments in response to greater use of winter cover crops, buffer strips, and wastewater treatment. To help answer the question about how much nutrient reduction is needed, all model-based assessments need to report the input reduction together with the change in the nutrient output concentration and load.

Numerous modelling studies have looked at the projected changes in the stream water nutrient concentrations of the River Thames with the modelled response to climate change alone, suggesting only a relatively small change in nitrate concentrations [36]. Consideration of the co-evolution of climate and land cover change also suggests that the stream water nitrate changes will be small [67]. This is likely because of the large store of groundwater nitrate which continues to move to the stream network irrespective of land cover change [68]. Much bigger changes are projected for SRP because the simulated land cover change has a much greater impact [10]. This emphasizes the need to consider long-term changes due to the legacy of pollutants in groundwater, soils, and bed sediments to determine when changes in the stream water concentrations might occur [4,9,68] and whether N and P can be reduced together by targeting the same sources. A previous study of the River Kennet, which is a tributary of the Thames, has suggested stream water NO_3^--N reductions may occur through denitrification because of higher temperatures [9], and further work is needed to quantify the importance of this N transformation pathway across a range of catchment types and river lengths.

There is also a need to consider both stream water concentrations and loads, as concentration may change in a different direction to load due to changes in flow. Concentration is important for biological impacts, whereas the load describes delivery to lake, coastal, and estuarine areas. Load will almost always increase with increased flow, but concentration is important especially during ecologically sensitive periods such as summer low flows. Model-based assessments also need to account for exhaustion of the N and P supply, which is evident from detailed studies of nutrient transport, and at the continental and global scale, nutrient retention in lakes and reservoirs is also important [13,69].

The model projections presented are a first step towards quantifying how proposed land use change will affect stream water N and P concentrations given projected climate change. Given that the results suggest that the 'environmental' land use changes proposed here reduce N and P concentrations and remain effective despite climate change, then a next step is to explore more specific land-management practices (e.g., Nature Based Solutions), and which of these might be most effective, or whether only large-scale land use change will result in change. This next step will need to consider that N and P differ in their availability to crops dependent on the applied form (i.e., fertilizers, manures, and wastewater applications) with P typically being less available than the supplied N. More nuanced assessments will also need to consider crop N and P requirements and yield response, predominant transport pathways, and the legacy N and P that has built in groundwaters (N) and catchment soils and stream and lake-bed sediments (P) that can provide a significant internal loading to freshwaters.

5.4. Will Reductions in Nutrient Concentrations Lead to Better Freshwater Ecology?

Where nutrients have been reduced, a commensurate reduction in stream and lake water algal blooms or improvement in plant diversity has not always been observed [70]. This has been ascribed to local factors typically, including internal loading of P that maintains high P concentrations in the reach or lake water column [71], long residence times of NO_3^--N stored in groundwater-dominated catchments [68,72], and the complexity of aquatic ecology in which many factors, including planned and inadvertent species introductions, affect the ecosystem. Reach-scale and lake light and temperature regimes have also been found to be as, if not more, important for primary production than nutrient concentrations in some cases [7,73]. Thus, whilst moves to more sustainable production will likely reduce freshwater N and P concentrations, the ecological response is far less predictable, as demonstrated here where the modelled lake nutrient and chl-a concentrations are far more variable than those in the stream water. The result from the IJsselmeer study is an interesting one, namely that the reduction in primary productivity due to NO_3^--N reduction will provide less food for birds, which emphasizes the need for a whole ecosystem view.

6. Conclusions

The simulations of the observed stream water flow, NO_3^--N, TP, and SRP and lake chl-a concentrations were satisfactory and sufficient to further explore future changes in the mean annual flows and concentrations (Objective 1). Building models that adequately represent the flow and concentration extremes for rivers and lakes remains challenging due to issues around adequate characterization of the model inputs, the representation of catchment water and pollutant storage and transport, and the complex and multiple physiochemical and biological interactions in lake ecosystems. The under-representation of high flow events in water quality monitoring networks often means there are insufficient data to constrain water quality models.

The projected effects of future climate and land use change on stream flow and stream and lake N, P, and chl-a concentrations differ between the northern and southern sites. Considering climate change alone, then in the north and mid-latitudes of Europe, the projected increases in temperatures are balanced to some extent by increased precipitation, leading to relatively small effects on water flows. In the south, at low altitudes, increased temperatures and lower precipitation act in the same direction to reduce water flows considerably. In the case of Lake Beyşehir in neighbouring Turkey, this may even lead to the lake drying up in the foreseeable future, and this effect would far outweigh any nutrient-related problems. In general, in the north and mid-latitudes of Europe, the effects of climate change alone on nutrient concentrations and loads are rather small except where point sources dominate, and reduced flows lead to less dilution of sewage inputs or in those catchments where soil bound P transport during storms is substantial and may therefore increase in the future (Objective 2). The effects due to large-scale land use change are generally much larger, with environmentally focused land use change generally reducing N and P concentrations (Objective 3). Thus, there is considerable potential to reduce the nutrient concentrations and loads across Europe despite projected climate change. Adoption of land use practices, equivalent to the sustainable production land use scenarios, would mitigate climate-induced changes even in some southern European sites at higher elevations. More accurate quantification of nutrient transport during periods of intense rainfall in large catchments is needed to reduce uncertainty in nutrient concentration and load simulation.

Modelled lake chl-a changes were not generally proportional to the changes in nutrient inputs (Objective 4). Ecological change can be less than the nutrient change (e.g., chlorophyll at Lake Beyşehir and the Orlík Reservoir in the Czech Republic) or greater due to complex food web interactions (e.g., at the IJsselmeer). Of all the projections in this study, the lake chlorophyll concentrations have the greatest uncertainty, and thus, the current emphasis to understand better the aquatic ecosystem response to multiple drivers of change is appropriate. Given the socio-economic implications of a shift to more sustainable agriculture practices, a better quantified understanding of how reductions in N and P translate to an ecosystem response seems imperative.

Supplementary Materials: The following supporting information can be downloaded at: https://www.mdpi.com/article/10.3390/w14050829/s1, Table S1. Description of land use (including wastewater treatment), water use and nitrogen deposition changes under the scenarios: LU1 = SRES A1, World market; LU2 = SRES A2, National enterprise, LU3 = SRES B1, Global sustainability, LU4 = SRES B2, Local stewardship. Nitrogen deposition was modelled based on European Monitoring and Evaluation Programme data and a chemical transport model developed at the Meteorological Synthesizing Centre–West (MSC-W). Wet and dry forms and land cover where accounted for in the catchment nitrogen deposition estimate. Further details are given in Jackson-Blake et al., 2008. Table S2. Summary of GCM-RCM model combinations and land use scenarios consisdered at each study area. For some study areas, there are two rows to note that the climate model combinsations and land use scenarios varied when considering different aspects of water quality. This was done to make best use of existing model set-ups and simulations to integrate these within the study and increase the number of model simulations overall. Figure S1. Mean air temperatures, present (1981–2010) and future (2031–2060), at the study areas based on runs with three different climate models: ECHAM5-KNMI (blue); SMHIRCA-BCM (red); and HadRM3P-HadCM3Q0 (orange). At some sites, observed temperatures are used instead of modelled values (purple). Figure S2. Mean annual precipitation, present (1981–2010) and future (2031–2060), at the study sites based on runs with three different climate models: ECHAM5-KNMI (blue); SMHIRCA-BCM (red); and HadRM3P-HadCM3Q0 (orange). At some sites, observed precipitation was used instead of modelled values (purple). Figure S3. The percentage change in stream water nitrate load at each of the study areas between the baseline (1981–2010) and future (2031–2060) periods due to climate change (climate) and climate and land cover change (LU1–4). Blank columns indicate that GCM-RCM combination was not used for a study site. A '0' notes a percentage change of zero. Figure S4. The percentage change in stream water total phosphorus load at each of the study areas between the baseline (1981–2010) and future (2031–2060) periods due to climate change (climate) and climate and land cover change (LU1–4). Blank columns indicate that GCM-RCM combination was not used for a study site. A '0' notes a percentage change of zero. Figure S5. The percentage change in stream water soluble reactive phosphorus load at each of the study areas between the baseline (1981–2010) and future (2031–2060) periods due to climate change (climate) and climate and land cover change (LU1–4). Blank columns indicate that GCM-RCM combination was not used for a study site. A '0' notes a percentage change of zero.

Author Contributions: Conceptualization, A.J.W. and R.A.S.; methodology, R.-M.C., M.E.L., S.G., S.J.H., V.H., J.H., L.A.J.-B., A.L., E.P., J.L.R., K.R., M.S., D.T., P.G.W., D.P. and D.S.; software, A.J.W., L.A.J.-B., A.L., K.R. and P.G.W.; formal analysis, R.-M.C., M.E.L., S.G., S.J.H., V.H., J.H., L.A.J.-B., A.L., E.P., J.L.R., K.R., M.S., D.T., P.G.W., D.P. and D.S.; writing—original draft preparation, A.J.W. and R.A.S.; writing—review and editing, R.-M.C., M.E.L., S.G., S.J.H., V.H., J.H., L.A.J.-B., A.L., E.P., J.L.R., K.R., M.S., D.T., P.G.W., D.P. and D.S.; visualization, A.J.W. and R.A.S.; project administration, A.J.W. and R.A.S. All authors have read and agreed to the published version of the manuscript.

Funding: This research was funded by the European Union's Seventh Programme for research and technological development including demonstration activities under grant number 244121. The FIRI2021 HYDRO-RI platform was supported by the Academy of Finland (grant 346165).

Institutional Review Board Statement: Not applicable.

Informed Consent Statement: Not applicable.

Data Availability Statement: The data presented in this study are available on request from the authors responsible for each case study, though some of the data may be held by third parties and permission would need to be sought to obtain those data. Further details about the data used and detailed reports about the model applications, including the data used and the sources, are given in individual case study project reports, and these are cited in Table 2.

Conflicts of Interest: The authors declare no conflict of interest. The funders had no role in the design of the study; in the collection, analyses, or interpretation of data; in the writing of the manuscript; or in the decision to publish the results.

References

1. Smith, V.H.; Tilman, G.D.; Nekola, J.C. Eutrophication: Impacts of excess nutrient inputs on freshwater, marine, and terrestrial ecosystems. *Environ. Pollut.* **1999**, *100*, 179–196. [CrossRef]
2. Robins, P.E.; Skov, M.W.; Lewis, M.J.; Gimenez, L.; Davies, A.G.; Malham, S.K.; Neill, S.P.; McDonald, J.E.; Whitton, T.A.; Jackson, S.E.; et al. Impact of climate change on UK estuaries: A review of past trends and potential projections. *Estuar. Coast. Shelf Sci.* **2016**, *169*, 119–135. [CrossRef]
3. Carpenter, S.R.; Caraco, N.F.; Correll, D.L.; Howarth, R.W.; Sharpley, A.N.; Smith, V.H. Nonpoint pollution of surface waters with phosphorus and nitrogen. *Ecol. Appl.* **1998**, *8*, 559–568. [CrossRef]
4. Ockenden, M.C.; Hollaway, M.J.; Beven, K.J.; Collins, A.L.; Evans, R.; Falloon, P.D.; Forber, K.J.; Hiscock, K.M.; Kahana, R.; Macleod, C.J.A.; et al. Major agricultural changes required to mitigate phosphorus losses under climate change. *Nat. Commun.* **2017**, *8*, 161. [CrossRef]

5. Whitehead, P.G.; Wilby, R.L.; Battarbee, R.W.; Kernan, M.; Wade, A.J. A review of the potential impacts of climate change on surface water quality. *Hydrol. Sci. J.* **2009**, *54*, 101–123. [CrossRef]
6. Michalak, A.M. Study role of climate change in extreme threats to water quality. *Nature* **2016**, *535*, 349–350. [CrossRef]
7. Bowes, M.J.; Loewenthal, M.; Read, D.S.; Hutchins, M.G.; Prudhomme, C.; Armstrong, L.K.; Harman, S.A.; Wickham, H.D.; Gozzard, E.; Carvalho, L. Identifying multiple stressor controls on phytoplankton dynamics in the River Thames (UK) using high-frequency water quality data. *Sci. Total Environ.* **2016**, *569*, 1489–1499. [CrossRef]
8. Withers, P.J.A.; Neal, C.; Jarvie, H.P.; Doody, D.G. Agriculture and eutrophication: Where do we go from here? *Sustainability* **2014**, *6*, 5853–5875. [CrossRef]
9. Wilby, R.L.; Whitehead, P.G.; Wade, A.J.; Butterfield, D.; Davis, R.J.; Watts, G. Integrated modelling of climate change impacts on water resources and quality in a lowland catchment: River Kennet, UK. *J. Hydrol.* **2006**, *330*, 204–220. [CrossRef]
10. Crossman, J.; Whitehead, P.G.; Futter, M.N.; Jin, L.; Shahgedanova, M.; Castellazzi, M.; Wade, A.J. The interactive responses of water quality and hydrology to changes in multiple stressors, and implications for the long-term effective management of phosphorus. *Sci. Total Environ.* **2013**, *454*, 230–244. [CrossRef]
11. Mack, L.; Andersen, H.E.; Beklioglu, M.; Bucak, T.; Couture, R.M.; Cremona, F.; Ferreira, M.T.; Hutchins, M.G.; Mischke, U.; Molina-Navarro, E.; et al. The future depends on what we do today—Projecting Europe's surface water quality into three different future scenarios. *Sci. Total Environ.* **2019**, *668*, 470–484. [CrossRef] [PubMed]
12. Reder, K.; Barlund, I.; Voss, A.; Kynast, E.; Williams, R.; Malve, O.; Florke, M. European scenario studies on future in-stream nutrient concentrations. *Trans. Asabe* **2013**, *56*, 1407–1417.
13. Beusen, A.H.W.; Bouwman, A.F.; Van Beek, L.P.H.; Mogollon, J.M.; Middelburg, J.J. Global riverine N and P transport to ocean increased during the 20th century despite increased retention along the aquatic continuum. *Biogeosciences* **2016**, *13*, 2441–2451. [CrossRef]
14. Sinha, E.; Michalak, A.M.; Balaji, V. Eutrophication will increase during the 21st century as a result of precipitation changes. *Science* **2017**, *357*, 405–408. [CrossRef] [PubMed]
15. Kronvang, B.; Behrendt, H.; Andersen, H.E.; Arheimer, B.; Barr, A.; Borgvang, S.A.; Bouraoui, F.; Granlund, K.; Grizzetti, B.; Groenendijk, P.; et al. Ensemble modelling of nutrient loads and nutrient load partitioning in 17 European catchments. *J. Environ. Monit.* **2009**, *11*, 572–583. [CrossRef]
16. Hurkmans, R.; Terink, W.; Uijlenhoet, R.; Torfs, P.; Jacob, D.; Troch, P.A. Changes in Streamflow Dynamics in the Rhine Basin under Three High-Resolution Regional Climate Scenarios. *J. Clim.* **2010**, *23*, 679–699. [CrossRef]
17. Wade, A.J.; Skeffington, R.A.; Couture, R.M.; Erlandsson, M.; Groot, S.; Halliday, S.J.; Harelzak, V.; Hejzlar, J.; Jackson-Blake, L.A.; Lepisto, A.; et al. *The REFRESH Common Modelling Framework for the Demonstration Catchments*; University College London: London, UK, 2013; p. 27. Available online: http://www.refresh.ucl.ac.uk/webfm_send/1968 (accessed on 10 January 2022).
18. Schelske, C.L.; Stoermer, E.F.; Fahnenstiel, G.L.; Haibach, M. Phosphorus enrichment, silica utilization, and biogeochemical silica depletion in the Great Lakes. *Can. J. Fish. Aquat. Sci.* **1986**, *43*, 407–415. [CrossRef]
19. Harelzak, V.; Groot, S.; Duel, H. *Final Report on the Biophysical Modelling of Lake Ijsselmeer*; University College London: London, UK, 2014; p. 44. Available online: http://www.refresh.ucl.ac.uk/webfm_send/2389 (accessed on 10 January 2022).
20. Rankinen, K.; Granlund, K.; Futter, M.N.; Butterfield, D.; Wade, A.J.; Skeffington, R.; Arvola, L.; Veijalainen, N.; Huttunen, I.; Lepisto, A. Controls on inorganic nitrogen leaching from Finnish catchments assessed using a sensitivity and uncertainty analysis of the INCA-N model. *Boreal Environ. Res.* **2013**, *18*, 373–386.
21. Etheridge, J.R.; Lepisto, A.; Granlund, K.; Rankinen, K.; Birgand, F.; Burchell, M.R. Reducing uncertainty in the calibration and validation of the INCA-N model by using soft data. *Hydrol. Res.* **2014**, *45*, 73–88. [CrossRef]
22. Lepistö, A.; Etheridge, J.R.; Granlund, K.; Kotamäki, N.; Malve, O.; Rankinen, K.; Varjopuro, R. *Report on the Biophysical Catchment-Scale Modelling of Yläneenjoki–Pyhäjärvi Demonstration Site*; University College London: London, UK, 2014; p. 67. Available online: http://www.refresh.ucl.ac.uk/webfm_send/2161 (accessed on 10 January 2022).
23. Couture, R.M.; Tominaga, K.; Starrfelt, J.; Moe, S.J.; Kaste, Ø.; Wright, R.; Farkas, C.; Engebretsen, A. *Report on the Catchment-scale Modelling of the Vansjø-Hobøl and Skuterud Catchments, Norway*; University College London: London, UK, 2014; p. 59. Available online: http://www.refresh.ucl.ac.uk/webfm_send/2209 (accessed on 10 January 2022).
24. Couture, R.M.; Tominaga, K.; Starrfelt, J.; Moe, S.J.; Kaste, O.; Wright, R.F. Modelling phosphorus loading and algal blooms in a Nordic agricultural catchment-lake system under changing land-use and climate. *Environ. Sci.-Processes Impacts* **2014**, *16*, 1588–1599. [CrossRef]
25. Couture, R.-M.; Moe, S.J.; Lin, Y.; Kaste, Ø.; Haande, S.; Lyche Solheim, A. Simulating water quality and ecological status of Lake Vansjø, Norway, under land-use and climate change by linking process-oriented models with a Bayesian network. *Sci. Total Environ.* **2018**, *621*, 713–724. [CrossRef]
26. Jackson-Blake, B.M.; Dunn, S.M.; Hershkovitz, Y.; Sample, J.; Helliwell, R.C.; Balana, B. *Biophysical Catchment-Scale Modelling in the River Dee Catchment, Scotland*; University College London: London, UK, 2014; p. 111. Available online: http://www.refresh.ucl.ac.uk/webfm_send/2163 (accessed on 10 January 2022).
27. Jackson-Blake, L.A.; Wade, A.J.; Futter, M.N.; Butterfield, D.; Couture, R.M.; Cox, B.A.; Crossman, J.; Ekholm, P.; Halliday, S.J.; Jin, L.; et al. The Integrated Catchment model of phosphorus dynamics (INCA-P): Description and demonstration of new model structure and equations. *Environ. Model. Softw.* **2016**, *83*, 356–386. [CrossRef]

28. Hejzlar, J.; Jarošík, J.; Kopáček, J. *River Vltava Modelling, Final Report*; University College London: London, UK, 2014; p. 32. Available online: http://www.refresh.ucl.ac.uk/webfm_send/2387 (accessed on 10 January 2022).
29. Erlandsson, M.; Wade, A.J.; Riera, J.L.; Puig, M.; Skeffington, R.A.; Halliday, S.J. *River Arbúcies Biophysical Modelling, Final Report*; University College London: London, UK, 2014; p. 67. Available online: http://www.refresh.ucl.ac.uk/webfm_send/2255 (accessed on 10 January 2022).
30. Erlandsson, M.; Wade, A.J.; Hershkovitz, Y.; Papadaki, C.; Manolaki, P.; Papastergiadou, E. *River Louros Modeling, Final Report*; University College London: London, UK, 2014; p. 53. Available online: http://www.refresh.ucl.ac.uk/webfm_send/2162 (accessed on 10 January 2022).
31. Beklioğlu, M.; Bucak, T.; Erdoğan, S.; Çakıroğlu, A.I.; Trolle, D.; Andersen, H.E.; Thodsen, H.; Elliott, J.A. *Lake Beyşehir Modelling: Final Report*; University College London: London, UK, 2014; p. 30. Available online: http://www.refresh.ucl.ac.uk/webfm_send/2158 (accessed on 10 January 2022).
32. Bucak, T.; Trolle, D.; Andersen, H.E.; Thodsen, H.; Erdogan, S.; Levi, E.E.; Filiz, N.; Jeppesen, E.; Beklioglu, M. Future water availability in the largest freshwater Mediterranean lake is at great risk as evidenced from simulations with the SWAT model. *Sci. Total Environ.* **2017**, *581*, 413–425. [CrossRef] [PubMed]
33. Bucak, T.; Trolle, D.; Tavsanoglu, U.N.; Cakiroglu, A.I.; Ozen, A.; Jeppesen, E.; Beklioglu, M. Modeling the effects of climatic and land use changes on phytoplankton and water quality of the largest Turkish freshwater lake: Lake Beysehir. *Sci. Total Environ.* **2017**, *621*, 802–816. [CrossRef] [PubMed]
34. Skarbøvik, E.; Haande, S.; Bechmann, M. *Overvåking Vansjø/Morsa 2011–2012*; Resultater fra overvåkingen I perioden oktober 2011 til oktober 2012; Bioforsk: Ås, Norway, 2013; Volume 8, p. 212.
35. Skarbøvik, E.; Bechmann, M.E. *Some Characteristics of the Vansjø-Hobøl (Morsa) Catchment*; Bioforsk Soil and Environment: Ås, Norway, 2010; p. 44.
36. Jin, L.; Whitehead, P.G.; Futter, M.N.; Lu, Z.L. Modelling the impacts of climate change on flow and nitrate in the River Thames: Assessing potential adaptation strategies. *Hydrol. Res.* **2012**, *43*, 902–916. [CrossRef]
37. Ovezikoglou, V.; Ladakis, M.; Dassenakis, M.; Skoullos, M. The fluctuation of nutrients and organic carbon in the waters of some rivers in the Western Greece. In Proceedings of the 8th International Conference on Environmental Science and Technology, Lemnos Island, Greece, 8–10 September 2003; pp. 628–632.
38. Mackereth, F.J.H.; Heron, J.; Talling, J.F. Water Analysis: Some revised methods for limnologists. *Freshw. Biol. Assoc. Sci. Publ.* **1978**, *36*, 117.
39. Jespersen, A.M.; Christoffersen, K. Measurements of chlorophyll-a from phytoplankton using ethanol as extraction solvent. *Arch. Hydrobiol.* **1987**, *109*, 445–454.
40. Wade, A.J.; Durand, P.; Beaujouan, V.; Wessel, W.W.; Raat, K.J.; Whitehead, P.G.; Butterfield, D.; Rankinen, K.; Lepisto, A. A nitrogen model for European catchments: INCA, new model structure and equations. *Hydrol. Earth Syst. Sci.* **2002**, *6*, 559–582. [CrossRef]
41. Arnold, J.G.; Moriasi, D.N.; Gassman, P.W.; Abbaspour, K.C.; White, M.J.; Srinivasan, R.; Santhi, C.; Harmel, R.D.; van Griensven, A.; Van Liew, M.W.; et al. SWAT: Model use, calibration, and validation. *Trans. Asabe* **2012**, *55*, 1491–1508. [CrossRef]
42. Dunn, S.M.; McDonnell, J.J.; Vache, K.B. Factors influencing the residence time of catchment waters: A virtual experiment approach. *Water Resour. Res.* **2007**, *43*, W06408. [CrossRef]
43. Hamilton, D.P.; Schladow, S.G. Prediction of water quality in lakes and reservoirs. 1. Model description. *Ecol. Model.* **1997**, *96*, 91–110. [CrossRef]
44. Schladow, S.G.; Hamilton, D.P. Prediction of water quality in lakes and reservoirs. 2. Model calibration, sensitivity analysis and application. *Ecol. Model.* **1997**, *96*, 111–123. [CrossRef]
45. Elliott, J.A.; Irish, A.E.; Reynolds, C.S. Modelling phytoplankton dynamics in fresh waters: Affirmation of the PROTECH approach to simulation. *Freshw. Rev.* **2010**, *3*, 75–96. [CrossRef]
46. Janse, J.H. A model of nutrient dynamics in shallow lakes in relation to multiple stable states. *Hydrobiologia* **1997**, *342*, 1–8.
47. Kotamaki, N.; Patynen, A.; Taskinen, A.; Huttula, T.; Malve, O. Statistical dimensioning of nutrient loading reduction: LLR assessment tool for lake managers. *Environ. Manag.* **2015**, *56*, 480–491. [CrossRef] [PubMed]
48. Saloranta, T.M.; Andersen, T. MyLake—A multi-year lake simulation model code suitable for uncertainty and sensitivity analysis simulations. *Ecol. Model.* **2007**, *207*, 45–60. [CrossRef]
49. Cole, T.M.; Wells, S.A. *CE-QUAL-W2: A Two-Dimensional, Laterally Averaged, Hydrodynamic and Water Quality Model, 3.7*; Department of Civil and Environmental Engineering, Portland State University: Portland, OR, USA, 2011.
50. Jackson-Blake, L.A.; Starrfelt, J. Do higher data frequency and Bayesian auto-calibration lead to better model calibration? Insights from an application of INCA-P, a process-based river phosphorus model. *J. Hydrol.* **2015**, *527*, 641–655. [CrossRef]
51. Jolliff, J.K.; Kindle, J.C.; Shulman, I.; Penta, B.; Friedrichs, M.A.M.; Helber, R.; Arnone, R.A. Summary diagrams for coupled hydrodynamic-ecosystem model skill assessment. *J. Mar. Syst.* **2009**, *76*, 64–82. [CrossRef]
52. van der Linden, P.; Mitchell, J.F.B. *ENSEMBLES: Climate Change and Its Impacts: Summary of Research and Results from the ENSEMBLES Project*; Met Office Hadley Centre: Exeter, UK, 2009; p. 160.
53. Nakicenovic, N.; Alcamo, J.; Davis, G.; de Vries, H.J.M.; Fenhann, J.; Gaffin, S.; Gregory, K.; Grubler, A.; Jung, T.Y.; Kram, T.; et al. *IPCC Special Report: Emissions Scenarios*; Cambridge University Press: Cambridge, UK, 2000; p. 570.

54. Jackson-Blake, L.A.; Dunn, S.M.; Helliwell, R.C.; Skeffington, R.A.; Stutter, M.I.; Wade, A.J. How well can we model stream phosphorus concentrations in agricultural catchments? *Environ. Model. Softw.* **2015**, *64*, 31–46. [CrossRef]
55. Kumar, D.; Kodra, E.; Ganguly, A.R. Regional and seasonal intercomparison of CMIP3 and CMIP5 climate model ensembles for temperature and precipitation. *Clim. Dyn.* **2014**, *43*, 2491–2518. [CrossRef]
56. Skeffington, R.A.; Wade, A.J.; Couture, R.M.; Erlandsson, M.; Groot, S.; Halliday, S.J.; Harelzak, V.; Hejzlar, J.; Jackson-Blake, L.A.; Lepisto, A.; et al. *Integrated Catchment Biophysical Modelling: Synthesis Report*; University College London: London, UK, 2014; p. 44. Available online: http://www.refresh.ucl.ac.uk/webfm_send/2383 (accessed on 10 January 2022).
57. Arnell, N.W.; Halliday, S.J.; Battarbee, R.W.; Skeffington, R.A.; Wade, A.J. The implications of climate change for the water environment in England. *Prog. Phys. Geogr.* **2015**, *39*, 93–120. [CrossRef]
58. Wade, A.J.; Palmer-Felgate, E.J.; Halliday, S.J.; Skeffington, R.A.; Loewenthal, M.; Jarvie, H.P.; Bowes, M.J.; Greenway, G.M.; Haswell, S.J.; Bell, I.M.; et al. Hydrochemical processes in lowland rivers: Insights from in situ, high-resolution monitoring. *Hydrol. Earth Syst. Sci.* **2012**, *16*, 4323–4342. [CrossRef]
59. Howden, N.J.K.; Burt, T.P.; Worrall, F.; Whelan, M.J.; Bieroza, M. Nitrate concentrations and fluxes in the River Thames over 140 years (1868–2008): Are increases irreversible? *Hydrol. Processes* **2010**, *24*, 2657–2662. [CrossRef]
60. UNEP. *A Snapshot of the World's Water Quality: Towards a Global Assessment*; United Nations Environment Programme: Nairobi, Kenya, 2016; p. 162.
61. Shoda, M.E.; Sprague, L.A.; Murphy, J.C.; Riskin, M.L. Water-quality trends in US rivers, 2002 to 2012: Relations to levels of concern. *Sci. Total Environ.* **2019**, *650*, 2314–2324. [CrossRef]
62. Stalnacke, P.; Aakeroy, P.A.; Blicher-Mathiesen, G.; Iital, A.; Jansons, V.; Koskiaho, J.; Kyllmar, K.; Lagzdins, A.; Pengerud, A.; Povilaitis, A. Temporal trends in nitrogen concentrations and losses from agricultural catchments in the Nordic and Baltic countries. *Agric. Ecosyst. Environ.* **2014**, *198*, 94–103. [CrossRef]
63. Longphuirt, S.N.; O'Boyle, S.; Stengel, D.B. Environmental response of an Irish estuary to changing land management practices. *Sci. Total Environ.* **2015**, *521*, 388–399. [CrossRef]
64. Cheng, P.; Li, X.Y.; Su, J.J.; Hao, S.N. Recent water quality trends in a typical semi-arid river with a sharp decrease in streamflow and construction of sewage treatment plants. *Environ. Res. Lett.* **2018**, *13*, 014026. [CrossRef]
65. Chen, B.H.; Chang, S.X.; Lam, S.K.; Erisman, J.W.; Gu, B.J. Land use mediates riverine nitrogen export under the dominant influence of human activities. *Environ. Res. Lett.* **2017**, *12*, 094018. [CrossRef]
66. Sharpley, A.; Jarvie, H.P.; Buda, A.; May, L.; Spears, B.; Kleinman, P. Phosphorus Legacy: Overcoming the Effects of Past Management Practices to Mitigate Future Water Quality Impairment. *J. Environ. Qual.* **2013**, *42*, 1308–1326. [CrossRef]
67. Bussi, G.; Janes, V.; Whitehead, P.G.; Dadson, S.J.; Holman, I.P. Dynamic response of land use and river nutrient concentration to long-term climatic changes. *Sci. Total Environ.* **2017**, *590*, 818–831. [CrossRef]
68. Jackson, B.M.; Browne, C.A.; Butler, A.P.; Peach, D.; Wade, A.J.; Wheater, H.S. Nitrate transport in Chalk catchments: Monitoring, modelling and policy implications. *Environ. Sci. Policy* **2008**, *11*, 125–135. [CrossRef]
69. Bowes, M.J.; Jarvie, H.P.; Halliday, S.J.; Skeffington, R.A.; Wade, A.J.; Loewenthal, M.; Gozzard, E.; Newman, J.R.; Palmer-Felgate, E.J. Characterising phosphorus and nitrate inputs to a rural river using high-frequency concentration-flow relationships. *Sci. Total Environ.* **2015**, *511*, 608–620. [CrossRef] [PubMed]
70. O'Hare, M.T.; Baattrup-Pedersen, A.; Baumgarte, I.; Freeman, A.; Gunn, I.D.M.; Lazar, A.N.; Sinclair, R.; Wade, A.J.; Bowes, M.J. Responses of Aquatic Plants to Eutrophication in Rivers: A Revised Conceptual Model. *Front. Plant Sci.* **2018**, *9*, 451. [CrossRef]
71. Powers, S.M.; Bruulsema, T.W.; Burt, T.P.; Chan, N.I.; Elser, J.J.; Haygarth, P.M.; Howden, N.J.K.; Jarvie, H.P.; Lyu, Y.; Peterson, H.M.; et al. Long-term accumulation and transport of anthropogenic phosphorus in three river basins. *Nat. Geosci.* **2016**, *9*, 353–356. [CrossRef]
72. Gu, B.J.; Zhu, Y.M.; Chang, J.; Peng, C.H.; Liu, D.; Min, Y.; Luo, W.D.; Howarth, R.W.; Ge, Y. The role of technology and policy in mitigating regional nitrogen pollution. *Environ. Res. Lett.* **2011**, *6*, 014011. [CrossRef]
73. Richardson, J.; Miller, C.; Maberly, S.C.; Taylor, P.; Globevnik, L.; Hunter, P.; Jeppesen, E.; Mischke, U.; Moe, S.J.; Pasztaleniec, A.; et al. Effects of multiple stressors on cyanobacteria abundance vary with lake type. *Glob. Change Biol.* **2018**, *24*, 5044–5055. [CrossRef]

water

MDPI

Article

Improving Environmental DNA Sensitivity for Dreissenid Mussels by Targeting Tandem Repeat Regions of the Mitochondrial Genome

Nathaniel T. Marshall [1,2], Henry A. Vanderploeg [3] and Subba Rao Chaganti [1,*]

1 Cooperative Institute for Great Lakes Research (CIGLR), University of Michigan, 4840 South State Road, Ann Arbor, MI 48108, USA; nathaniel.marshall@stantec.com
2 Stantec Consulting Ltd., 1500 Lake Shore Dr, Columbus, OH 43204, USA
3 Great Lakes Environmental Research Laboratory, National Oceanic and Atmospheric Administration, 4840 South State Road, Ann Arbor, MI 48108, USA; henry.vanderploeg@noaa.gov
* Correspondence: chaganti@umich.edu

Abstract: The recent genetic revolution through the analysis of aquatic environmental DNA (eDNA) has become a powerful tool for improving the detection of rare and/or invasive species. For the majority of eDNA studies, genetic assays are designed to target mitochondrial genes commonly referred to as "barcode" regions. However, unlike the typical structure of an animal mitochondrial genome, those for the invasive zebra and quagga mussels are greatly expanded with large extended tandem repeat regions. These sections of repeated DNA can appear hundreds of times within the genome compared to a single copy for the mitochondrial barcode genes. This higher number of target copies per mitochondrial genome presents an opportunity to increase eDNA assay sensitivity for these species. Therefore, we designed and evaluated new eDNA assays to target the extended repeat sections for both zebra and quagga mussels. These assays lower the limit of detection of genomic DNA by 100-fold for zebra mussels and 10-fold for quagga mussels. Additionally, these newly developed assays provided longer durations of detection during degradation mesocosm experiments and greater sensitivity for eDNA detection from water samples collected across western Lake Erie compared to standard assays targeting mitochondrial genes. This work illustrates how understanding the complete genomic structure of an organism can improve eDNA analysis.

Keywords: environmental DNA; quantitative PCR; dreissenid mussels; mitochondrial genome; aquatic invasive species

Citation: Marshall, N.T.; Vanderploeg, H.A.; Chaganti, S.R. Improving Environmental DNA Sensitivity for Dreissenid Mussels by Targeting Tandem Repeat Regions of the Mitochondrial Genome. *Water* **2022**, *14*, 2069. https://doi.org/10.3390/w14132069

Academic Editor: George Arhonditsis

Received: 22 April 2022
Accepted: 23 June 2022
Published: 28 June 2022

1. Introduction

The closely related dreissenid bivalve zebra mussel (*Dreissena polymorpha*, referred to herein as ZM) and quagga mussel (*D. rostriformis*, referred to herein as QM) are two of the most widely distributed freshwater aquatic invasive species [1,2]. Dreissenid mussels are ecosystem engineers, causing extensive environmental alterations within invaded habitats, such as the sequestration of nutrients, changes to the benthic habitat and structure, increased water clarity via prolific filtration, and changes to the composition and biomass of phytoplankton [1,3,4]. Within North America, the first appearance of both species was reported in the Laurentian Great Lakes, with the discovery of ZMs in 1986 from Lake St. Clair [5] and QMs in 1989 from Lake Erie [6]. Since then, these species have continued their invasional spread to inland water bodies via recreational boating [7]. Furthermore, new invasion pathways have been identified through the aquatic pet and plant trade [8,9], thus posing additional risk.

The early detection of new dreissenid invasions is critical for implementing successful eradication and management strategies [10]. Additionally, monitoring population densities and the invasional front along a waterbody are important for estimating dreissenid-related ecological impacts [11,12]. As a result, many studies have aimed to develop high-sensitivity molecular methods for the detection of new invasions of dreissenid mussels [13]. Initially, molecular methods were aimed at detecting dreissenid veliger larvae from zooplankton tow samples [14]. However, with the recent genetic revolution through the analysis of aquatic environmental DNA (eDNA—the genetic material released from urine, waste, mucus, or sloughed cells), molecular methods are now routinely implemented for detecting aquatic invasive species from free-floating DNA found within the water column [15]. The detection and semi-quantification of dreissenid infestations using eDNA have shown promise in both lacustrine [16–18] and riverine habitats [18–22].

Molecular methods for the analysis of eDNA can involve targeted species-specific approaches using the conventional polymerase chain reaction (cPCR) or quantitative PCR (qPCR), or passive approaches utilizing metabarcoding for the detection of general biodiversity [23]. Targeted species-specific assays tend to be more sensitive than broad metabarcoding approaches [22]. Furthermore, cPCR has been found to be a fast and cheap method for the early detection of dreissenid veligers [24]; however, cPCR is typically less sensitive than qPCR [25] and is limited in its quantification output [22]. While qPCR provides the highest molecular sensitivity for eDNA analysis, it has failed to detect ZMs within Lake Superior from locations of low abundance [26]. This suggests that current qPCR assays are potentially unreliable for the detection of expanding invasions within large water bodies. Therefore, it is desirable to have a qPCR assay that targets the most abundant DNA fragment hypothesized to be within the environment to ultimately increase the probability of successful collection in the field and successful detection in the laboratory.

Over the past decade, macrobial eDNA assays have been typically designed for mitochondrial (mt) genes, as the mt-genome is expected to be in higher concentrations than the nuclear (nu) genome within a cell [27] and thus are thought to be found in higher concentrations within the environment. A meta-review of more than 500 published qPCR assays for eDNA analysis has shown that ~98% of assays for the detection of macrobial organisms target a mt-gene, with ~77% of those assays targeting just two genes (cytochrome oxidase I (COI) and cytochrome *b* (Cyt *b*)) (data is summarized from [28]). Similarly, of the 29 dreissenid-specific molecular assays, ~72% are designed to target a mt-gene, with ~76% of those targeting COI (Table S1). However, these commonly targeted mt-genes appear as only a single copy within a mt-genome, and thus multi-copy genes (e.g., ribosomal nu-DNA genes) may provide higher sensitivity for eDNA detection. Targeting multi-copy genes within molecular microbial studies has provided increased sensitivity with qPCR analysis [29,30]. Yet, only a few macrobial eDNA studies have compared the sensitivity between single-copy and multi-copy gene targets, with the results suggesting increased sensitivity for multi-copy gene assays for fish [31–33] and bivalves [34]. Thus, the development of the most sensitive eDNA assay requires knowledge about the genomic structure (e.g., the gene location and the number of copies per genome) of the taxa of interest.

Recently, genome sequencing has been completed for the entire nu- and mt-genomes of ZMs [35] and for the entire mt-genome of QMs [36]. This new breadth of information provides tremendous insight into the genomic composition and structure of these two species. Unlike the typical structure of other animal mt-genomes, which are ~14–20 kilobases (kb) in length and homoplasmic, dreissenid mt-genomes appear more similar to many plant mt-genomes, which display frequent gene rearrangements, greatly expanded repetitive regions, encode various open reading frames of unknown function, and can be heteroplasmic [35,36]. Of particular interest for eDNA applications, both dreissenid mussels display largely expanded mt-genomes (ZM: ~67 kb and QM: ~46 kb) composed of long extended tandem repeat regions (three repeat regions totaling >50 kb and seven repeat regions totaling > 30 kb, respectively) [35,36]. These extended regions can be repeated hundreds of times per mt-genome, compared to only a single copy for any of the coding mt-genes (e.g., 16S, COI, or Cyt *b*).

In light of this new whole mt-genomic data, we aimed to develop species-specific qPCR assays that target highly repetitive regions of non-coding mt-DNA. We hypothesized these highly repetitive non-coding mt-DNA sections would be found in higher concentrations within tissue and eDNA water samples compared to any nu- or mt- coding gene region and thus provide higher sensitivity for eDNA sampling. We tested this hypothesis with newly developed qPCR assays across (1) log dilutions of tissue genomic DNA (gDNA), (2) mesocosm experiments consisting of varying abundances of mussels, and (3) water samples collected from Lake Erie with known dreissenid populations. This study demonstrates the potential advantages of increasing the levels of detection and quantification from eDNA when leveraging whole genomic datasets for primer development.

2. Methods

2.1. Estimation of Gene Copy Number per Genome

To estimate the number of copies present within the mt- or nu-genome for each gene region with a developed molecular assay (Table S1), in silico PCR was performed to calculate the number of PCR products of expected size within the mt- and nu-genomes. The number of copies for each nu-gene within the nu-genome (within each of the 16 chromosomes) was calculated using the National Center for Biotechnology Information (NCBI) tool Primer-BLAST (https://www.ncbi.nlm.nih.gov/tools/primer-blast, accessed on 15 October 2021 [37]) against the ZM-assembled nu-genome (BioProject: PRJNA533175). Note that this process does not distinguish functional genes from pseudogenes within the ZM's genome but rather identifies the number of regions within the genome that would result in PCR amplification. The number of copies for each mt-gene within the mt-genome was calculated using in silico PCR (http://insilico.ehu.eus/user_seqs/, accessed on 15 October 2021) with the assembled mt-genome for each species (ZM BioProject: PRJNA533175, and QMs BioProject: PRJNA666063–accession MW080914).

2.2. Assay Design

Within the ZM's and QM's whole mt-genomes, the large tandem repeat sections for each species were identified using the program Tandem Repeat Finder [38]. Potential primers for the tandem repeat sections were identified using the PrimerQuest Tool from Integrated DNA Technologies (IDT; https://www.idtdna.com/PrimerQuest/Home/Index, accessed on 15 October 2021). Assays were further inspected for primer-dimer and heterodimer formation using the OligoAnalyzer Tool from IDT (https://www.idtdna.com/calc/analyzer, accessed on 15 October 2021). One mt tandem repeat assay was chosen for in vitro laboratory validation for each of the two species (Table 1).

Table 1. Species-specific and/or dreissenid-specific primer pairs (F-Forward and R-Reverse) used for quantitative PCR analysis of environmental DNA for (**A**) zebra mussel (ZM) and (**B**) quagga mussel (QM). The genomic origin (mitochondrial (mt) or nuclear (nu)) is listed for each gene, as well as the length (base pairs), primer efficiency (%), and R^2 for each primer set. Limit of detection (LOD) and limit of quantification (LOQ) are listed in ng of tissue extraction per reaction and were estimated for each qPCR assay using the qPCR_LOD_Calc R script [39].

Species	Assay	Zebra Mussel Primer	Length (bps)	Efficiency	R2	LOD	LOQ	Source
A. ZM	mt-tr285	F: GTTTTCCAGTTCTTCTGTCG R: CTCTCACTTTTTTCCCCTATCCCTC	97	96.83	0.990	2.2×10^{-6}	1.2×10^{-4}	Present Study
	mt-COI	F: TAGAGCTAAGGGCACCTGGAA R: AGCCCATGAGTGGTGACAAT	73	90.63	0.990	2.5×10^{-4}	3.3×10^{-3}	[40]
	mt-16S	F: TGGGGCAGTAAGAAGAAAAAAAATAA R: CATCGAGGTCGCAAACCG	141	91.00	0.995	2.5×10^{-4}	2.4×10^{-3}	[16]
	nu-H2B	F: CGCGCGCTCCACTGACAAGA R: CACCAGGCAGCAGGAGACGC	251	88.36	0.999	2.2×10^{-6}	4.1×10^{-5}	[19]
B. QM	mt-tr258	F: TCGGTTCAACGGGATTCCC R: CCCCCTTACAAGATTTTCGATTT	232	102.32	0.995	9.3×10^{-6}	2.8×10^{-4}	Present Study
	mt-COI	F: GGAAACTGGTTGGTCCCGAT R: GGCCCTGAATGCCCCATAAT	188	97.16	0.998	9.8×10^{-5}	3.5×10^{-4}	[40]

Table 1. *Cont.*

Species	Assay	Zebra Mussel Primer	Length (bps)	Efficiency	R2	LOD	LOQ	Source
B. QM	mt-16S	F: TGGGGCAGTAAGAAGAAAAAAATAA R: CATCGAGGTCGCAAACCG	141	101.15	0.996	5.7×10^{-5}	1.1×10^{-3}	[16]
	nu-H2B	F: CGCGCGCTCCACTGACAAGA R: CACCAGGCAGCAGGAGACGC	251	95.89	0.994	5.7×10^{-5}	1.8×10^{-3}	[19]

2.3. Assay Validation

The two primer pairs were tested for cross-amplification of non-target organisms in silico using the Primer-BLAST tool. The primer pairs were further tested for cross-amplification between the sister taxa by performing qPCR using 1 ng per reaction of gDNA. Total gDNA was extracted from the foot tissue of adult ZMs, or QMs collected from Lake Erie using a Qiagen EZDNA extraction kit following the manufacturer's protocol. DNA extractions were diluted to 10 ng/µL by quantification on a nanodrop lite with subsequent dilutions using DNase-free water.

To determine the sensitivity of each mtTR assay and previously developed species-specific COI assays [40] (Table 1), serial dilutions of gDNA (ranging from 10 ng to 0.00001 ng per reaction) were amplified using each assay for either the ZM's or QM's DNA. Each of the serial dilutions was amplified in triplicates. PCR reactions were run using 2× Fast Plus EvaGreen qPCR Master Mix (CAT #31014, Biotium, Fremont, CA, USA) on an Applied Biosystems QunatStudio Flex 6 Real-Time PCR System. Reactions consisted of 20 µL in volume and included 10 µL 2× Master Mix, 0.5 µL forward and reverse primers at a 10 mM concentration, 6.5 µL diH2O, and 2.5 µL of the sample template. Cycling began with 10 min at 94 °C followed by 40 cycles of 94 °C for 15 s and 60 °C for 60 s. A negative PCR control was run with each plate of samples. Melt curve analysis was completed after each qPCR run for each assay. Melt curves were inspected for a single peak corresponding to the same temperature as the peak from the positive control samples included in each run.

Quantitative PCR assays were evaluated for sensitivity based on the qPCR efficiency, the limit of quantification (LOQ)—defined as the lowest concentration of the target that can be accurately quantified with a coefficient of variance below a threshold of ≤35%—and the limit of detection (LOD)—defined as the lowest concentration of DNA that can be detected in 95% of replicates [39]. PCR amplification efficiency for each assay was calculated following the Minimum Information for Publication of Quantitative Real-time PCR Experiments (MIQE Guidelines) [41]. Using the calibrated standard curves for serial dilutions of gDNA (described above), PCR efficiency was calculated as $10^{-1/\text{slope}} - 1$, where the slope was calculated from the logarithm of the initial template concentration plotted against the Cq (quantification cycle) for each dilution. The LOD and LOQ were estimated for each qPCR assay using the qPCR_LOD_Calc R script [39].

2.4. Comparisons of Assay Sensitivity

To assess the differences in sensitivity and duration of the detection between COI and mtTR assays, water samples previously collected and processed from mesocosm experiments were further analyzed here [34]. Briefly, three abundance treatments of three replicates (a total of nine mesocosms) were chosen for analysis. These abundance treatments consisted of 2, 12, and 48 mussels, with each mesocosm consisting of equal numbers of ZMs and QMs (i.e., 1, 6, and 24 mussels, respectively). Each tank was covered with saran wrap for the duration of the experiment. The organisms were maintained within the tanks to release eDNA for a duration of 24 h, after which they were removed by a gloved hand. A new glove was worn for each tank. The time after removing the mussels from each tank was defined as time 0. Water samples collected at 0 h and 144 h were analyzed in this study. The water samples were collected after mixing the water within each mesocosm by stirring with a gloved hand for ~3–5 s. Further description of the water collection and eDNA extraction can be found in [34].

Additionally, detection and assay sensitivity were compared from water samples collected across six sites within western Lake Erie (WLE, Figure 1). Triplicate water samples of 500 mL were collected from ~30 cm below the surface in August of 2019. Water samples were filtered on a 47-mm-diameter glass microfiber filter GF/C (nominal pore size 1.2 µm; GE Healthcare Life Science, Westborough, MA, USA), and eDNA was extracted following a modified Zymo extraction protocol [34]. Water samples were analyzed with the species-specific COI [40] and mtTR assays described above, as well as two genus-specific assays targeting the mt 16S gene and the nu H2B gene regions (Table 1). The 16S gene assay (Dre_16S) was originally designed with a hydrolysis probe [16] but was adopted as a non-probe assay for comparisons in this study. The other three assays (ZM-specific COI and QM-specific COI [40] and the genus-specific H2B (Dre_H2B) [19]) were designed as non-probe assays with EvaGreen-based dye analysis.

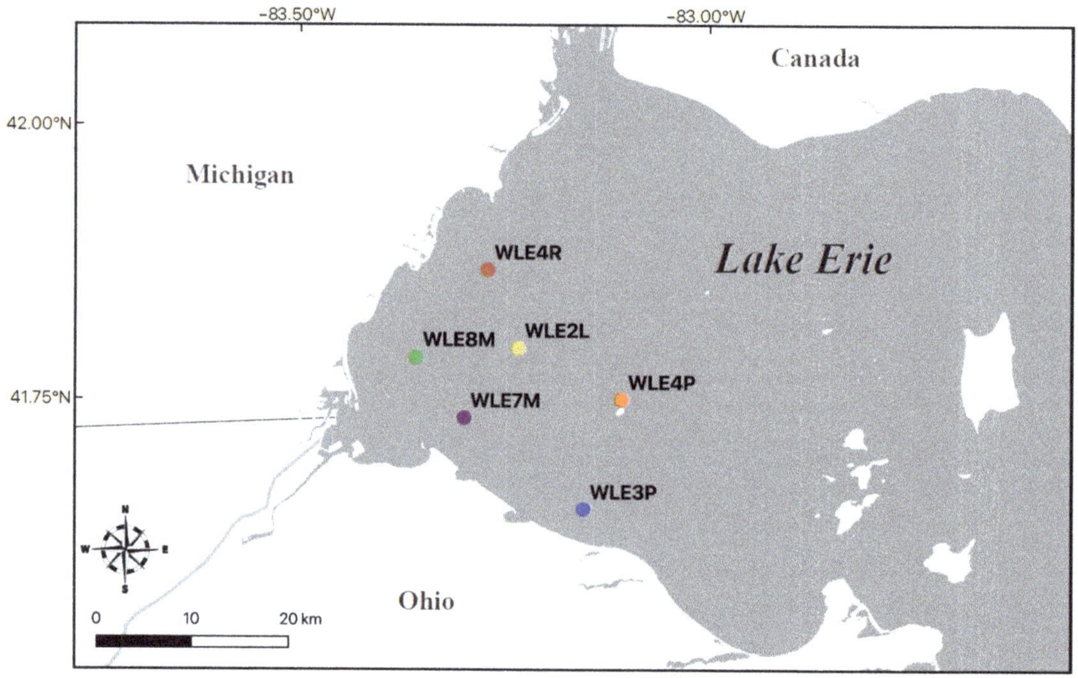

Figure 1. Map of western Lake Erie displaying the six sampling sites for eDNA water collection.

PCR conditions followed those listed for the assay validation. Positive standards were constructed for each PCR run using gDNA from ZMs and QMs that was log-diluted from 1 ng to 0.01 ng per reaction. The efficiency of each qPCR run was evaluated by comparing the positive standards to the calibrated standard curve.

The standardized Cq values (40-Cq) from the mesocosm and Lake Erie eDNA samples were compared between the COI and mtTR assays for ZMs and QMs using a Student's *t*-test. Additionally, to determine if the COI and mtTR assays provided similar eDNA quantification estimates within the mesocosm and Lake Erie eDNA samples, the concentration of gDNA was quantified from the standard curve for each assay. Linear regression was used to compare estimates of concentrations of gDNA between COI and mtTR assays for both ZMs and QMs.

3. Results

3.1. Copy Number per Genome

The copy number of nu-genes ranged from 6–30 copies per haploid genome, with ribosomal genes displaying the highest copy number (30 copies for 18S and 20 copies for 28S; Table S1). Copies of the 18S gene were found on half of the chromosomes (Chr 2, 3, 4, 5, 7, 9, 12, and 14), while copies of the 28S gene were found on seven of the chromosomes (Chr 2, 3, 4, 5, 6, 9, and 14). Note that these results suggest the presence of pseudogenes for both 18S and 28S within the ZM's nu-genome, which is common for nu-ribosomal genes among eukaryotes [42]. The two histone genes (H2B and H1) each had 19 copies spread across two chromosomes (Chr 2 and 5), while the MetRS gene had six copies found only on chromosome 5. The gene coding regions within the mt-genome (i.e., 16S, COI, and Cyt *b*) all displayed the expected single copy per mt-genome (Table S1). The targeted extended repeat regions (Dpo_tr285 and Dro_tr258) within the available mt-genome reference sequences exhibited 115 and 51 copies per mt-genome, respectively (Table 1).

3.2. Assay Specificity

For the newly designed primers for the mt-repeat regions, Primer-BLAST yielded no results for the cross-amplification of any non-target DNA within the NCBI database. Furthermore, no evidence of cross-amplification between sister taxa was observed during the qPCR trials. The melting curve for both primers displayed a single distinct peak with qPCR analysis (ZM = 76 °C and QM = 78 °C).

3.3. Assay Sensitivity—Standard Curves

Standard curve analysis using gDNA from the foot tissue estimated high qPCR efficiencies and R^2 values for all six assays tested in this study (Table 1). Across the first five standard dilutions (the standards that displayed 100% detection with both the COI and mtTR assays), the mtTR displayed increased sensitivity, as demonstrated by a lower Cq value for both of the species. The ZM's mtTR assay displayed a shift of 7.05 ± 0.62 Cq values (Figure 2A), while the QM's mtTR assay displayed a shift of 4.17 ± 0.25 Cq values (Figure 2B).

Additionally, for ZMs, the calculated LOD was determined to be two orders of magnitude lower for the mtTR assay, and LOQ was determined to be one order of magnitude lower for the mtTR assay compared to the COI assay (Table 1). For QMs, the calculated LOD was determined to be one order of magnitude lower for the mtTR assay compared to the COI assay (Table 1).

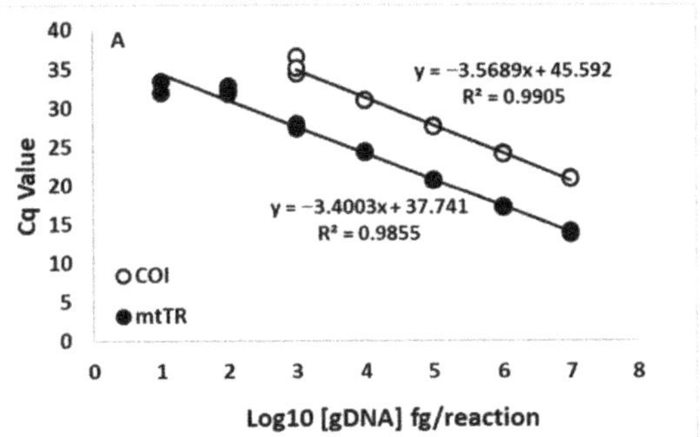

Figure 2. *Cont.*

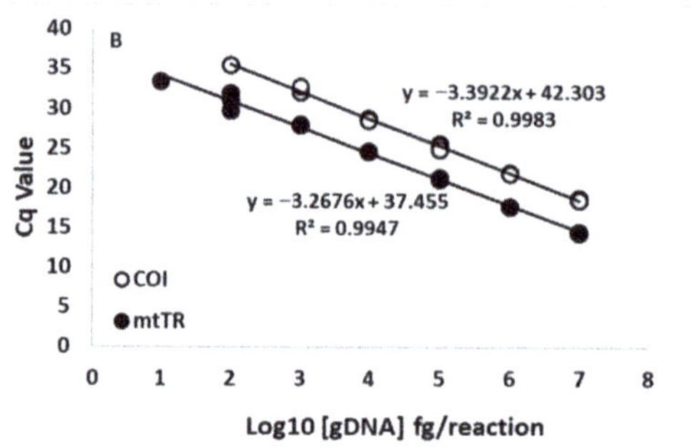

Figure 2. Standard curve analysis of log10 [genomic DNA] (fg per reaction) for (**A**) zebra mussels (*Dreissena polymorpha*) and (**B**) quagga mussels (*D. rostriformis*) for a mitochondrial cytochrome oxidase I (COI) assay (Blackman et al. 2020) (open circles) and a newly developed mitochondrial tandem repeat (mtTR) assay (filled circles).

3.4. Assay Sensitivity—Mesocosm Experiments

Across all of the mussel abundances at time 0 h, the mtTR assays displayed higher standardized Cq values than the COI assays for both of the species (all of the abundances' standardized Cq values—ZM: COI = 10.53 ± 1.79, mtTR = 16.20 ± 1.54, $p < 0.001$ ***; QM: COI = 10.80 ± 3.47, mtTR = 16.23 ± 3.34, $p = 0.006$ **) (Figure 3A,C). The estimated concentration of the eDNA log10 (fg/uL), calculated based on standard curves, did not differ across the mtTR and COI assays for either of the species (all of the abundances' eDNA concentrations—ZM: COI = 4.52 ± 0.50, mtTR = 4.10 ± 0.45, $p = 0.10$; QM: COI = 3.86 ± 1.02, mtTR = 4.19 ± 1.02, $p = 0.53$) (Figure 3B,D). At the time of 144 h of the eDNA degradation, the COI assay was detected in 44.44% (4/9) of the samples for ZMs and 33.33% (3/9) of the samples for QMs, while the mtTR assay was detected in 89% (8/9) and 100% (9/9) of the samples for ZMs and QMs, respectively.

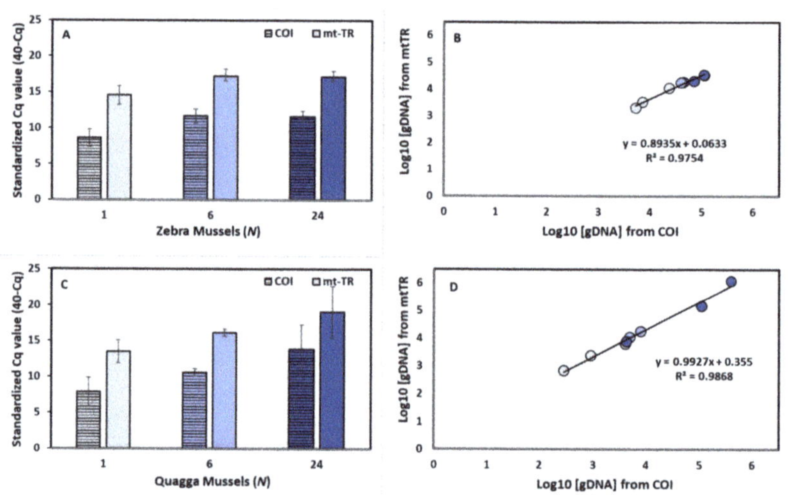

Figure 3. Standardized Cq values (±standard deviation) for (**A**) zebra mussels (*Dreissena polymorpha*) and (**C**) quagga mussels (*D. rostriformis*) for a mitochondrial cytochrome oxidase I (COI) assay (Blackman et al., 2020) (lined bar), and a newly developed mitochondrial tandem repeat (mtTR) assay (filled bar) across the three mesocosm tanks. The relationship of quantified genomic DNA (log10 fg/µL) between the COI and mtTR assay for (**B**) zebra mussels and (**D**) quagga mussels.

3.5. Assay Sensitivity—Western Lake Erie

Across the six sampling sites, the standardized Cq values were higher for the mtTR assay compared to the COI for both of the species (all of the sites' standardized Cq values—ZM: COI = 8.41 ± 4.02, mtTR = 14.12 ± 3.67, $p < 0.001$ ***; QM: COI = 13.93 ± 2.15, mtTR = 19.13 ± 1.79, $p < 0.001$ ***) (Figure 4A,C). The estimated concentration of the eDNA log10 (fg/uL), calculated based on tissue extraction standard curves, did not differ across the mtTR and COI for either of the species (all of sites' standardized eDNA concentrations—ZM: COI = 3.84 ± 1.36, mtTR = 3.50 ± 1.08, $p = 0.40$; QM: COI = 4.78 ± 0.63, mtTR = 5.08 ± 0.55, $p = 0.15$) (Figure 4B,D). For both the mtTR and COI assays, ZMs displayed a lower concentration of eDNA compared to QMs across Lake Erie (Figures 5, S1 and S2). Further, ZMs displayed lower eDNA concentrations as sites moved further offshore (Figures 4A and S1). The single-copy genus specific assay (mt-16S) displayed lower standardized Cq values to the QM's mtTR assay (Figures 5, S1 and S2), while the multi-copy H2B gene assay displayed similar standardized Cq values (Figures 5, S1 and S2).

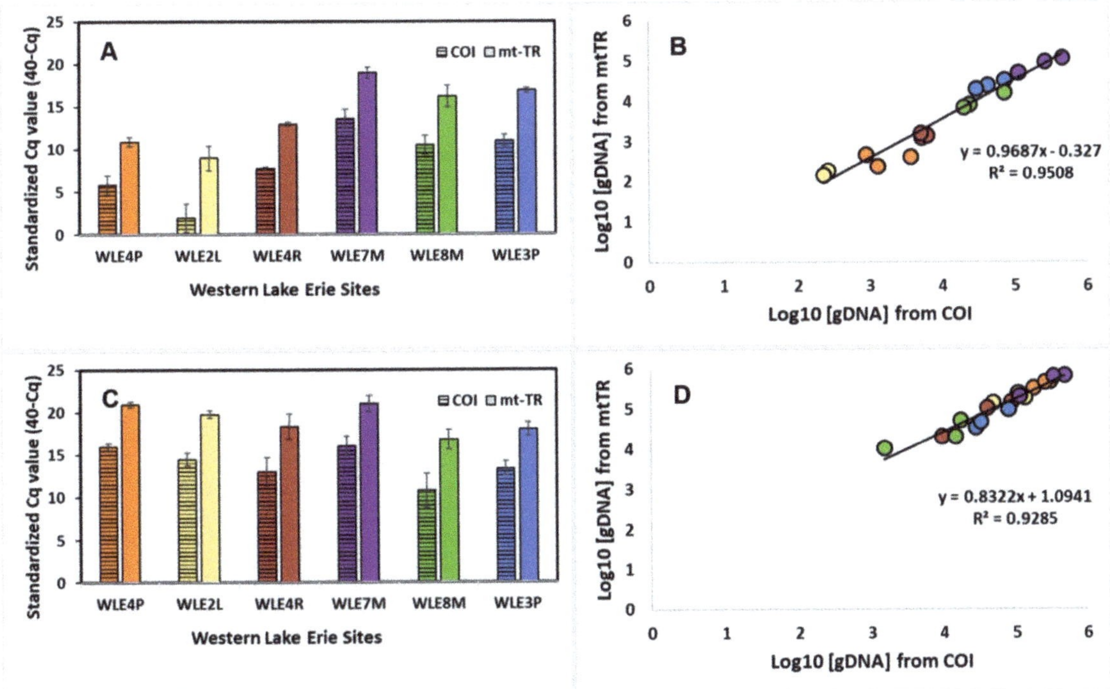

Figure 4. Standardized Cq values (±standard deviation) for (**A**) zebra mussels (*Dreissena polymorpha*) and (**C**) quagga mussels (*D. rostriformis*) for a mitochondrial cytochrome oxidase I (COI) assay (Blackman et al. 2020) (lined bar) and a newly developed mitochondrial tandem repeat (mtTR) assay (filled bar) across the six western Lake Erie sampling sites. The relationship of quantified genomic DNA (log10 of fg/μL) between the COI and mtTR assay for (**B**) zebra mussels and (**D**) quagga mussels.

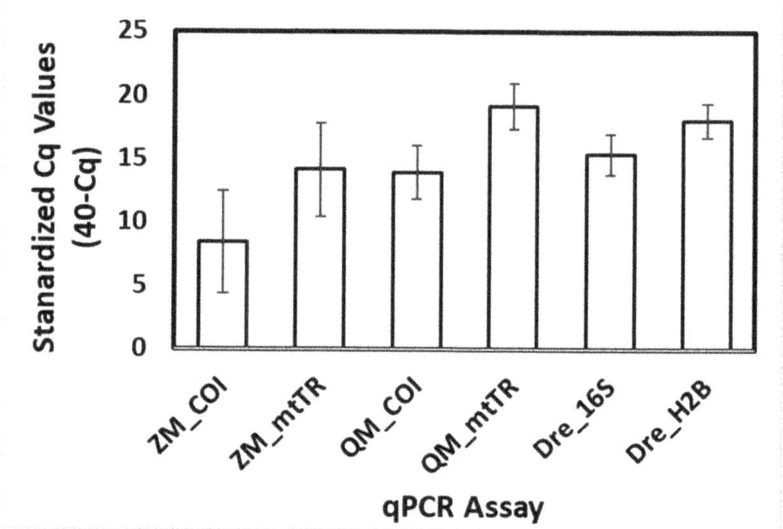

Figure 5. Mean standardized Cq values (±standard deviation) for each of the six eDNA assays across all water samples collected from western Lake Erie.

4. Discussion

As eDNA becomes a standardized sampling methodology for the detection and management of invasive species [10,43], targeting the most abundant genomic fragment can increase the sensitivity of an eDNA assay. Here we demonstrate proof of this concept by taking advantage of the unique structure of the mt-genome of dreissenid mussels. By targeting the tandem repeat regions of the mt-genome, we demonstrated higher sensitivity for the detection of dreissenid eDNA compared to the standard single-copy mt-gene regions (e.g., 16S and COI). Accordingly, the LOD from the tissue gDNA was estimated to be one to two orders of magnitude lower for the mt tandem repeat sections compared to the single-copy COI assays for either of the species. The higher number of copies per cell for these repeat regions allows for a lower total number of cells to be collected within the environment and still achieve a positive detection. While the number of mt-genomes per cell is highly dependent upon the type of tissue, the concentration of the mt-genome is always expected to outnumber the concentration of the nu-genome [44]. Therefore, by targeting a highly repeated section of mt-DNA, we designed species-specific assays targeting the theoretically most abundant DNA region found within a dreissenid mussel cell.

Standard operating procedures for eDNA qPCR assays are beginning to be outlined [28,39,45], and these procedures typically recommend the use of synthetic standards for the analysis of LOD and LOQ [45]. However, while synthetic standards are important for evaluating the efficacy of an eDNA assay in the laboratory, the ability to successfully collect and detect a genetic target from the environment is reliant on the inherent number of target copies found within a cell for the species of interest. Therefore, synthetic standards do not provide the full context of LOD and LOQ in terms of realized copies for various target genes expected to be found within the environment. For example, a qPCR targeting a multi-copy gene for the anthrax pathogen *Bacillus anthracis* was found to have a similar LOD to the established single-copy assays using synthetic standards; yet the multi-copy gene significantly improved its detection by lowering the Cq threshold by > two cycles [29]. Similarly, our newly developed mtTR assays lowered Cq values by seven and four cycles for ZMs and QMs, respectively. Therefore, these mtTR regions can increase detection rates for eDNA of low population sizes when the concentration of cellular material may be below the detection threshold for standard single-copy gene assays.

While the degradation of the mt-genome is typically slower compared to the nu-genome [33,34,44], it is not known the purpose of these mtTR regions [36] or if they degrade at similar rates to coding regions within the mt-genome. We tested samples collected from a previous mesocosm experiment for the detection of these repeat regions after six days of degradation. While the tandem repeat sections fell below the LOQ for both QMs and ZMs, these two markers still provided high detection rates after six days of degradation, compared to limited detections with the COI assays. This suggests that the high concentrations of the mtTR regions allow for longer time frames of positive detections post- eDNA release. Thus, the longer timeframe allows for an increased detection probability. However, this also presents a potential risk of a false-positive detection relating to the legacy-genomic material or to eDNA transport within a flowing system. Therefore, when an eDNA study focuses on recent presence, such as studies evaluating the success of an eradication event, eDNA analysis may benefit from targeting a faster degrading gene region. In such cases, the use of eRNA may improve the assessments of recent presence, as eRNA has been shown to degrade at expeditated rates [34]. However, because the mtTR assays target non-coding DNA, these mtTR regions are not expected to be translated into RNA within the cell and thus will not be useful with eRNA analysis.

The water samples from Lake Erie were found to have much larger standardized Cq values for the mtTR assay compared to the COI assay for both ZMs and QMs. Additionally, for majority of the sites, the QM mtTR assay provided higher standardized Cq values than the genus-specific 16S gene assay, which amplifies both the ZM's and QM's eDNA. This suggests that in some cases, the species-specific mtTR assays can even be more sensitive than a genus-specific single-copy assay. However, the multi-copy genus-specific H2B gene assay provided similarly high standardized Cq values, further suggesting that multi-copy targets greatly improve sensitivity [34]. If monitoring programs are not particularly interested in distinguishing between the two species, the genus-specific H2B target may provide better sensitivity than a species-specific assay to detect low-density dreissenid populations. However, the designed primers in this study allow high sensitivity at the species level, providing improved detection when species-specific information is necessary. For example, in this study, large variations in eDNA concentration was found for ZM, with the highest concentrations occurring in sites nearshore, while QM's eDNA concentration displayed a much smaller variation across the sites. These patterns of eDNA concentration between the species are not possible with a genus-specific assays.

Unlike single-copy genes, it is unknown how much variation occurs in the number of mtTR repeats within individuals, between individuals, and between populations. Currently, only one mt-genome has been sequenced for both ZMs and QMs; thus, it is not possible to estimate a population and spatial variation within this non-coding region. It is hypothesized that the mt-genome displays some levels of heteroplasmy, whereas the number of repeats can differ between individuals and even between cells within an individual [36]. Therefore, while both mtTR assays displayed increasing DNA concentrations with the mussel abundance across the mesocosms, the unknown level of heteroplasmy may result in misleading abundance estimates when quantifying eDNA with these mtTR assays. Additionally, no mt-genomes have been sequenced for other closely related *Dreissena* taxa, and thus it is not known if these mtTR assays will cross-amplify with other sister taxa in co-occurring habitats. However, these assays show a clear specificity against the two tested species, with a clear distinction between the ZM's and QM's mtTR regions. Continued investigations into the mt-genome structure across geographical populations and between dreissenid taxa will improve the evaluation and interpretation of quantified eDNA from mtTR assays in the future.

While these extended mt-genome repeat regions are rare within the animal kingdom, this study provides proof of concept to increase the efficiency and effectiveness of eDNA by incorporating whole genome information. Additionally, mt tandem repeat regions have recently been discovered in the parasitic worm *Schistosoma haematobium* [46] and are common among plants [47]. Thus, a similar approach for assay development may be beneficial for improving eDNA detection for a wide range of taxa. Likewise, other studies have shown increased assay sensitivity when targeting a multi-copy nu-gene [31–34], suggesting that gene copy number should be considered in eDNA assay development. It is important to note that the majority of developed assays target the COI gene region due to the wealth of information available to properly evaluate species-specificity [28,45]; thus, the development of assays targeting other gene regions may be limited until more in-depth genomic databases are developed. This study expresses the need to understand the unique cellular aspects (including the genomic structure) of an organism to develop the most efficient and effective eDNA methodology for detection efforts.

Supplementary Materials: The following are available online at https://www.mdpi.com/article/10.3390/w14132069/s1, Figure S1: The Standardized Cq values (±standard deviation) across water samples collected from six sites in western Lake Erie for ZM_COI, ZM_mtTR, QM_COI, QM_mtTR, Dre_16S, and Dre_H2B. Figure S2: The Standardized Cq values (±standard deviation) for each of the six eDNA assays from triplicate water samples collected at WLE4P, WLE2L, WLE4R, WLE7M, WLE8M, and WLE3P. Table S1: Gene regions across the nuclear (nu-) and mitochondrial (mt-) genomes used for developing diagnostic molecular assays targeting zebra mussel (*Dreissena polymorpha*), quagga mussel (*D. rostriformis*), or both [48–56].

Author Contributions: N.T.M., H.A.V. and S.R.C. were involved in developing the experimental design. N.T.M. and S.R.C. conducted the experiments. Data analysis was performed by N.T.M. and S.R.C. provided the feedback for the analysis. The first draft was written by N.T.M. All authors have read and agreed to the published version of the manuscript.

Funding: Funding was awarded to the Cooperative Institute for Great Lakes Research (CIGLR) through the NOAA Cooperative Agreement with the University of Michigan (NA17OAR4320152). The CIGLR contribution number is 1195 and the NOAA GLERL contribution number is 2007.

Data Availability Statement: Corresponding to the standard curves of genomic DNA and the qPCR Cq values for the mesocosm and Lake Erie eDNA samples are uploaded to GitHub https://github.com/ntmarshall406/Dreissenid_mtTR_eDNA_Assays, accessed on 15 October 2021.

Acknowledgments: We thank the staff of CIGLR and NOAA GLERL for providing opportunities and assistance with the Lake Erie sample collection. Specifically, we thank Kent Baker for coordinating and Paul Den Uyl for the sample collection during the NOAA GLERL and CIGLR HABs monitoring in western Lake Erie.

Conflicts of Interest: The authors declare no conflict of interest.

References

1. Nalepa, T.F.; Schloesser, D.W. (Eds.) *Quagga and Zebra Mussels: Biology, Impacts, and Control*; CRC Press: Boca Raton, FL, USA, 2013; pp. 9–32.
2. Marshall, N.T.; Stepien, C.A. The family Dreissenidae. In *Freshwater Mollusk Families of the World*; Cummings, K., Lydeard, C., Eds.; JHU Press: Baltimore, MD, USA, 2019; pp. 193–196.
3. Vanderploeg, H.A.; Nalepa, T.F.; Jude, D.J.; Mills, E.L.; Holeck, K.T.; Liebig, J.R.; Grigorovich, I.A.; Ojaveer, H. Dispersal and emerging ecological impacts of Ponto-Caspian species in the Laurentian Great Lakes. *Can. J. Fish. Aquat. Sci.* **2002**, *59*, 1209–1228. [CrossRef]
4. Higgins, S.N.; Zanden, M.J.V. What a difference a species makes: A meta–analysis of dreissenid mussel impacts on freshwater ecosystems. *Ecol. Monogr.* **2010**, *80*, 179–196. [CrossRef]
5. Carlton, J.T. The zebra mussel *Dreissena polymorpha* found in North America in 1986 and 1987. *J. Great Lakes Res.* **2008**, *34*, 770–773. [CrossRef]
6. May, B.; Marsden, J.E. Genetic identification and implications of another invasive species of dreissenid mussel in the Great Lakes. *Can. J. Fish. Aquat. Sci.* **1992**, *49*, 1501–1506. [CrossRef]
7. De Ventura, L.; Weissert, N.; Tobias, R.; Kopp, K.; Jokela, J. Overland transport of recreational boats as a spreading vector of zebra mussel *Dreissena polymorpha*. *Biol. Invasions* **2016**, *18*, 1451–1466. [CrossRef]
8. Snyder, M.R.; Stepien, C.A.; Marshall, N.; Scheppler, H.B.; Black, C.L.; Czajkowski, K.P. Detecting aquatic invasive species in bait and pond stores with targeted environmental (e)DNA high-throughput sequencing metabarcode assays: Angler, retailer, and manager implications. *Biol. Conserv.* **2020**, *245*, 108430. [CrossRef]
9. Patoka, J.; Patoková, B. Hitchhiking Exotic Clam: *Dreissena polymorpha* (Pallas, 1771) transported via the ornamental plant trade. *Diversity* **2021**, *13*, 410. [CrossRef]

10. Sepulveda, A.; Smith, D.; O'Donnell, K.; Owens, N.; White, B.; Richter, C.; Merkes, C.; Wolf, S.; Rau, M.; Neilson, M.; et al. Using structured decision making to evaluate potential management responses to detection of dreissenid mussel (*Dreissena* spp.) environmental DNA. *Manag. Biol. Invasions* **2022**, *13*, 344–368. [CrossRef]
11. Ginn, B.K.; Bolton, R.; Coulombe, D.; Fleischaker, T.; Yerex, G. Quantifying a shift in benthic dominance from zebra (*Dreissena polymorpha*) to quagga (*Dreissena rostriformis bugensis*) mussels in a large, inland lake. *J. Great Lakes Res.* **2018**, *44*, 271–282. [CrossRef]
12. Larson, C.E.; Barge, J.T.; Hatzenbuhler, C.L.; Hoffman, J.C.; Peterson, G.S.; Pilgrim, E.M.; Wiechman, B.; Rees, C.B.; Trebitz, A.S. Invasive *Dreissena* mussel coastal transport from an already invaded estuary to a nearby archipelago detected in DNA and zooplankton surveys. *Front. Mar. Sci.* **2022**, *9*, 818738. [CrossRef]
13. Feist, S.M.; Lance, R.F. Advanced molecular-based surveillance of quagga and zebra mussels: A review of environmental DNA/RNA (eDNA/eRNA) studies and considerations for future directions. *NeoBiota* **2021**, *66*, 117–159. [CrossRef]
14. Frischer, M.E.; Hansen, A.S.; Wyllie, J.A.; Wimbush, J.; Murray, J.; Nierzwicki-Bauer, S.A. Specific amplification of the 18S rRNA gene as a method to detect zebra mussel (*Dreissena polymorpha*) larvae in plankton samples. *Hydrobiologia* **2002**, *487*, 33–44. [CrossRef]
15. Sepulveda, A.J.; Nelson, N.M.; Jerde, C.L.; Luikart, G. Are environmental DNA methods ready for aquatic invasive species management? *Trends Ecol. Evol.* **2020**, *35*, 668–678. [CrossRef]
16. Gingera, T.; Bajno, R.; Docker, M.; Reist, J. Environmental DNA as a detection tool for zebra mussels *Dreissena polymorpha* (Pallas, 1771) at the forefront of an invasion event in Lake Winnipeg, Manitoba, Canada. *Manag. Biol. Invasions* **2017**, *8*, 287–300. [CrossRef]
17. Amberg, J.J.; Merkes, C.M.; Stott, W.; Rees, C.B.; Erickson, R.A. Environmental DNA as a tool to help inform zebra mussel, *Dreissena polymorpha*, management in inland lakes. *Manag. Biol. Invasions* **2019**, *10*, 96–110. [CrossRef]
18. Marshall, N.; Stepien, C.A. Invasion genetics from eDNA and thousands of larvae: A targeted metabarcoding assay that distinguishes species and population variation of zebra and quagga mussels. *Ecol. Evol.* **2019**, *9*, 3515–3538. [CrossRef]
19. Peñarrubia, L.; Alcaraz, C.; De Vaate, A.B.; Sanz, N.; Pla, C.; Vidal, O.; Viñas, J. Validated methodology for quantifying infestation levels of dreissenid mussels in environmental DNA (eDNA) samples. *Sci. Rep.* **2016**, *6*, 39067. [CrossRef]
20. De Ventura, L.; Kopp, K.; Seppälä, K.; Jokela, J. Tracing the quagga mussel invasion along the Rhine river system using eDNA markers: Early detection and surveillance of invasive zebra and quagga mussels. *Manag. Biol. Invasions* **2017**, *8*, 101–112. [CrossRef]
21. Shogren, A.J.; Tank, J.L.; Egan, S.P.; Bolster, D.; Riis, T. Riverine distribution of mussel environmental DNA reflects a balance among density, transport, and removal processes. *Freshw. Biol.* **2019**, *64*, 1467–1479. [CrossRef]
22. Blackman, R.C.; Ling, K.K.S.; Harper, L.R.; Shum, P.; Hänfling, B.; Lawson-Handley, L. Targeted and passive environmental DNA approaches outperform established methods for detection of quagga mussels, *Dreissena rostriformis bugensis* in flowing water. *Ecol. Evol.* **2020**, *10*, 13248–13259. [CrossRef]
23. Marshall, N.T.; Stepien, C.A. Macroinvertebrate community diversity and habitat quality relationships along a large river from targeted eDNA metabarcode assays. *Environ. DNA* **2020**, *2*, 572–586. [CrossRef]
24. Johansson, M.L.; Lavigne, S.Y.; Ramcharan, C.W.; Heath, D.D.; MacIsaac, H.J. Detecting a spreading non-indigenous species using multiple methodologies. *Lake Reserv. Manag.* **2020**, *36*, 432–443. [CrossRef]
25. Xia, Z.; Johansson, M.L.; Gao, Y.; Zhang, L.; Haffner, G.D.; MacIsaac, H.J.; Zhan, A. Conventional versus real-time quantitative PCR for rare species detection. *Ecol. Evol.* **2018**, *8*, 11799–11807. [CrossRef] [PubMed]
26. Trebitz, A.S.; Hatzenbuhler, C.L.; Hoffman, J.C.; Meredith, C.S.; Peterson, G.S.; Pilgrim, E.M.; Barge, J.T.; Cotter, A.M.; Wick, M. *Dreissena* veligers in western Lake Superior—Inference from new low-density detection. *J. Great Lakes Res.* **2019**, *45*, 691–699. [CrossRef]
27. Robin, E.D.; Wong, R. Mitochondrial DNA molecules and virtual number of mitochondria per cell in mammalian cells. *J. Cell. Physiol.* **1988**, *136*, 507–513. [CrossRef]
28. Thalinger, B.; Deiner, K.; Harper, L.R.; Rees, H.C.; Blackman, R.C.; Sint, D.; Traugott, M.; Goldberg, C.S.; Bruce, K. A validation scale to determine the readiness of environmental DNA assays for routine species monitoring. *Environ. DNA* **2021**, *3*, 823–836. [CrossRef]
29. Braun, P.; Nguyen, M.D.-T.; Walter, M.C.; Grass, G. Ultrasensitive Detection of *Bacillus anthracis* by Real-Time PCR Targeting a Polymorphism in Multi-Copy 16S rRNA Genes and Their Transcripts. *Int. J. Mol. Sci.* **2021**, *22*, 12224. [CrossRef]
30. Shan, J.; Jia, Y.; Teulières, L.; Patel, F.; Clokie, M.R.J. Targeting multicopy prophage genes for the increased detection of *Borrelia burgdorferi* sensu lato (s.l.), the causative agents of Lyme disease, in blood. *Front. Microbiol.* **2021**, *12*, 464. [CrossRef]
31. Minamoto, T.; Uchii, K.; Takahara, T.; Kitayoshi, T.; Tsuji, S.; Yamanaka, H.; Doi, H. Nuclear internal transcribed spacer-1 as a sensitive genetic marker for environmental DNA studies in common carp *Cyprinus carpio*. *Mol. Ecol. Resour.* **2017**, *17*, 324–333. [CrossRef]
32. Dysthe, J.C.; Franklin, T.W.; McKelvey, K.S.; Young, M.K.; Schwartz, M.K. An improved environmental DNA assay for bull trout (*Salvelinus confluentus*) based on the ribosomal internal transcribed spacer I. *PLoS ONE* **2018**, *13*, e0206851. [CrossRef]
33. Jo, T.; Arimoto, M.; Murakami, H.; Masuda, R.; Minamoto, T. Estimating shedding and decay rates of environmental nuclear DNA with relation to water temperature and biomass. *Environ. DNA* **2020**, *2*, 140–151. [CrossRef]

34. Marshall, N.T.; Vanderploeg, H.A.; Chaganti, S.R. Environmental (e)RNA advances the reliability of eDNA by predicting its age. *Sci. Rep.* **2021**, *11*, 2769. [CrossRef] [PubMed]
35. McCartney, M.A.; Auch, B.; Kono, T.; Mallez, S.; Zhang, Y.; Obille, A.; Becker, A.; Abrahante, J.E.; Garbe, J.; Badalamenti, J.P.; et al. The genome of the zebra mussel, *Dreissena polymorpha*: A resource for comparative genomics, invasion genetics, and biocontrol. *G3 Genes Genom. Genet.* **2022**, *12*, jkab423. [CrossRef] [PubMed]
36. Calcino, A.; Baranyi, C.; Wanninger, A. Heteroplasmy and repeat expansion in the plant-like mitochondrial genome of a bivalve mollusc. *bioRxiv* **2020**, 1–32. [CrossRef]
37. Ye, J.; Coulouris, G.; Zaretskaya, I.; Cutcutache, I.; Rozen, S.; Madden, T.L. Primer-BLAST: A tool to design target-specific primers for polymerase chain reaction. *BMC Bioinform.* **2012**, *13*, 134. [CrossRef]
38. Benson, G. Tandem repeats finder: A program to analyze DNA sequences. *Nucleic Acids Res.* **1999**, *27*, 573–580. [CrossRef]
39. Klymus, K.E.; Merkes, C.M.; Allison, M.J.; Goldberg, C.S.; Helbing, C.C.; Hunter, M.E.; Jackson, C.A.; Lance, R.F.; Mangan, A.M.; Monroe, E.M.; et al. Reporting the limits of detection and quantification for environmental DNA assays. *Environ. DNA* **2019**, *2*, 271–282. [CrossRef]
40. Blackman, R.; Benucci, M.; Donnelly, R.; Hänfling, B.; Harper, L.; Sellers, G.; Lawson-Handley, L. Simple, sensitive and species-specific assays for detecting quagga and zebra mussels (*Dreissena rostriformis bugensis* and *D. polymorpha*) using environmental DNA. *Manag. Biol. Invasions* **2020**, *11*, 218–236. [CrossRef]
41. Bustin, S.A.; Benes, V.; Garson, J.A.; Hellemans, J.; Huggett, J.; Kubista, M.; Mueller, R.; Nolan, T.; Pfaffl, M.W.; Shipley, G.L.; et al. The MIQE Guidelines: Minimum Information for Publication of Quantitative Real-Time PCR Experiments. *Clin. Chem.* **2009**, *55*, 611–622. [CrossRef]
42. Li, Y.; Yang, R.-H.; Jiang, L.; Hu, X.-D.; Wu, Z.-J.; Yao, Y.-J. rRNA pseudogenes in filamentous ascomycetes as revealed by genome data. *G3 Genes Genom. Genet.* **2017**, *7*, 2695–2703. [CrossRef]
43. Morisette, J.; Burgiel, S.; Brantley, K.; Daniel, W.; Darling, J.; Davis, J.; Franklin, T.; Gaddis, K.; Hunter, M.; Lance, R.; et al. Strategic considerations for invasive species managers in the utilization of environmental DNA (eDNA): Steps for incorporating this powerful surveillance tool. *Manag. Biol. Invasions* **2021**, *12*, 747–775. [CrossRef] [PubMed]
44. Jo, T.; Takao, K.; Minamoto, T. Linking the state of environmental DNA to its application for biomonitoring and stock assessment: Targeting mitochondrial/nuclear genes, and different DNA fragment lengths and particle sizes. *Environ. DNA* **2022**, *4*, 271–283. [CrossRef]
45. Langlois, V.S.; Allison, M.J.; Bergman, L.C.; To, T.A.; Helbing, C.C. The need for robust qPCR-based eDNA detection assays in environmental monitoring and species inventories. *Environ. DNA* **2021**, *3*, 519–527. [CrossRef]
46. Kinkar, L.; Gasser, R.; Webster, B.; Rollinson, D.; Littlewood, D.; Chang, B.; Stroehlein, A.; Korhonen, P.; Young, N. Nanopore sequencing resolves elusive long tandem-repeat regions in mitochondrial genomes. *Int. J. Mol. Sci.* **2021**, *22*, 1811. [CrossRef] [PubMed]
47. Wynn, E.L.; Christensen, A.C. Repeats of unusual size in plant mitochondrial genomes: Identification, incidence and evolution. *G3 Genes Genom. Genet.* **2019**, *9*, 549–559. [CrossRef] [PubMed]
48. Williams, M.R.; Stedtfeld, R.D.; Engle, C.; Salach, P.; Fakher, U.; Stedtfeld, T.; Dreelin, E.; Stevenson, R.J.; Latimore, J.; Hashsham, S.A. Isothermal amplification of environmental DNA (eDNA) for direct field-based monitoring and laboratory confirmation of *Dreissena* sp. *PLoS ONE* **2017**, *12*, e0186462. [CrossRef]
49. Hoy, M.S.; Kelly, K.; Rodriguez, R.J. Development of a molecular diagnostic system to discriminate *Dreissena polymorpha* (zebra mussel) and *Dreissena bugensis* (quagga mussel). *Mol. Ecol. Resour.* **2010**, *10*, 190–192. [CrossRef]
50. Ram, J.L.; Karim, A.S.; Acharya, P.; Jagtap, P.; Purohit, S.; Kashian, D.R. Reproduction and genetic detection of veligers in changing *Dreissena* populations in the Great Lakes. *Ecosphere* **2011**, *2*, 1–16. [CrossRef]
51. Ardura, A.; Zaiko, A.; Borrell, Y.J.; Samuiloviene, A.; Garcia-Vazquez, E. Novel tools for early detection of a global aquatic invasive, the zebra mussel *Dreissena polymorpha*. *Aquatic. Conserv.* **2017**, *27*, 165–176. [CrossRef]
52. Mahon, A.R.; Barnes, M.A.; Senapati, S.; Feder, J.L.; Darling, J.A.; Chang, H.C.; Lodge, D.M. Molecular detection of invasive species in heterogeneous mixtures using a microfluidic carbon nanotube platform. *PLoS ONE* **2011**, *6*, e17280. [CrossRef]
53. Bronnenhuber, J.E.; Wilson, C.C. Combining species-specific COI primers with environmental DNA analysis for targeted detection of rare freshwater species. *Conserv. Genet. Resour.* **2013**, *5*, 971–975. [CrossRef]
54. Egan, S.P.; Barnes, M.A.; Hwang, C.T.; Mahon, A.R.; Feder, J.L.; Ruggiero, S.T.; Tanner, C.E.; Lodge, D.M. Rapid invasive species detection by combining environmental DNA with light transmission spectroscopy. *Conserv. Lett.* **2013**, *6*, 402–409. [CrossRef]
55. Egan, S.P.; Grey, E.; Olds, B.; Feder, J.L.; Ruggiero, S.T.; Tanner, C.E.; Lodge, D.M. Rapid molecular detection of invasive species in ballast and harbor water by integrating environmental DNA and light transmission spectroscopy. *Environ. Sci. Technol.* **2015**, *49*, 4113–4121. [CrossRef] [PubMed]
56. Sepulveda, A.J.; Amberg, J.J.; Hanson, E. Using environmental DNA to extend the window of early detection for dreissenid mussels. *Manag. Biol. Invasion.* **2019**, *10*, 342. [CrossRef]

water

MDPI

Article

Ecological Quality Assessment of Greek Lowland Rivers with Aquatic Macrophytes in Compliance with the EU Water Framework Directive

Konstantinos Stefanidis [1,2,*], Georgios Dimitrellos [1], Maria Sarika [3], Dionysios Tsoukalas [1] and Eva Papastergiadou [1,*]

1 Department of Biology, University of Patras, University Campus, 26504 Rio, Greece
2 Hellenic Centre for Marine Research, Institute of Marine Biological Resources and Inland Waters, 46.7 km of Athens—Sounio Ave., Anavyssos, 19013 Attiki, Greece
3 Department of Biology, National and Kapodistrian University of Athens, 15701 Athens, Greece
* Correspondence: kstefani@upatras.gr (K.S.); evapap@upatras.gr (E.P.); Tel.: +30-229-107-6439 (K.S.)

Abstract: Aquatic macrophytes are one of the four biological quality elements (BQE) used for assessing the ecological status of inland waters according to the EU Water Framework Directive (WFD 2000/60). With this article, we present the methodological approach for the implementation of a WFD compliant macrophyte index to the riverine systems of Greece. In addition to the definition and harmonization of the ecological quality class boundaries, the results from the pilot application of the index and the ecological classification of the monitored river reaches are also presented. Aquatic plants and environmental parameters were sampled from 93 river reaches between 2012 and 2015. A multivariate analysis with optimal scaling (MVAOS) was conducted to define the main stressor gradient and to identify the least disturbed sites and the reference conditions that are required for the derivation of the ecological quality classes. The Macrophyte Biological Index IBMR for Greek rivers ($IBMR_{GR}$) was calculated for all the sites and the boundaries for the five quality classes were derived according to the methodology proposed by the Mediterranean Geographic Intercalibration Group (MedGIG). The main findings showed that the hydromorphological modifications were the main environmental stressors that correlated strongly with the $IBMR_{GR}$, whereas physicochemical stressors were of lesser importance. More specifically, the first principal component explained 51% of the total variance of the data, representing a moderately strong gradient of hydromorphological stress, whereas the second component explained 22.5%, representing a weaker gradient of physicochemical stress. In addition, the ecological assessment showed that almost 60% of the sites failed the WFD target of the "Good" ecological quality class, which agrees with classification assessments based on other BQEs for Greece and many Mediterranean countries. Overall, this work provides a first assessment of the ecological classification of Greek rivers with the BQE of aquatic macrophytes with significant implications for ecological monitoring and decision making within the frame of the WFD implementation.

Keywords: aquatic macrophytes; Water Framework Directive; rivers; ecological quality; ecological monitoring; Eastern Mediterranean

Citation: Stefanidis, K.; Dimitrellos, G.; Sarika, M.; Tsoukalas, D.; Papastergiadou, E. Ecological Quality Assessment of Greek Lowland Rivers with Aquatic Macrophytes in Compliance with the EU Water Framework Directive. *Water* **2022**, *14*, 2771. https://doi.org/10.3390/w14182771

Academic Editor: Yongjiu Cai

Received: 27 July 2022
Accepted: 29 August 2022
Published: 6 September 2022

1. Introduction

Aquatic macrophytes are aquatic photosynthetic organisms easily seen with the naked eye and include vascular plants, mosses, liverworts and macro-algal growths [1]. They have been widely used as bioindicators in freshwater habitats because certain species and communities are known to respond to environmental changes caused by anthropogenic perturbations such as eutrophication, acidification and hydromorphological alteration [2–6]. Several studies have investigated the role of anthropogenic disturbances in shaping the structure and functioning of macrophyte communities [7–10] revealing various and complex responses of diversity and community indices to gradients of hydromorphological features and nutrients. It is well known that numerous/multiple human activities such as agriculture, aquaculture, urban infrastructure and settlements, alterations in the hydromorphology and flow regime, significantly influence the abundance, structure and the extent of the macrophyte communities [11,12]. Naturally, aquatic macrophytes were recognized as an important tool for biomonitoring and assessment of freshwater ecosystems and were adopted as one of the four biological quality elements (BQEs) that are used for the ecological classification of streams and rivers in Europe, following the implementation of the Water Framework Directive (WFD 2000/60) [13].

The goal of the Water Framework Directive [14] is to restore or maintain good ecological state of freshwater systems of all EU member states. Thus, the WFD provides very detailed guidelines for the implementation of the ecological monitoring and the assessment of all European inland and coastal waters, including rivers and streams. The ecological monitoring and assessment involve the monitoring of biological, hydromorphological and physicochemical quality elements. For running waters (rivers and streams) the goal of the "Good" ecological status is defined in Annex V of the WFD and refers to terms of quality assessed with the use of biological communities, based mainly on diatoms, benthic invertebrates, fish and aquatic macrophytes. Basically, the ecological status is derived by comparing the biological community of a certain site with the respective community that would be expected in environmental conditions with no or minimal anthropogenic impact. These conditions are known as reference conditions and can be defined using different approaches [15].

Today, numerous biological assessment systems based on macrophytes have been developed and used by EU members [1]. Most of these systems are based on indices that consider the species composition of macrophyte assemblages and species indicator values that reflect the tolerance to a certain disturbance (e.g., organic pollution) [16,17]. The Macrophyte Biological Index for Rivers (IBMR) is one of these indices originally developed for France [17] and adopted by other EU members (e.g., Portugal, Italy, Cyprus and Greece) [1,13]. The calculation of the IBMR is based on indicator taxa that belong to various macrophyte groups, such as macroalgae (e.g., Characeae), aquatic bryophytes (e.g., *Fontinalis* sp.), truly aquatic macrophytes (e.g., *Potamogeton* sp.) and emergent vascular species (e.g., *Polygonum* sp.) [1,17]. In Greece, the $IBMR_{GR}$ is the national assessment method for classification of ecological quality of rivers with the use of macrophytes that has been intercalibrated during the Mediterranean Geographic Intercalibration Group exercise (MedGIG) [1] and it has been implemented during the first round of the National monitoring program (2012–2015) and the second phase which started on 2018 and is still running [18]. The $IBMR_{GR}$ list of indicator species included new species that are characteristic of the Greek rivers to adjust the index in the local conditions. Earlier studies on the aquatic macrophyte communities of the Greek riverine ecosystems have focused on specific rivers examining mostly associations between plant communities and environmental gradients [19,20]. A more recent study by Stefanidis et al. [2] examined the biodiversity patterns of aquatic macrophytes across environmental gradients at a larger spatial scale covering multiple catchments and sampling sites, which are included in the current study.

Driven by the lack of a national ecological assessment method for Greek rivers with the use of aquatic macrophytes, the current study is the first ever that focuses on the development and use of a macrophyte index as an official WFD-compliant national method. More specifically, this paper describes the methodological approach for the implementation of the IBMR index to the riverine systems of Greece, including the multi-step procedure for the definition and harmonization of the ecological quality class boundaries. It also provides a first ever detailed overview of the ecological classification of the riverine ecosystems of Greece based on macrophytes that resulted from the implementation of this pilot monitoring scheme during the first phase of the ecological monitoring (2012–2015) within the frame of the WFD 2000/60.

2. Materials and Methods

2.1. Field Surveys

Field surveys were conducted in a total of 93 stream sites (Figure 1), which are part of the National Monitoring Network [21], covering all the biogeographic regions of mainland Greece [22]. Samplings were conducted once between April and August of 2014 and 2015 following standardized protocols from the MedGIG [1]. Macrophytes were sampled from both banks and the channel when feasible, by wading upstream for a 100 m long section of the river reach. The abundance of each species was assessed using a 5-point cover scale where 1 stands for rare plants with cover ranging from 1 to 5%, 2 for occasional plants with cover 6–25%, 3 for frequent plants and cover 26–50%, 4 for abundant plants with cover between 51 and 75% and 5 for dominant plants with cover between 76 and 100%. Most species were identified in the field, but some specimens were collected and transferred to the laboratory for further identification. A complete list with the identified plant taxa can be found in the supplementary material (Table S1).

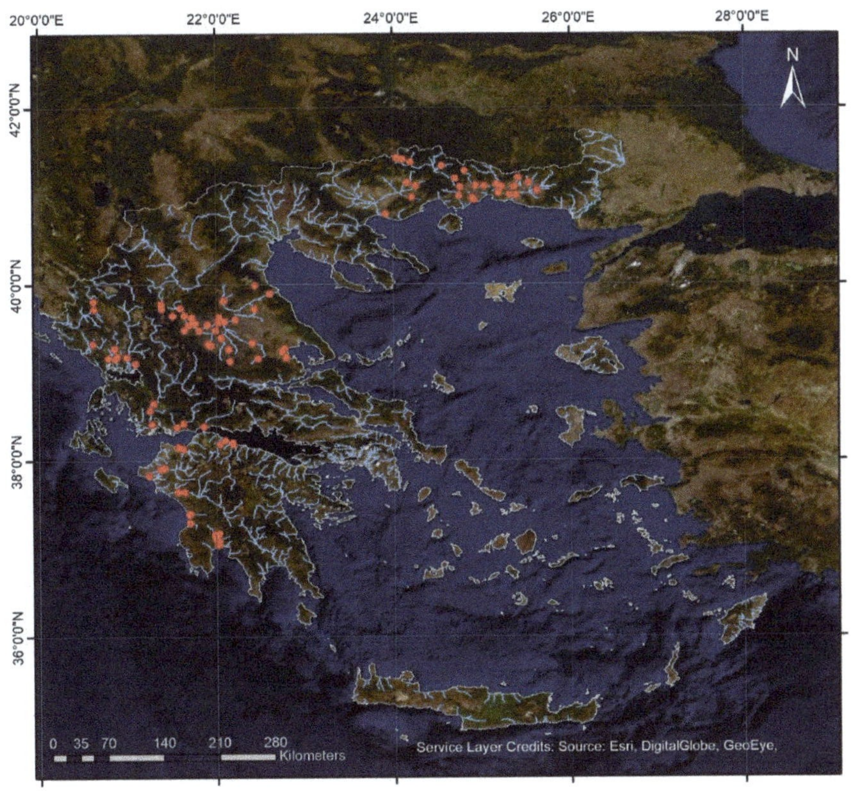

Figure 1. Location of sampling sites (n = 93) of the National Monitoring network, across streams and rivers of mainland Greece.

Water was sampled and transferred to the laboratory for the chemical quantification of orthophosphates, nitrogen species (nitrate, nitrite, and ammonium concentrations in water), total inorganic nitrogen and total phosphorus following the analytical procedures according to APHA [23]. Electrical conductivity, water temperature, dissolved oxygen, and pH were measured in site with a portable multi-meter probe. In parallel with the macrophyte and water sampling, hydromorphological characteristics such as channel cross section alteration, water abstraction, presence of dykes, hydrological alteration, etc., were recorded as described by Feio et al. [15] (see Table 1). Both physicochemical and hydromorphological variables were used as proxies of anthropogenic stressors (e.g., nutrient pollution, acidification, hydromorphological modifications) in order to define the main stressors that influence the macrophyte index.

Table 1. Environmental variables that were considered as potential stressor indicators.

	Stressor	Description
Hydromorphological	Channel profile/cross section alteration	Degree of channel profile modification present at the site/cross section alteration
	Channel morphology	Degree of the morphological modification of the channel present at the site
	Local habitat alteration	Alteration of instream habitats
	Stream hydrology	Degree of the hydrological alteration present at the site
	Upstream dams influence	Effect of upstream dams
	Water abstraction	Effect of water abstraction at the site
	Dykes (flood protection)	Effect of dykes for flood protection
Physicochemical	pH	Sorensen scale
	Conductivity	Conductivity [mS/cm]
	Ammonium	Ammonia concentration in the water [mg/L NH_4^+]
	Nitrate	Nitrate concentration in the water [mg/L NO_3^-]
	Total nitrogen	Total Nitrogen [mg/L TN]
	Total phosphorus	Concentration of total phosphorus in the water [mg/L TP]
	Orthophosphates	Concentration of Orthophosphates in the water [mg/L PO_4^{3-}]
	DO	Concentration of dissolved oxygen [mg/L]
Land use	Urbanization	Urban and industrial areas in immediate vicinity of site
	Agriculture	Agriculture at the immediate vicinity of site

2.2. Statistical Treatment of Environmental Pressure Data—Identification of the Main Stress Gradient

We applied a multivariate analysis with optimal scaling (MVAOS) for the entire dataset of environmental variables (Table 1) to identify those with the highest contribution in the variance of the data. Although there are many other methods for handling multivariate data analysis in water sciences [24,25], we used MVAOS because it allows us to extend the concept of the principal component analysis (PCA) to ordinal variables by transforming them to scale variables [26]. Basically, the MVAOS procedure transforms the ordinal variables and then a PCA is conducted. After omitting those variables with a correlation coefficient $r < 0.7$ with the first two principal components (PC1 and PC2), a second PCA was conducted with the remainder variables. Thus, the MVAOS PCA was used for dimensionality reduction and for providing an environmental gradient as a proxy for stressor gradient. Furthermore, the PC1 scores for the sites were used to identify the less disturbed or unstressed sites which represented the reference sites according to the guidelines provided by the MedGIG. Unstressed sites were defined as those with a PC1 value less than the 25th percentile of the total scores. The MVAOS procedure was applied with the "Gifi" package [27]. PCA were then performed on the transformed variables with the "FactoMineR" package [28]. All analyses were done in R environment [29].

2.3. Development and Implementation of the IBMR$_{GR}$ Index

The IBMR$_{GR}$ index was calculated for all sites according to the following formula [17]:

$$\text{IBMR} = \frac{\sum_i (E_i K_i CS_i)}{\sum_i (E_i K_i)} \quad (1)$$

where E_i the coefficient of ecological amplitude for a given species i, K_i the scale of cover and CS_i is the species-specific score that indicates tolerance to organic pollution.

Sites with only one or two scoring macrophyte species were excluded from further analysis leaving a total number of 79 from the initial 93 sites. In order to distinguish how well the IBMR$_{GR}$ index responds to the pressure a simple linear regression was conducted between the macrophyte index and the PC1 scores derived from the PCA on the most important stressor variables.

Then, the index is normalized between 0 and 1 as follows [30]:

$$IBRM_{NORM} = \frac{I - SI_5}{USI_{75} - SI_5} \quad (2)$$

where I is the index value at a given site, SI_5 is the 5th percentile of the stressed sites and USI_{75} is the 75th percentile of the unstressed sites. Normalized values larger than 1 are set to 1 and lower than 0 are set to 0.

To determine the boundary of the index between the High and Good ecological quality class the 25th percentile of the unstressed sites was used. For the boundaries between the other quality classes the guidelines from the Common Implementation Strategy [31] were followed and the 25th percentile value was divided to 4 so each quality class (Good, Moderate, Poor and Bad) has the same range as the others.

3. Results and Discussion

3.1. Determination of the Least Disturbed/Unstressed Sites

The results from the first MVAOS PCA including all the environmental data showed that the variables with the highest correlation with the PC1 were the channel profile/cross section alteration, channel morphology, habitat alteration, stream hydrology, water abstraction and agriculture, whereas ammonium and phosphate concentrations correlated strongly with the PC2 (Table 2). A second MVAOS PCA was conducted keeping only these nine variables and the results showed that the PC1 acts as a gradient of hydromorphological stressors whereas PC2 clearly shows a strong relationship with three physicochemical variables (ammonium, nitrate, and phosphate) (Figure 2). The first component explained 51% of the total variance of the data whereas the second component explained 22.5%. Thus, the PC1 represents a moderately strong gradient of hydromorphological stress and the PC2 a weaker gradient of physicochemical stress. Then, using the PC1 scores of the sites, that indicate their position along the first component, we distinguished the sites that are less affected by the hydromorphological stressors (those that are positioned at the left side of the biplot in Figure 2). To do so we defined as less disturbed sites (or unstressed) those with a PC1 score less than the 25th percentile of the total PC1 scores. Sites with PC1 score between the 25th and 75th percentile were considered as moderately stressed and those with a PC1 score >75th percentile were the highly stressed sites. Based on this discrimination, Figure 2 shows that unstressed sites are clearly separated as they are placed at the far left in the biplot. Moderately stressed sites are distributed along axis 1 and highly stressed sites are placed on the right part of the plot with several points also showing a high correlation with the second axis that represents a gradient of physicochemical stress. The results showed that the examined variables were responsible for a substantial portion of the variance of the dataset, indicating the importance of hydro-morphological disturbances in the macrophyte communities of the investigated stream sites. To further show the clear discrimination of the sites between these three groups of stress intensity we performed a Kruskal-Wallis test for the IBMR$_{GR}$ values among the stress level and we found that the macrophyte index medians are significantly different ($p \leq 0.001$). Figure 3 illustrates these differences between the unstressed, moderately stressed and highly stressed sites.

Table 2. Correlations between the environmental variables and the first two principal components. Values > 0.7 (in bold) indicate which variables are retained for further analysis.

Stressor	PC1	PC2
Channel profile/cross section alteration	**0.882**	−0.082
Channel morphology	**0.872**	−0.099
Habitat alteration	**0.848**	0.012
Stream hydrology	**0.877**	−0.093
Dams influence	−0.137	0.35
Water abstraction	**0.822**	−0.048
Dykes	0.619	−0.282
DO	−0.436	−0.225
pH	−0.168	−0.21
Electrical conductivity	0.596	−0.109
Ammonium	0.407	**0.799**
Nitrate	0.283	**0.834**
Phosphate	0.268	**0.82**
Urbanization	0.325	0.167
Agriculture	**0.868**	−0.189

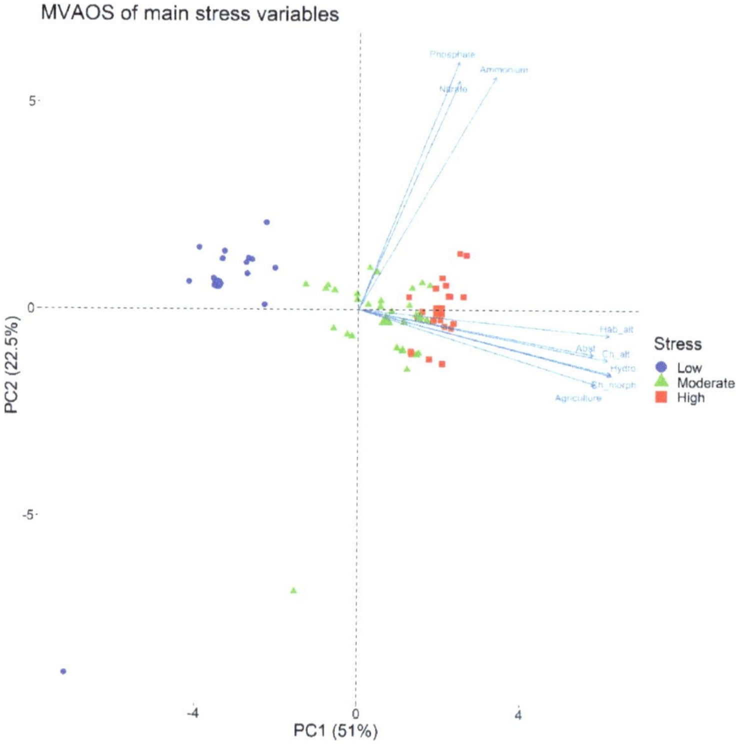

Figure 2. A MVAOS principal component analysis biplot with the nine most important environmental stressors. The position of the sites along the principal component 1 indicates the level or perturbation (stressed, moderately stressed and unstressed).

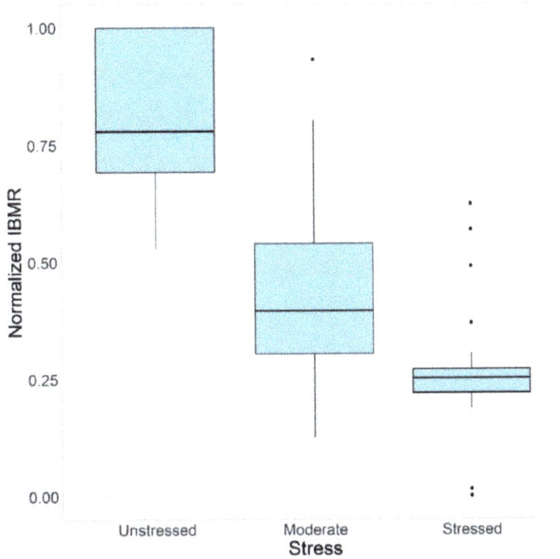

Figure 3. Boxplots of the normalized $IBMR_{GR}$ index among the unstressed, moderately stressed and stressed river reaches.

A simple Spearman correlation analysis between the $IBMR_{GR}$ index and the PC1 scores revealed a significant negative correlation (r = −0.81) which indicates a decline in the macrophyte index with a simultaneous increase of the hydromorphological stress. A linear regression showed a rather strong negative relationship ($R^2 = 0.63$) between the biological indicator ($IBMR_{GR}$) and the stressor (PC1 scores) (Figure 4).

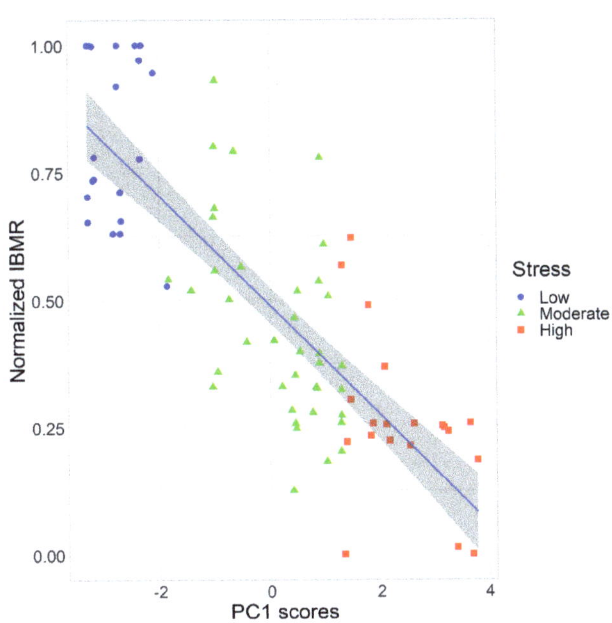

Figure 4. Linear regression between the normalized $IBMR_{GR}$ index and the scores of PC1. Higher PC1 values indicate higher levels of stress. Shaded area represents the 95% confidence intervals.

3.2. Definition of the Ecological Class Boundaries of the $IBMR_{GR}$

The next step was the definition of the ecological class boundaries of the $IBMR_{GR}$ index, so the index is WFD compliant. For this purpose the index was first normalized [30] to a range from 0 to 1. The boundary value of the normalized index between the High and the Good ecological quality class is the 25th percentile of the unstressed sites (0.69). The boundaries between the other quality classes are derived by dividing the 25th percentile of the unstressed sites (0.69) to the four remaining quality classes. Thus, the boundary between Good and Moderate is 0.518, between Moderate and Poor is 0.345 and between Poor and Bad is 0.173 (Figure 5).

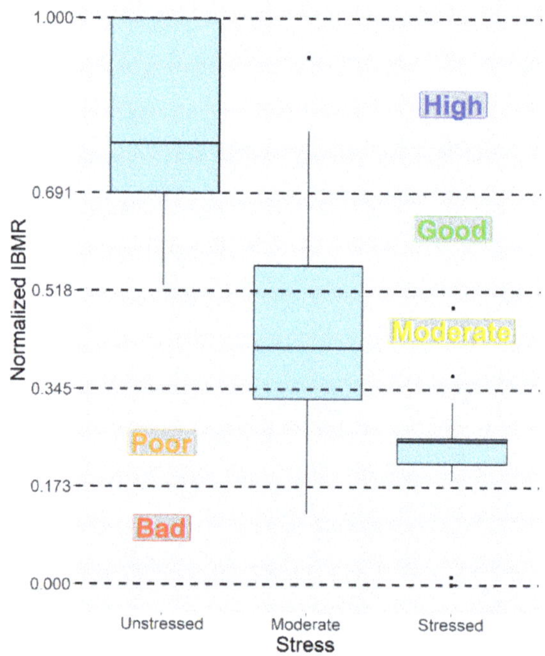

Figure 5. Ecological quality class boundaries of the normalized $IBMR_{GR}$ index.

3.3. Ecological Classification of the River Sites

The implementation of the index classified 15 sites as Good and 19 sites as High ecological quality classes. Thus, 45 from the total 79 sites failed the target of the Good quality class having either Bad, Poor or Moderate with the majority of them (27) classified as Poor (Figure 6). These results are generally in agreement with classification schemes based on other BQEs (e.g., benthic invertebrates and diatoms) [18,32] that have shown that a large share of rivers and streams in Greece is classified with less than Good ecological quality. This pattern is generally found in the whole of Europe where 40 to 50% of water bodies have failed the target of the Good ecological quality [33,34]. Currently there is a long discussion on why European freshwaters have not improved after the implementation of the measures proposed by the River Basin Management Plans and the Programmes of Measures [34,35], but deciphering the causes is a quite complex matter and beyond the scope of this study.

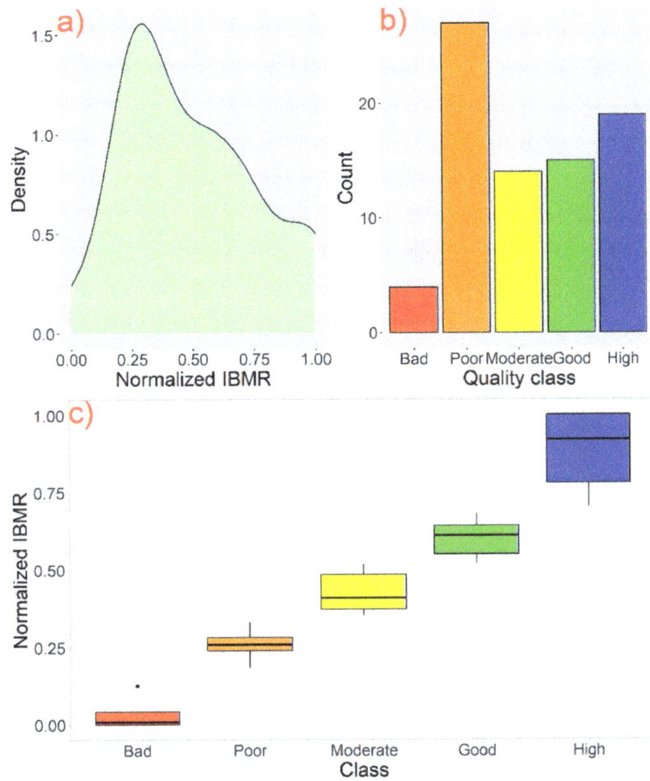

Figure 6. (**a**) Distribution of the normalized $IBMR_{GR}$ and (**b**) histogram of the ecological quality classes for all the studied reaches. Both plots show a high frequency of low values of normalized $IBMR_{GR}$ and Poor ecological quality. More than half of the river reaches have failed the Good ecological quality threshold. Subplot (**c**) shows boxplots of the normalized IBMR per class of ecological quality.

Since the WFD requires from the member states to differentiate the water bodies to types and establish type-specific reference conditions [36], ecological status assessment must be fulfilled for each type separately. Member states that share the same eco-region use a harmonized typology. For the Mediterranean region there are currently six intercalibration river types which are described as R-M1: Small, medium altitude Mediterranean streams with strong seasonal flow; R-M2: Small-medium lowland Mediterranean streams; R–M3: Large Mediterranean streams with strong seasonal flow; R-M4: Small–medium Mediterranean mountain streams with strong seasonal flow; R-M5: Small lowland temporary streams with temporary flow and VL: Very large rivers [32].

A first attempt to assess the ecological quality of the sites per intercalibration river type was made only for those types that were represented by a sufficient number of sites. The total of the 93 sites that were considered initially for the ecological assessment with the use of macrophytes are classified into three distinct river types, R-M1, R-M2 and R-M3. The majority (52) are characterized as R-M2, 35 sites as R-M1 and only 6 sites are classified as R-M3. Thus, the statistical analysis that involves the definition of a stressor gradient and quality class boundaries was only feasible for the types R-M1 and R-M2.

The methodology is the same as described in Sections 2.2 and 2.3. A MVAOS PCA was conducted separately for each type to identify the main stressor gradient and to distinguish the less disturbed or unstressed sites (Figure 7). Then the quality class boundaries are defined as described previously (Table 3).

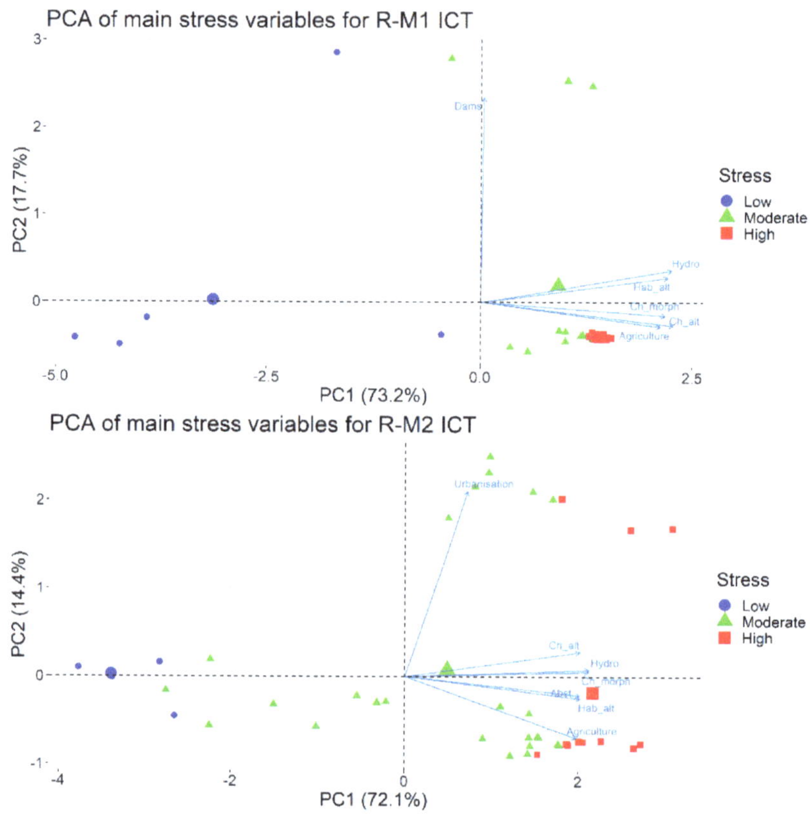

Figure 7. PCA biplots for sites belonging to ICT R-M1 (**top**) and R-M2 (**bottom**). PC1 components indicate a strong gradient of hydromorphological stressors. Unstressed sites (red dots) are positioned on the left part of the plot which, means low correlation with the stressor gradient.

Table 3. Quality class boundaries of the normalized IBMR defined for two MedGIG intercalibration river types (R-M1 and R-M2).

Type	**Ecological Quality Class Boundaries**									
	High		**Good**		**Moderate**		**Poor**		**Bad**	
	MIN	*MAX*	*MIN*	*MAX*	*MIN*	*MAX*	*MIN*	*MAX*	*MIN*	*MAX*
R-M1	>0.705	1	>0.529	≤0.705	>0.352	≤0.529	>0.176	≤0.352	0	≤0.176
R-M2	>0.754	1	>0.567	≤0.754	>0.378	≤0.567	>0.189	≤0.378	0	≤0.189

The linear regressions between the $IBMR_{GR}$ and the PC1 scores for each river type (R-M1 and R-M2) showed relatively good relationships with R^2 of 0.692 and 0.632 respectively (Figure 8). Most of the R-M1 sites (9 of 31) were classified as Bad, whereas only 6 sites met the target of the Good ecological quality class. Nine sites were not classified because only two or less macrophyte species with a specific score CS were recorded. For R-M2 sites, 22 of the 47 assessed sites had a Good or High quality class, 10 were classified as Moderate and 15 as Bad and Poor. In general, R-M2 sites showed better ecological quality conditions than the R-M1 sites.

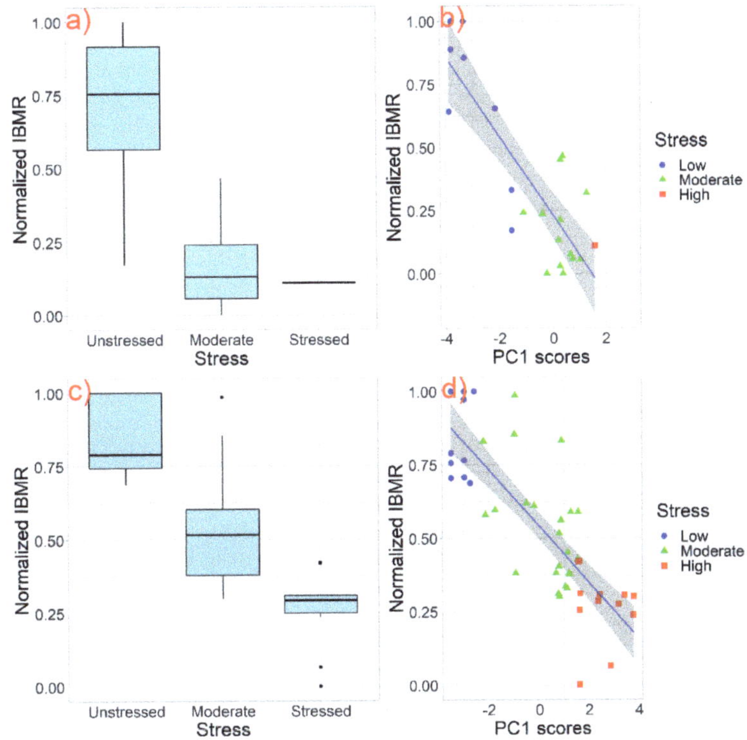

Figure 8. Boxplots of normalized $IBMR_{GR}$ index per stress level for R-M1 sites (**a**) and R-M2 sites (**c**). Linear regressions between normalized $IBMR_{GR}$ and PC1 scores of the main hydromorphological stress gradient as shown for R-M1 sites (**b**) and R-M2 sites (**d**).

3.4. The Use of the $IBMR_{GR}$ Index for Ecological Classification of Greek Running Waters—Strengths and Potential Caveats

The assessment of riverine ecological quality with the IBMR is a quite common method applied in several EU member states [13,37]. Although the index was primarily built to reflect the macrophyte responses to organic and trophic pollution gradients [17], in our case it was shown to correlate positively with a gradient of hydromorphological alterations. In Greece, hydromorphological modifications, such as bank and channel resectioning and realignment, are common stressors that occur in many river courses [38]. On many occasions, they are attributed to expansion of the agricultures [38] and other human activities and may co-occur with other perturbations, including point-sources of organic pollution [21]. This finding corroborates the results of previous studies done in different types of running water-bodies [39,40]. It is highly likely that hydromorphologically perturbed sites are also polluted. In our case, the PCA of the main stressor metrics showed that some of the most impaired sites were also related with the second PC component, which represents a gradient of ammonium, nitrate and phosphate concentrations in water. Still, we should bear in mind that the presented results derive from the pilot application of the index in rivers in Greece, and, as such the total number of tested sites is considered relatively small to exclude ultimate conclusions. An overall assessment of all the sites, regardless their typology, revealed rather promising results but further testing on discrete river types requires extensive collection of field data not only on macrophytes but mostly on environmental descriptors of key stressors. Here, we were able to test the $IBMR_{GR}$ index on two MedGIG types (R-M1 and R-M2) and we found relatively strong relationships between the index and the stressor gradient. These two types refer to mid-altitude or lowland streams which are characterized by seasonal flows and show diverse aquatic plant communities [2]. For other river types, such as very large rivers and temporary streams, the application of the index might be problematic. For instance, macrophyte sampling at very large rivers or rivers of the type R-M3 might be less effective due to limitations associated with the large channel dimensions (depth and width) or muddy and clay substrate conditions. In addition, large rivers are more likely to be regulated and present frequent hydropeaking which although it plays a major role in shaping riparian plant communities [40–42], it may not be considered as a key stressor by the ecological assessment schemes. This is not a problem only for macrophytes but for other BQEs as well. Especially for countries where it is difficult to define sites with "reference conditions" different and additional approaches might be optimal [43,44]. Low land large rivers for example are more likely to be degraded and present a short gradient of pollution which makes it very challenging to identify sites with near natural conditions in terms of organic and nutrient pollution. For temporary streams, indicator species during low flow and dry conditions might be absent [4], which makes non-feasible the estimation of the index [45]. If there is not a sufficient number of indicator species the index will produce an unreliable result [1]. In our case, several sites were excluded from the ecological assessment because of a low number of indicator species. To deal with this issue, indicator species lists should be adjusted to include more new species that are characteristic to the local conditions [1,46]. Overall, the results of our research show that the macrophyte index $IBMR_{GR}$ can be used as a reliable indicator for the biological assessment of water quality and therefore it is recommended for future use in river management planning. Despite these promising results, there is still a need for expanding the stressor gradient, including additional monitoring sites. Future studies could then allow a comparative assessment between indices from different BQEs for the same water bodies [47] which would provide useful insights about the implementation of the ecological monitoring and assessment of the riverine systems, and promote the knowledge exchange towards to an enhanced management of rivers and riparian zones [48].

4. Conclusions

With this article we presented the first ever detailed overview of the implementation of a biological index based on macrophytes for the ecological assessment of the riverine ecosystems of Greece according to the guidelines of the WFD 2000/60. We showed that the $IBMR_{GR}$ index for Greek rivers relates strongly with the main stressor gradient and can distinguish the monitoring sites according to the stress level. We also found that the hydromorphological modifications were the main environmental stressors that correlated strongly with the $IBMR_{GR}$, whereas physicochemical stressors were of lesser importance. From a management perspective this means that mitigation and restoration measures should prioritize the recovery of the hydromorphologial functionality in order to improve the ecological quality and the overall ecosystem status. Furthermore, the results from the ecological classification showed that almost 40% of the sites had Good or High ecological quality, which is close to what other classification schemes have shown. Still, a type-specific classification such as the WFD dictates was only possible for two Mediterranean river types, mostly because of the small number of sites for the other types that did not allow for reliable statistical treatments. Nevertheless, the overall results are promising and indicate that the $IBMR_{GR}$ index is an efficient and applicable index for the rivers of Greece. However, future research is needed to explore the feasibility of application to additional river types and to strengthen the relationship between the index and the stressor gradient by collecting more, and possibly new metrics that could expand the range of the gradient.

Supplementary Materials: The following are available online at https://www.mdpi.com/article/10.3390/w14182771/s1. Table S1: A checklist with the plant species found at the studied sites.

Author Contributions: Conceptualization, E.P., K.S.; methodology, E.P., K.S.; data analysis, E.P., K.S., D.T.; visualization, K.S.; writing—original draft preparation, K.S., E.P.; writing—review and editing,

K.S., E.P., G.D., M.S., D.T.; data curation, E.P., G.D., M.S., D.T.; funding acquisition, E.P. All authors have read and agreed to the published version of the manuscript.

Funding: This research was funded by European and National grants from the Hellenic Centre for Marine Research under the "Monitoring of ecological quality of Greek rivers for the Implementation of Article 8 of WFD 2000/60/EE: samplings and analyses of aquatic macrophytes" research project. The publication of this article has been financed by the Research Committee of the University of Patras, funding number 826/30.08.2022).

Institutional Review Board Statement: Not applicable.

Informed Consent Statement: Not applicable.

Data Availability Statement: Derived data supporting the findings of this study are available from the corresponding author (E.P.) on request.

Acknowledgments: We are grateful to the staff of Patras Laboratory of Ecology for their invaluable help during the field work and collection of the samples. We would also like to express our gratitude to the Research Committee of the University of Patras for the support on managing the project and related activities.

Conflicts of Interest: The authors declare no conflict of interest.

References

1. Aguiar, F.C.; Segurado, P.; Urbanič, G.; Cambra, J.; Chauvin, C.; Ciadamidaro, S.; Dörflinger, G.; Ferreira, J.; Germ, M.; Manolaki, P.; et al. Comparability of River Quality Assessment Using Macrophytes: A Multi-Step Procedure to Overcome Biogeographical Differences. *Sci. Total Environ.* **2014**, *476–477*, 757–767. [CrossRef] [PubMed]
2. Stefanidis, K.; Oikonomou, A.; Papastergiadou, E. Responses of Different Facets of Aquatic Plant Diversity along Environmental Gradients in Mediterranean Streams: Results from Rivers of Greece. *J. Environ. Manag.* **2021**, *296*, 113307. [CrossRef] [PubMed]
3. Stefanidis, K.; Papastergiadou, E. Linkages between Macrophyte Functional Traits and Water Quality: Insights from a Study in Freshwater Lakes of Greece. *Water* **2019**, *11*, 1047. [CrossRef]
4. Manolaki, P.; Guo, K.; Vieira, C.; Papastergiadou, E.; Riis, T. Hydromorphology as a Controlling Factor of Macrophytes Assemblage Structure and Functional Traits in the Semi-Arid European Mediterranean Streams. *Sci. Total Environ.* **2020**, *703*, 134658. [CrossRef]
5. Tremp, H.; Kohler, A. The Usefulness of Macrophyte Monitoring-Systems, Exemplified on Eutrophication and Acidification of Running Waters. *Acta Bot. Gall.* **1995**, *142*, 541–550. [CrossRef]
6. Szoszkiewicz, K.; Ciecierska, H.; Kolada, A.; Schneider, S.C.; Szwabińska, M.; Ruszczyńska, J. Parameters Structuring Macrophyte Communities in Rivers and Lakes—Results from a Case Study in North-Central Poland. *Knowl. Manag. Aquat. Ecosyst.* **2014**, *145*, 08. [CrossRef]
7. O'Hare, M.T.; Baattrup-Pedersen, A.; Baumgarte, I.; Freeman, A.; Gunn, I.D.; Lázár, A.N.; Sinclair, R.; Wade, A.J.; Bowes, M.J. Responses of Aquatic Plants to Eutrophication in Rivers: A Revised Conceptual Model. *Front. Plant Sci.* **2018**, *9*, 451. [CrossRef]
8. Baattrup-Pedersen, A.; Göthe, E.; Riis, T.; O'Hare, M.T. Functional Trait Composition of Aquatic Plants Can Serve to Disentangle Multiple Interacting Stressors in Lowland Streams. *Sci. Total Environ.* **2016**, *543*, 230–238. [CrossRef]
9. Stromberg, J.C.; Lite, S.J.; Dixon, M.D. Effects of Stream Flow Patterns on Riparian Vegetation of a Semiarid River: Implications for a Changing Climate. *River Res. Appl.* **2010**, *26*, 712–729. [CrossRef]
10. Son, D.; Cho, H.; Lee, E.J. Determining Factors for the Occurrence and Richness of Submerged Macrophytes in Major Korean Rivers. *Aquat. Bot.* **2018**, *150*, 82–88. [CrossRef]
11. Gecheva, G.; Pall, K.; Todorov, M.; Traykov, I.; Gribacheva, N.; Stankova, S.; Birk, S. Anthropogenic Stressors in Upland Rivers: Aquatic Macrophyte Responses. a Case Study from Bulgaria. *Plants* **2021**, *10*, 2708. [CrossRef] [PubMed]
12. Peternel, A.; Gaberščik, A.; Zelnik, I.; Holcar, M.; Germ, M. Long-Term Changes in Macrophyte Distribution and Abundance in a Lowland River. *Plants* **2022**, *11*, 401. [CrossRef] [PubMed]
13. Birk, S.; Willby, N. Towards Harmonization of Ecological Quality Classification: Establishing Common Grounds in European Macrophyte Assessment for Rivers. *Hydrobiologia* **2010**, *652*, 149–163. [CrossRef]
14. European Commission. Directive of the European Parliament and of the Council 2000/60/EC. Establishing a Framework for Community Action in the Field of Water Policy. *Off. J. Eur. Parliam.* **2000**, *L327*, 1–82.
15. Feio, M.J.; Aguiar, F.C.; Almeida, S.F.P.; Ferreira, J.; Ferreira, M.T.; Elias, C.; Serra, S.R.Q.; Buffagni, A.; Cambra, J.; Chauvin, C.; et al. Least Disturbed Condition for European Mediterranean Rivers. *Sci. Total Environ.* **2014**, *476–477*, 745–756. [CrossRef]
16. Demars, B.O.L.; Potts, J.M.; Trémolières, M.; Thiébaut, G.; Gougelin, N.; Nordmann, V. River Macrophyte Indices: Not the Holy Grail! *Freshw. Biol.* **2012**, *57*, 1745–1759. [CrossRef]
17. Haury, J.; Peltre, M.C.; Trémolières, M.; Barbe, J.; Thiébaut, G.; Bernez, I.; Daniel, H.; Chatenet, P.; Haan-Archipof, G.; Muller, S.; et al. A New Method to Assess Water Trophy and Organic Pollution—The Macrophyte Biological Index for Rivers (IBMR): Its Application to Different Types of River and Pollution. *Hydrobiologia* **2006**, *570*, 153–158. [CrossRef]

18. Skoulikidis, N.T.; Karaouzas, I.; Amaxidis, Y.; Lazaridou, M. Impact of EU Environmental Policy Implementation on the Quality and Status of Greek Rivers. *Water* **2021**, *13*, 1858. [CrossRef]
19. Manolaki, P.; Papastergiadou, E. Environmental Factors Influencing Macrophytes Assemblages in a Middle-Sized Mediterranean Stream. *River Res. Appl.* **2016**, *32*, 639–651. [CrossRef]
20. Manolaki, P.; Papastergiadou, E. The Impact of Environmental Factors on the Distribution Pattern of Aquatic Macrophytes in a Middle-Sized Mediterranean Stream. *Aquat. Bot.* **2013**, *104*, 34–46. [CrossRef]
21. Stefanidis, K.; Papaioannou, G.; Markogianni, V.; Dimitriou, E. Water Quality and Hydromorphological Variability in Greek Rivers: A Nationwide Assessment with Implications for Management. *Water* **2019**, *11*, 1680. [CrossRef]
22. Oikonomou, A.; Leprieur, F.; Leonardos, I.D. Biogeography of Freshwater Fishes of the Balkan Peninsula. *Hydrobiologia* **2014**, *738*, 205–220. [CrossRef]
23. APHA. *Standard Methods for the Examination of Water and Wastewater*, 23rd ed.; Baird, R.B., Eaton, A.D., Rice, E.W., Eds.; American Public Health Association, American Water Works Association, Water Environment Federation: Denver, CO, USA, 2017; ISBN 978-0875532875.
24. Najafzadeh, M.; Niazmardi, S. A Novel Multiple-Kernel Support Vector Regression Algorithm for Estimation of Water Quality Parameters. *Nat. Resour. Res.* **2021**, *30*, 3761–3775. [CrossRef]
25. Najafzadeh, M.; Homaei, F.; Farhadi, H. *Reliability Assessment of Water Quality Index Based on Guidelines of National Sanitation Foundation in Natural Streams: Integration of Remote Sensing and Data-Driven Models*; Springer: Amsterdam, The Netherlands, 2021; Volume 54, ISBN 0123456789.
26. Gifi, A. *Nonlinear Multivariate Analysis*; Wiley: Hoboken, NJ, USA, 1990; ISBN 978-0-471-92620-7.
27. Mair, P.; Leeuw, J. *De Gifi: Multivariate Analysis with Optimal Scaling*. R Package Version 0.3-9. 2019. Available online: https://cran.r-project.org/web/packages/Gifi/Gifi.pdf (accessed on 1 July 2022).
28. Lê, S.; Josse, J.; Husson, F. FactoMineR: An R Package for Multivariate Analysis. *J. Stat. Softw.* **2008**, *25*, 253–258. [CrossRef]
29. R Core Team. *R: A Language and Environment for Statistical Computing*; R Core Team: Vienna, Austria, 2020.
30. Böhmer, J.; Rawer-Jost, C.; Zenker, A. Multimetric Assessment of Data Provided by Water Managers from Germany: Assessment of Several Different Types of Stressors with Macrozoobenthos Communities. *Hydrobiologia* **2004**, *516*, 215–228. [CrossRef]
31. Pollard, P.; van de Bund, W. *Template for the Development of a Boundary Setting Protocol for the Purposes of the Intercalibration Exercise*; Version 1.2; CIS—Ecostat Working Group: Brussels, Belgium, 2005.
32. Stefanidis, K.; Oikonomou, A.; Stoumboudi, M.; Dimitriou, E.; Skoulikidis, N.T. Do Water Bodies Show Better Ecological Status in Natura 2000 Protected Areas than Non-Protected Ones?—The Case of Greece. *Water* **2021**, *13*, 3007. [CrossRef]
33. Masouras, A.; Karaouzas, I.; Dimitriou, E.; Tsirtsis, G.; Smeti, E. Benthic Diatoms in River Biomonitoring-Present and Future Perspectives within the Water Framework Directive. *Water* **2021**, *13*, 478. [CrossRef]
34. Lemm, J.U.; Venohr, M.; Globevnik, L.; Stefanidis, K.; Panagopoulos, Y.; Gils, J.; Posthuma, L.; Kristensen, P.; Feld, C.K.; Mahnkopf, J.; et al. Multiple Stressors Determine River Ecological Status at the European Scale: Towards an Integrated Understanding of River Status Deterioration. *Glob. Chang. Biol.* **2021**, *27*, 1962–1975. [CrossRef]
35. Birk, S.; Chapman, D.; Carvalho, L.; Spears, B.M.; Andersen, H.E.; Argillier, C.; Auer, S.; Baattrup-Pedersen, A.; Banin, L.; Beklioğlu, M.; et al. Impacts of Multiple Stressors on Freshwater Biota across Spatial Scales and Ecosystems. *Nat. Ecol. Evol.* **2020**, *4*, 1060–1068. [CrossRef]
36. European Commision. Common Implementation Strategy for the Water Framework Directive (2000/60/EC). Guidance Document n. 10. River and Lakes—Typology, Reference Conditions and Classification Systems. 2003. Available online: https://circabc.europa.eu/sd/a/dce34c8d-6e3d-469a-a6f3-b733b829b691/Guidance%20No%2010%20-%20references%20conditions%20inland%20waters%20-%20REFCOND%20(WG%202.3).pdf (accessed on 15 July 2022).
37. Wiegleb, G.; Gebler, D.; van de Weyer, K.; Birk, S. Comparative Test of Ecological Assessment Methods of Lowland Streams Based on Long-Term Monitoring Data of Macrophytes. *Sci. Total Environ.* **2016**, *541*, 1269–1281. [CrossRef]
38. Stefanidis, K.; Latsiou, A.; Kouvarda, T.; Lampou, A.; Kalaitzakis, N.; Gritzalis, K.; Dimitriou, E. Disentangling the Main Components of Hydromorphological Modifications at Reach Scale in Rivers of Greece. *Hydrology* **2020**, *7*, 22. [CrossRef]
39. Vukov, D.; Ilić, M.; Ćuk, M.; Igić, R. The Effect of Hydro-Morphology and Habitat Alterations on the Functional Diversity and Composition of Macrophyte Communities in the Large River. *Front. Environ. Sci.* **2022**, *10*, 863508. [CrossRef]
40. Lozanovska, I.; Rivaes, R.; Vieira, C.; Ferreira, M.T.; Aguiar, F.C. Streamflow Regulation Effects in the Mediterranean Rivers: How Far and to What Extent Are Aquatic and Riparian Communities Affected? *Sci. Total Environ.* **2020**, *749*, 141616. [CrossRef]
41. Lozanovska, I.; Bejarano, M.D.; Martins, M.J.; Nilsson, C.; Ferreira, M.T.; Aguiar, F.C. Functional Diversity of Riparian Woody Vegetation Is Less Affected by River Regulation in the Mediterranean Than Boreal Region. *Front. Plant Sci.* **2020**, *11*, 857. [CrossRef]
42. Bejarano, M.D.; Jansson, R.; Nilsson, C. The Effects of Hydropeaking on Riverine Plants: A Review. *Biol. Rev.* **2018**, *93*, 658–673. [CrossRef]
43. Lazaridou, M.; Ntislidou, C.; Karaouzas, I.; Skoulikidis, N.; Birk, S. Harmonization of the Assessment Method for Classifying the Ecological Quality Status of Very Large Greek Rivers. *Knowl. Manag. Aquat. Ecosyst.* **2018**, *419*, 50. [CrossRef]
44. Kelly, M.G.; Chiriac, G.; Soare-Minea, A.; Hamchevici, C.; Birk, S. Defining Ecological Status of Phytobenthos in Very Large Rivers: A Case Study in Practical Implementation of the Water Framework Directive in Romania. *Hydrobiologia* **2019**, *828*, 353–367. [CrossRef]

45. Rodrigues, C.; Alves, P.; Bio, A.; Vieira, C.; Guimarães, L.; Pinheiro, C.; Vieira, N. Assessing the Ecological Status of Small Mediterranean Rivers Using Benthic Macroinvertebrates and Macrophytes as Indicators. *Environ. Monit. Assess.* **2019**, *191*, 596. [CrossRef]
46. Schneider, S. Macrophyte Trophic Indicator Values from a European Perspective. *Limnologica* **2007**, *37*, 281–289. [CrossRef]
47. Karaouzas, I.; Smeti, E.; Kalogianni, E.; Skoulikidis, N.T. Ecological Status Monitoring and Assessment in Greek Rivers: Do Macroinvertebrate and Diatom Indices Indicate Same Responses to Anthropogenic Pressures? *Ecol. Indic.* **2019**, *101*, 126–132. [CrossRef]
48. Urbanič, G.; Politti, E.; Rodríguez-González, P.M.; Payne, R.; Schook, D.; Alves, M.H.; Anđelković, A.; Bruno, D.; Chilikova-Lubomirova, M.; Di Lonardo, S.; et al. Riparian Zones—From Policy Neglected to Policy Integrated. *Front. Environ. Sci.* **2022**, *10*, 868527. [CrossRef]

water

MDPI

Article

Ecological Responses of Meiofauna to a Saltier World—A Case Study in the Van Uc River Continuum (Vietnam) in the Dry Season

Hien Thanh Nguyen [1,2,*], Lucie Gourdon [2], Hoi Van Bui [2], Duong Thanh Dao [2], Huong Mai [2], Hao Manh Do [3], Thanh Vu Nguyen [4] and Sylvain Ouillon [2,5]

1 Faculty of Ecology and Biological Resources, Graduate University of Science and Technology, Vietnam Academy of Science and Technology, 18 Hoang Quoc Viet Str., Cau Giay, Hanoi 100000, Vietnam
2 Department of Water, Environment, Oceanography, University of Science and Technology of Hanoi, Vietnam Academy of Science and Technology, 18 Hoang Quoc Viet Str., Cau Giay, Hanoi 100000, Vietnam
3 Institute of Marine Environment and Resources, Vietnam Academy of Science and Technology, 246 Da Nang, Hai Phong 180000, Vietnam
4 Institute of Ecology and Biological Resources, Vietnam Academy of Science and Technology, 18 Hoang Quoc Viet Str., Cau Giay, Hanoi 100000, Vietnam
5 UMR LEGOS, IRD, CNES, CNRS, UPS, University of Toulouse, 14 Avenue Edouard Belin, 31400 Toulouse, France
* Correspondence: nguyen-thanh.hien@usth.edu.vn

Abstract: Increasing intensity of storms, typhoons, and sea level rise in conjunction with high water demand, especially for agriculture, in dry seasons in the Red River Delta may have led to seawater intruding deeper into the rivers' estuaries. Given that losses of agricultural productivity and shortages of freshwater resources are projected, a reliable early warning of salinity invasion is, therefore, crucially needed. To evaluate the impact of salinity variations on riverine ecosystems, distribution patterns of meiofauna were examined at 20 stations along the Van Uc River continuum in the dry season. Meiofaunal richness indices were higher in the estuary and slightly decreased upriver. Nematoda was the most dominant taxon in salty stations, while Rotifera was more abundant in the less salty ones. A multiple variate analysis showed a strong interplay among salinity, nutrients, and pore water conductivity, which shaped the meiofaunal distribution. The inclusion of pore water salinity, nutrients, and meiofaunal community structure indicated a greater extent of the saline ecosystem in the estuary, posing a greater risk of freshwater salinization. Our results highlight the potential role of meiofauna as bioindicators but also call for a reformation of salinity assessment for better freshwater conservation and management.

Keywords: salinity intrusion; meiofauna; community structure; bioindicator; ecotone

Citation: Nguyen, H.T.; Gourdon, L.; Bui, H.V.; Dao, D.T.; Mai, H.; Do, H.M.; Nguyen, T.V.; Ouillon, S. Ecological Responses of Meiofauna to a Saltier World—A Case Study in the Van Uc River Continuum (Vietnam) in the Dry Season. *Water* **2023**, *15*, 1278. https://doi.org/10.3390/w15071278

Academic Editors: Eva Papastergiadou and Kostas Stefanidis

Received: 11 February 2023
Revised: 18 March 2023
Accepted: 20 March 2023
Published: 24 March 2023

1. Introduction

River ecosystems are pivotal for humanity as they provide water for agriculture, industry, and power generation, as well as support high biodiversity values and many vital ecosystem services [1,2]. Nonetheless, given their strong economic potential, urbanization and industrialization rates in such river basins are often rapid, which increase competition for water resources and add more contaminants to aquatic environments [3]. Additionally, riverine ecosystems in low-lying areas are particularly susceptible to climate change, putting these habitats under serious threat [4]. Deterioration of water quality in both surface and groundwater are projected to be worsened in many regions, challenging the achievement of Sustainable Development Goals 6 (SDGs, clean water for all) and 11 (life below water). Hence, there is an urgent need for early warning systems and better management of rivers to provide multiple sustainable benefits [2,3].

As the third largest river in the Red River Delta, Vietnam, the Van Uc River has contributed to local livelihoods and the development of the city of Hai Phong, an important location in the northern key economic region of Vietnam [5]. Nevertheless, as with other rivers in the large coastal cities, the water quality degradation induced by both anthropogenic pollution and natural disturbance is a major concern to this densely populated river–sea continuum [6,7]. Nguyen et al. [8] emphasized a significant sea level rise trend in the Hai Phong coastal area, with a remarkably accelerated rate in the last 20 years (14.7 cm in comparison with 21.4 cm over 60 years). Seawater intrusion has recently been recorded moving further landward with longer retention [9–11]. Such events of extreme salinity intrusion in the northern region adversely impacted crop production and threatened freshwater resources [9,10,12].

It is worth highlighting that higher release of toxic heavy metals from the sediment to water environments, and increased heavy metal uptake by living aquatic organisms, have been observed, both induced by elevated salinity [13,14]. Additionally, as invertebrates with a freshwater affinity are sensitive to salinity change, their community structure will be altered by the intrusion of salt-tolerant or brackish species [15,16]. Those salt-tolerant species are poised to colonize new areas and expand their range, which can have severe impacts on ecological interactions and processes by altering the original components of the food web in a particular ecosystem [17]. The increase in salt concentration in the river, even in a small amount, has had results ranging from ecological mortality to sublethal effects, which, in turn, lead to biodiversity loss and reduction in related ecosystem services [18–20]. Hence, in the light of rapid population growth, increased demand for freshwater, and predicted climate change effects and biodiversity loss, a thorough assessment of salinity invasion impacts on the Van Uc River ecosystems is crucially needed to improve the city's risk mitigation and adaptation strategies.

The monitoring of salinity intrusion is often conducted using either direct water property measurements or remote sensing techniques [21]. Nonetheless, such approaches cannot reflect the integration of diverse environmental factors nor the long-term sustainability of river ecosystems [22]. In contrast, bioindicators are used to define the health of an ecosystem; they not only provide insights into their own response to environmental disturbance but also are capable of predicting how ecosystems might respond to future conditions [23,24]. Bioindicators are, therefore, among the proposed tools for monitoring ecological integrity and environmental pollution, with high applicability to water quality assessment [23,25]. Nonetheless, climate change impacts on freshwater environments are less well known compared to the terrestrial or marine realm [26]. Regarding salinity intrusion, numerous studies have focused on monitoring water quality changes and the hydrodynamics of both surface and groundwater in the saltwater-invaded area and its adjacent aquifer, yet few have addressed the response of organisms to saltwater intrusion [27,28].

Predominantly inhabiting the surface/groundwater interface, meiofauna, that is, small invertebrate species, have been widely used in many environmental quality assessments, owing to their benthic lifestyle, short generation time, and fast response to changing conditions [29–31]. Shifts in meiofaunal density, community structure, functioning traits, and other associated ecological indices have often been linked with or indicated for a wide range of environmental perturbations such as eutrophication, heavy metals, pesticides, seasonal variations, physical disturbance, and so on [31–35]. Importantly, the high sensitivity of meiofauna to salinity variations has been widely observed, including in polar, temperate, and tropical ecosystems [36–40], emphasizing their valuable role as a bioindicator for saline water invasion. However, only a few papers concern the composition of meiofaunal communities in Vietnam, and there remains a particular knowledge gap for meiofauna in the northern areas.

This study, therefore, aims to (i) investigate the changes in distribution patterns of meiofauna assemblages between different riverine habitats (upstream, freshwater/brackish ecotone, and downstream), and (ii) evaluate the shifts in community structure in relation to salinity variations along the Van Uc River.

2. Materials and Methods

2.1. Study Area

Located in the Red River Delta, the second largest of the important agriculture areas in Vietnam, the Van Uc River flows through the city of Hai Phong and finally meets the East Sea of Vietnam. The Van Uc River belongs to the Thai Binh River system and is one of nine distributaries of the Red River. It receives water and sediments inputs from both the Red River and the Thai Binh River via a complex network within the Red River Delta. The total river discharge through the Van Uc estuary was estimated at approximately $17.7 \times 109\ m^3/y$ (during the period 1989–2010), corresponding to 14.5% of the total water discharge from the Red–Thai Binh system into the Gulf of Tonkin [6]. The Van Uc sediment flux represents 14.4% of the total sediment flux of the Red–Thai Binh River to the Red River coastal area. Therefore, the Van Uc River is the third most important distributary of the Red River Delta in terms of water and sediment discharges (after the Day River and the Ba Lat River mouth). The Van Uc estuary is subjected to the Southeast Asian sub-tropical monsoon climate and experiences two distinct seasons, a wet season (May–October) and a dry season (November–April).

2.2. Sampling and Sample Processing

Samples were collected along the Van Uc River during the dry season in April 2021. Twenty sampling stations were selected following the salinity gradients, which presented downstream, brackish/freshwater ecotone, and upstream habitats (Figure 1, Table 1). It is worth noting that earlier in the dry season, the leading edge of saline water entering the river was estimated to reach 26–28 km upstream with a drastic change in salinity from brackish to freshwater occurred at 26 km (Figure 1, Km-26) from the sea [11]. In addition, the Van Uc River is characterized by a strong stratification of two distinct water masses during the dry season, with freshwater flowing seaward at the surface and seawater flowing landward near the bed [5]. Such typical estuarine circulation results from the combination of the longitudinal pressure gradient (a barotropic force, constant as a function of depth, and acting in a down-estuary direction) and the longitudinal density (salinity) gradient (a baroclinic force, increasing almost linearly with depth and acting in an up-estuary direction) [41]. Therefore, in order to track the salt front movement, the sediment samples were intensively taken both upriver and downriver from the station VU12 (Figure 1, Km-26) to assess the spatial changes in both environmental conditions and meiofaunal communities. The geographical coordinates and a brief description of the sampling stations are given in Table 1.

At each station, a Ponar grab sampler (sample area of 152 × 152 mm, at 4 m depth) was deployed three times to collect sediment. From the grab-sampling sediment, three subsamples for meiofauna and six others for environmental variable analyses, including granulometry and nutrient content, were then retrieved by inserting plexiglass corers (inner diameter, 3.4 cm, area 10 cm^2) down to 7 cm depth. All meiofauna samples were treated with 7% $MgCl_2$ to anesthetize organisms and then preserved in 4% formaldehyde solution while the other sediment samples were frozen until further analysis.

In the laboratory, the meiofauna samples were extracted by flotation with Ludox-TM50 (specific gravity of 1.18) and stained with Rose Bengal (0.5 $g\ L^{-1}$) before being enumerated and identified to the major taxon level under a stereomicroscope (EMZ-13TR, Meijitechno, San Jose, CA, USA). The following diversity indices were calculated for meiobenthos to assess their efficiency in describing environmental conditions [42]: the Margalef biodiversity (d), the Shannon index (H'), Pielou's evenness index (J'), and the Hill indices ($N1$, $N2$).

In addition, granulometric analyses were carried out in the laboratory using a Mastersizer 3000 (Malvern Panalytical, Malvern, Westborough, MA, USA). Organic matter in the sediment was estimated using the weight loss on ignition method and presented as the percentage of total matter (OM). Total phosphorous (TP) was measured using the vanado-molybdophosphoric acid colorimetric method, while total nitrogen (TN) was detected following the method of ISO 11464: 1994. Other environmental parameters, including salinity and dissolved oxygen, were measured in the surface water, pH and temperature were measured in situ for both surface water (Hanna HI98194; Cluj, Romania) and pore water in the sediment, and electrical conductivity was measured to evaluate the salinity of pore water in the sediment (ECs; Field Scout Direct Soil EC Meter and pH meter; Spectrum Technology, Inc., Aurora, IL, USA).

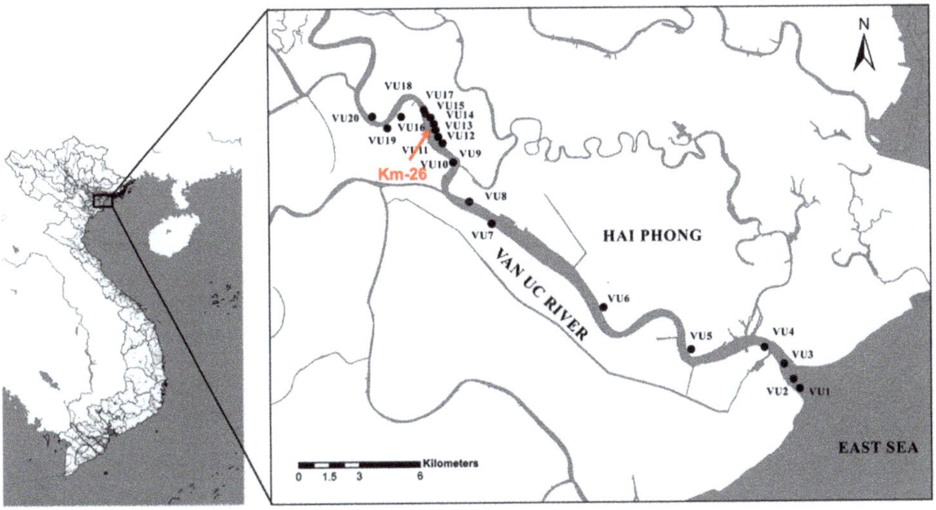

Figure 1. Map of the sampling sites along the Van Uc River.

Table 1. Description of sampling sites.

Habitat	Station	Geographical Location		Description
		Latitude	Longitude	
Downstream	VU1	20.678319	106.700849	Estuary, dense mangrove forest
Downstream	VU2	20.682121	106.698022	
Downstream	VU3	20.688487	106.693951	Mangrove forest, next to a freshwater outlet from agriculture irrigation channel
Downstream	VU4	20.695359	106.685127	Scattered distribution of mangroves
Downstream	VU5	20.69464	106.652553	
Downstream	VU6	20.712204	106.613992	Close to livestock farm (pig)
Ecotone	VU7	20.747069	106.565513	Close to rice field
Ecotone	VU8	20.756267	106.555787	Close to rice field
Ecotone	VU9	20.77239	106.548795	Close to rice field
Ecotone	VU10	20.772398	106.548794	Close to polychaeta farm
Ecotone	VU11	20.780523	106.544285	Close to rice field
Ecotone	VU12	20.783151	106.542376	Close to rice field (26 km from the estuary)
Ecotone	VU13	20.785923	106.541231	Close to polychaeta farm
Ecotone	VU14	20.788484	106.540525	Close to rice field
Ecotone	VU15	20.791128	106.53912	Riverbank under construction
Ecotone	VU16	20.792836	106.536795	Close to rice field
Ecotone	VU17	20.794296	106.536181	Close to polychaeta farm, on fertilized soil/mud in the dry season
Upstream	VU18	20.791533	106.525937	Industrial zone
Upstream	VU19	20.786903	106.519853	
Upstream	VU20	20.791663	106.513265	Close to rice field

2.3. Data Analysis

Principal component analysis (PCA) was applied to elucidate the interaction between the environmental variables. To determine the water suitability for irrigation use, electrical conductivity (EC) was classified into four classes according to Richards [43]: low salinity for irrigation purpose (EC < 250 μScm^{-1}), medium salinity (250 < EC < 750 $\mu S\ cm^{-1}$), high salinity (750 < EC < 2250 $\mu S\ cm^{-1}$), and very high salinity (2250 < EC < 5000 $\mu S\ cm^{-1}$). All environmental variables were normalized prior to the calculation of an environmental resemblance matrix. The spatial changes for each variable were tested by one-way PERMANOVA analysis (univariate, no transformation, Euclidean distance).

Differences among habitats were tested by a one-way ANOVA (analysis of variance, factor: habitat) performed on meiobenthic univariate variables (density, ecological indices). Homogeneity and normality of the dataset were checked using Kolmogorov–Smirnov tests. When required, the data were log (1 + x) transformed. A post hoc Tukey's test was applied when significant differences were detected by ANOVA.

To address the variations of meiofauna community structure, PERMANOVA and SIMPER analyses were performed. PERMANOVA tests were designed with one factor (habitat) and performed on a meiofaunal community structure resemblance matrix (no transformation, Bray–Curtis similarity). Pairwise tests were performed for the main factors and interactions when significant results were obtained. Monte Carlo tests were applied when the number of available permutations was <100. Similarity percentage (SIMPER) analysis was used to determine which taxa were responsible for observed differences between upstream and downstream stations.

The relationship between the environmental variables and the meiobenthic community structure was explored by carrying out a BIOENV analysis and a distance-based linear models (DistLM) routine [44] with forward selection of the independent variables and 999 permutations. Distance-based redundancy analysis (dbRDA) was then performed to visualize the model output from the DistLM, which showed the influence of environmental variables on meiofaunal communities.

All univariate and multivariate analyses were performed according to the procedures described by Clark and Warwick [45], using the PRIMER V6 software package [46] and the PERMANOVA+ add-on [44], except for the univariate analyses of meiofauna density and ecological indices (STATISTICA 7).

3. Results

3.1. Environmental Variables

The PCA revealed a strong salinity influence on the sample distribution pattern, with the first two principal components explaining 68.8% of the total variance (Figure 2). The first axis (PC1) was defined by pore water electrical conductivity, salinity, and dissolved oxygen and explained 44.1% of variance, while the second axis (PC2) was correlated with sediment grain size together with water pH and explained 24.7% of variance. Three groups of samples were formulated, representing three riverine habitats along a gradient of salinity (Figure 3), with the first one, the downstream/brackish ecosystem, characterized by oligohaline to mesohaline water and very high salinity for irrigation purposes (VU1–VU6), while the second group (ecotone) was defined by oligohaline water and high salinity for irrigation purposes (VU7–VU17). The last three stations represented upstream habitat (freshwater), although their ECs were still medium salinity for irrigation purposes. Univariate PERMANOVA (one-way, factor: habitat) analyses showed significant differences between the three habitats, except for in the granulometric analysis (D50). Salinity, pore water conductivity (ECs), dissolved oxygen (DO), and temperature significantly decreased from the estuary toward upstream, while nutrients (TN, TP, OM) were significantly higher in the ecotone and upstream habitats compared to downstream (Table 2).

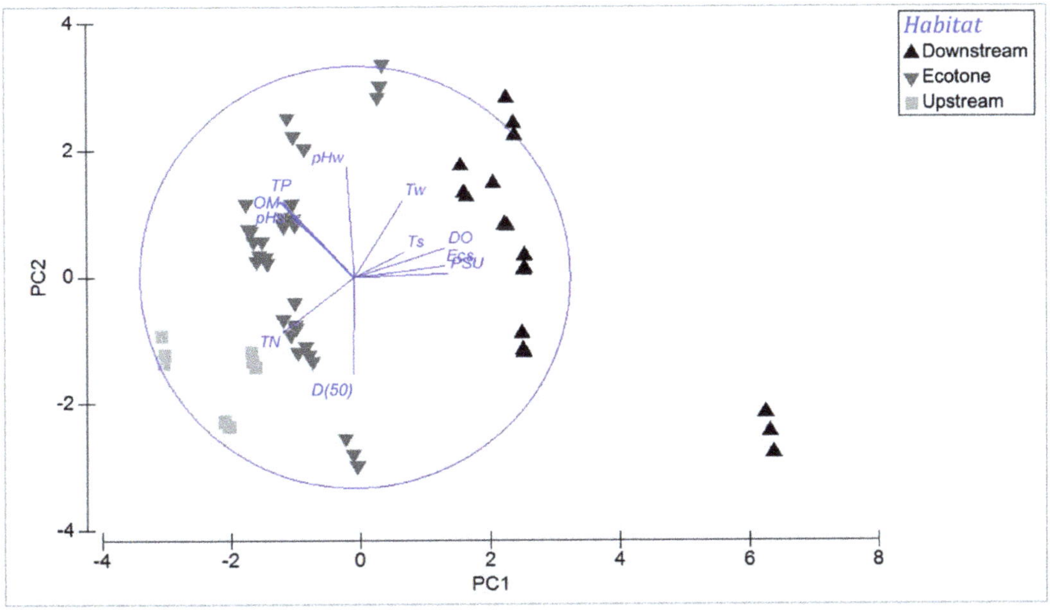

Figure 2. Principal component analysis based on environmental conditions in the Van Uc River (PC1 = 44.1% and PC2 = 24.7%). The circle displays correlation coefficients of the variables with the first two principal axes; the closer the projection vector is to the circle border, the higher the correlation is between the considered variable and its associated axis. T_W—water temperature, T_s—sediment temperature, DO—dissolved oxygen, pH_W—pH of water, pH_s—pH of sediment, $D_{(50)}$—median particle size, TN—total nitrogen, TP—total phosphorus, OM—organic matter, ECs—pore water electrical conductivity, PSU—practical salinity unit.

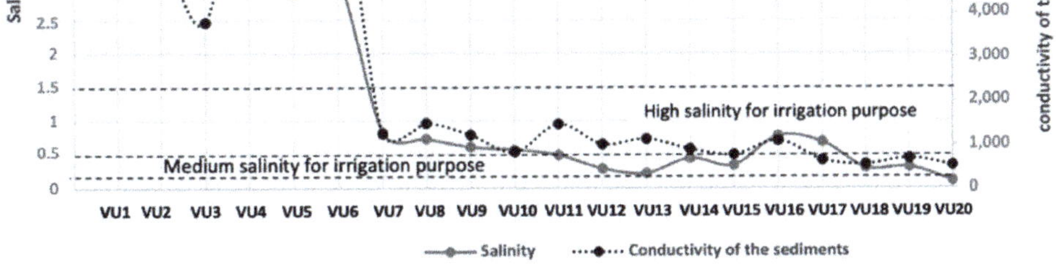

Figure 3. Salinity and sediment conductivity variations along the Van Uc River.

Table 2. Abiotic properties (mean ± SD, min.–max.) of sampling stations representing three habitats along the Van Uc River. T_W — water temperature, T_s — sediment temperature, DO — dissolved oxygen, pH_W — pH of water, pH_s — pH of sediment, $D_{(50)}$ — median particle size, TN — total nitrogen, TP — total phosphorus, OM — organic matter, ECs — pore water electrical conductivity, PSU—practical salinity unit.

Abiotic Factor	Downstream		Ecotone		Upstream		One-Way PERMANOVA			
	Mean ± SD	Min.–Max.	Mean ± SD	Min.–Max.	Mean ± SD	Min.–Max.	df	MS	Pseudo-F	*p*
PSU	4.07 ± 0.84	3.16–5.75	0.54 ± 0.19	0.22–0.83	0.24 ± 0.1	0.11–0.31	2	27.272	348.8	0.001
ECs (μScm^{-1})	6503 ± 2048	3800–9400	1064 ± 270	640–1480	583 ± 75	520–680	2	24.8	150.39	0.001
T_s (°C)	26.48 ± 0.49	25.8–27.1	26.23 ± 0.42	25.8–27.2	25.87 ± 0.43	25.4–26.4	2	5.0129	5.8345	0.007
T_w (°C)	25.94 ± 0.19	25.7–26.2	25.90 ± 0.16	25.6–26.2	25.44 ± 0.02	25.4–25.5	2	4.9071	5.6221	0.04
TN (%)	0.72 ± 0.24	0.39–1.04	2.13 ± 0.79	0.48–3.52	2.49 ± 0.09	2.36–2.57	2	16.92	38.335	0.001
TP (%)	0.14 ± 0.02	0.10–0.16	0.16 ± 0.01	0.14–0.18	0.16 ± 0.0005	0.16–0.161	2	5.3827	6.3609	0.006
OM (%)	3.39 ± 0.45	2.81–3.99	4.04 ± 0.48	2.77–4.66	3.95 ± 0.63	3.32–4.74	2	7.7696	10.19	0.001
$D_{(50)}$ (μm)	49.21 ± 27.27	22–98.9	56.21 ± 22.44	27.7–109	60.07 ± 15.61	38.6–76.7	2	0.8199	0.8148	0.45
DO (mgL^{-1})	5.59 ± 0.46	4.89–6.42	4.18 ± 0.23	3.88–4.73	3.22 ± 0.24	2.9–3.41	2	25.685	191.91	0.001
pH_w	7.78 ± 0.1	7.54–7.92	7.81 ± 0.08	7.64–8.05	7.67 ± 0.06	7.61–7.77	2	7.2054	9.2108	0.001
pH_s	7.2 ± 0.33	6.49–7.59	7.39 ± 0.12	7.2–7.72	7.38 ± 0.08	7.29–7.49	2	4.5527	5.201	0.016

3.2. Meiofauna Assemblage along the Van Uc River

The total meiofauna density (±SD) varied from 179 ± 27 inds/10 cm^2 (VU16) to 1454 ± 723 inds/10 cm^2 (VU14) (Figure 4). Higher variability of meiofaunal density was observed in the ecotone ecosystem compared to the downstream and upstream ones. Nevertheless, a one-way ANOVA test performed on the total meiofauna densities (under log transformation) showed that there were no significant differences among the three habitats (F = 2.482, p = 0.09).

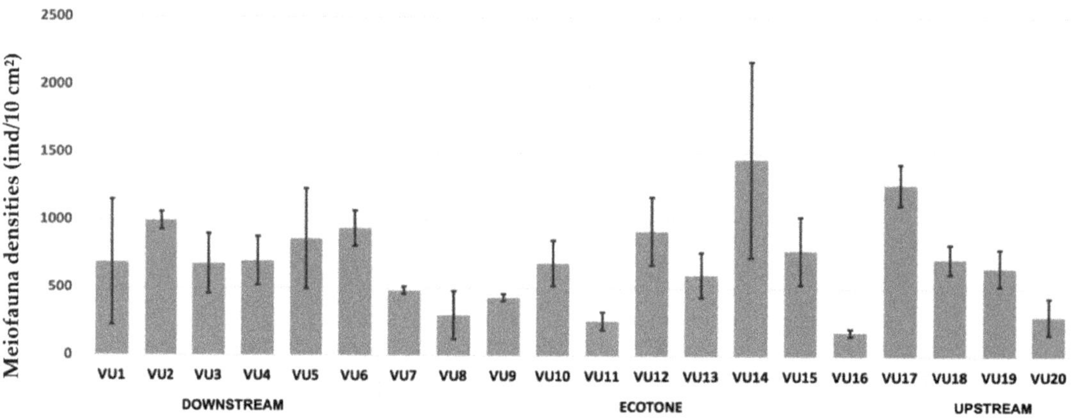

Figure 4. Meiofauna densities (mean ± SD/10 cm^2) along the Van Uc River.

A total of 23 taxa were identified along the Van Uc River during the sampling period, with Nematoda the most dominant taxon (72.6%), followed by Rotifera (19.1%), Copepoda (2.2%), nauplii (1.1%), and others (4.1%). The "others" category contained all meiofaunal taxa whose percentage of representation was less than 1%, including Bivalvia (0.89%), Foraminifera (0.83%), Oligochaeta (0.74%), Polychaeta (0.6%), Amphipoda (0.55%), Turbellaria (0.35%), Ciliophora (0.33%), Ostracoda (0.13%), and Sipuncula (0.11%), together with some taxa presented at a very low density (less than 0.1%) such as Insecta, Kinorhyncha, Gastrotricha, Acari, Nemertea, Cumacea, Bryozoa, Isopoda, Gastropoda, and Tanaidacea.

The contribution of the different meiofaunal taxa present in the samples for each sampling site during the dry season is shown in Figure 5. Nematodes were always the most abundant taxon, with the exception of VU14 and VU15, where Rotifera were predominant (with, respectively, 80.86% and 78.77%). Copepodes and nauplii were also represented by several individuals, changing, respectively, from 0.55% (VU13) and 0.32% (VU14) to 5.21% (VU9) and 5.52% (VU5). The representation of the "others" category changed from 0.33% (VU13) to 9.32% (VU20).

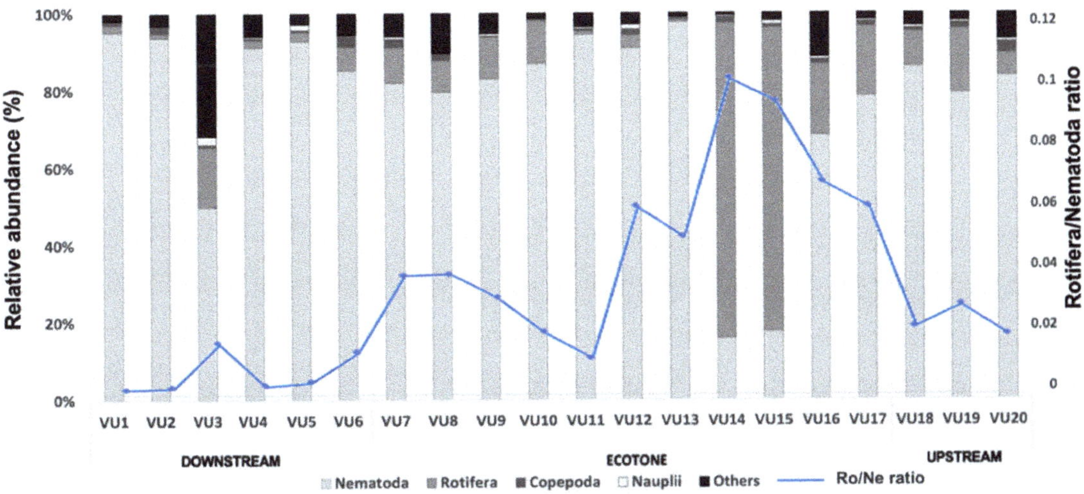

Figure 5. Proportional composition of meiofauna and Rotifera/Nematoda ratio along the Van Uc River.

The dynamics of nematodes and rotifers followed an inversely proportional relationship. When the percentage of Nematoda decreased, the percentage of Rotifera increased simultaneously and reciprocally. Moving inland, the Rotifera/Nematoda ratio in many stations increased as there were fewer nematodes, but more rotifers were found, especially in the ecotone, until nematodes took the lead again in the upstream habitat.

The significant changes in meiofaunal composition between stations were proven by the results of the one-way PERMANOVA test, which showed a pseudo-F of 3.03 and a *p*(perm)-value of 0.03 (Table 3).

Table 3. Results of one-way PERMANOVA test for meiofaunal community structure.

One-Way PERMANOVA						
Source	df	SS	MS	Pseudo-F	*p* (perm)	Permutation
Habitat	2	4133.4	2066.7	3.0274	0.003	998
Res	57	38,911	682.65			
Total	59	43,045				

Pairwise Test				
Groups	t	*p* (perm)	perms	*p* (MC)
Downstream, Ecotone	2.1901	0.001	999	0.001
Downstream, Upstream	1.4951	0.032	999	0.056
Ecotone, Upstream	1.1445	0.236	999	0.26

In addition, the SIMPER routine demonstrated a high level of dissimilarity of VU14 from VU15, mainly contributed by Rotifera (72–81%) and Nematodes (16–24%) (Table 4).

Table 4. Results of the SIMPER analysis.

Station	Average Similarity (%)	Taxa Contributions (%)							
		Nematodes	Rotifers	Polychaeta	Turbellaria	Nauplii	Copepods	Amphipoda	Bivalves
VU1	60.85	95.24							
VU2	92.88	95.86							
VU3	73.88	93.23	1.56	1.48					
VU4	68.43	73.42	17.07	1.98	2.74				
VU5	66.21	83.88	2.14			7.25	3.84		
VU6	75.45	80.79	10.33			2.58	2.88		
VU7	81.70	81.54	8.19				3.24	3.74	
VU8	63.80	91.18	5.44						
VU9	91.28	73.22	14.93				5.54		5.02
VU10	74.75	89.51	7.94						
VU11	82.46	97.21							
VU12	77.85	93.22					3.05		
VU13	80.13	99.14							
VU14	65.11	23.95	72.93						
VU15	70.26	16.82	81.50						
VU16	79.76	75.75	12.09						9.16
VU17	88.69	90.98	2.99				2.18		
VU18	82.48	93.34	2.53						
VU19	80.44	83.35	8.71				4.12		
VU20	65.94	82.13	3.72			2.77	3.06	5.30	

3.3. *Meiofaunal Ecological Indices*

The meiofaunal richness (S) ranged between 6 in the ecotone (VU13) and 18 in the downstream habitat (VU6), showing a slight decrease from the estuary toward upstream (Figure 6). Being among the species richness indices, the Margalef biodiversity index (d) was also highest in VU6 (1.804) and lowest in VU13 (0.682), which concurred with the changes in the number of meiofaunal groups. The meiofaunal assemblage in VU13 was not only low in diversity but also unequally distributed, with nematodes predominant, making for the lowest Pielou's evenness index score (J′) of 0.074 (VU13). The highest score for Pielou's evenness was obtained in VU16 (0.528). Furthermore, the maximum value of the Shannon index (H′) was recorded at location VU3 (1.186), while the minimum value of this index was obtained at location VU13 (0.126). Concerning Hill's indices, N1 changed from 1.136 to 3.274, whereas N2 changed from 1.043 to 2.272. Both maxima were reached at location VU4.

One-way ANOVA tests performed on the different meiofaunal ecological indices showed that only the Margalef biodiversity index was significantly different among the three habitats, while the other four did not differ significantly. Results obtained from a pairwise comparison indicated that the Margalef index was significantly higher downstream compared to the ecotone habitat ($p < 0.05$), but there were no significant differences between the downstream and upstream, and ecotone and upstream, habitats.

3.4. *Meiofauna in Relation to Environmental Variations*

The marginal and sequential tests performed using the DistLM routine indicated that salinity, ECs, nutrients (TN, TP), sediment temperature, and sediment grainsize (D50) were environmental variables that significantly affected the community structure of meiofauna along the Van Uc River (Table 5, $p < 0.05$). The ordination plot generated via analysis of the linear model based on distance (dbRDA) revealed that 84.5% of fitted variation and 36.1% of total variation in meiofauna assemblages could be explained by the first two axes (Figure 7a). dbRDA1 represented 26.8% of the total variation and 62.7% of the fitted variation in the meiofaunal assemblages, showing a positive correlation with salinity and sediment temperature. D50, TN, and TP were correlated with dbRDA2, which explained 8.2% of total variation and 22% of the fitted variation. The first axis represents the variation of assemblages in relation to the spatial changes in salinity, with the right side of the graph characterized by a brackish condition downstream with higher salinity. In contrast, ecotone samples were distributed on the left, with lower salinity values. The upstream samples were localized in the middle, but showed greatest similarity with the meiofauna in the ecotone. The second axis represents the spatial variation of meiofaunal assemblages in response to nutrient and sediment grain size changes. In our study, the increasing salinity was characterized by a greater diversity of meiofauna, with the predominance of nematodes, copepods, nauplii, amphipods, other shrimp-like crustaceans, and polychaetes (Figure 7b). In contrast, the left side of the graph, which presents ecotone and upstream habitats, is dominated by rotifers, insect larvae, some isopods, and acari. Nematodes were more abundant at the stations with higher salinity but lower nutrient contents, while rotifers favored the less saline condition and high OM and TN. Interestingly, copepods and nauplii mostly appeared in the direction of low TP and a coarser sediment grain size, whereas the insect larvae followed the same trend as rotifers.

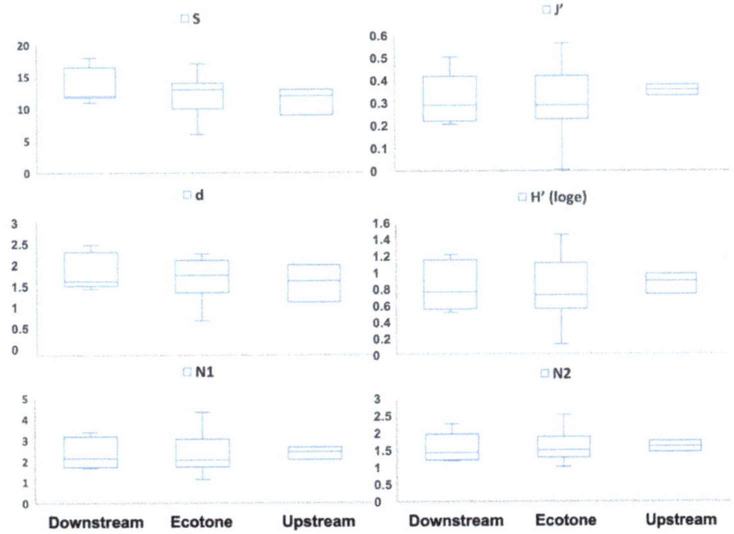

Figure 6. Meiofaunal diversity indices boxplots in downstream, ecotone, and upstream habitats along the Van Uc River in the dry season of 2021. Boxes display median, first and third quartiles, minimum, and maximum.

Table 5. DistLM analysis results showing correlations between environmental variables and meiofaunal community attributes along the Van Uc River. Values in red indicate significant correlations ($p < 0.05$).

Marginal Tests			Sequential Tests				
Variables	Pseudo-F	p	Variables	R^2	Pseudo-F	p	Res. df
pHw	0.80628	0.483	(+) pHw	0.013	0.80628	0.459	58
pHs	0.8369	0.498	(+) pHs	0.025	1.5032	0.207	57
DO	2.2615	0.072	(+) DO	0.056	1.0279	0.409	56
Ecs	3.0077	0.03	(+) Ecs	0.084	1.7092	0.142	55
PSU	4.3192	0.01	(+) PSU	0.152	4.2724	0.01	54
OM	1.6501	0.171	(+) OM	0.173	1.3829	0.231	53
TP	1.466	0.235	(+) TP	0.228	3.6862	0.009	52
TN	1.7807	0.131	(+) TN	0.287	4.2498	0.008	51
Tw	0.35302	0.842	(+) Tw	0.299	0.82954	0.451	50
Ts	2.4743	0.064	(+) Ts	0.371	5.5701	0.001	49
$D_{(50)}$	4.5717	0.006	(+) $D_{(50)}$	0.428	4.7905	0.002	48

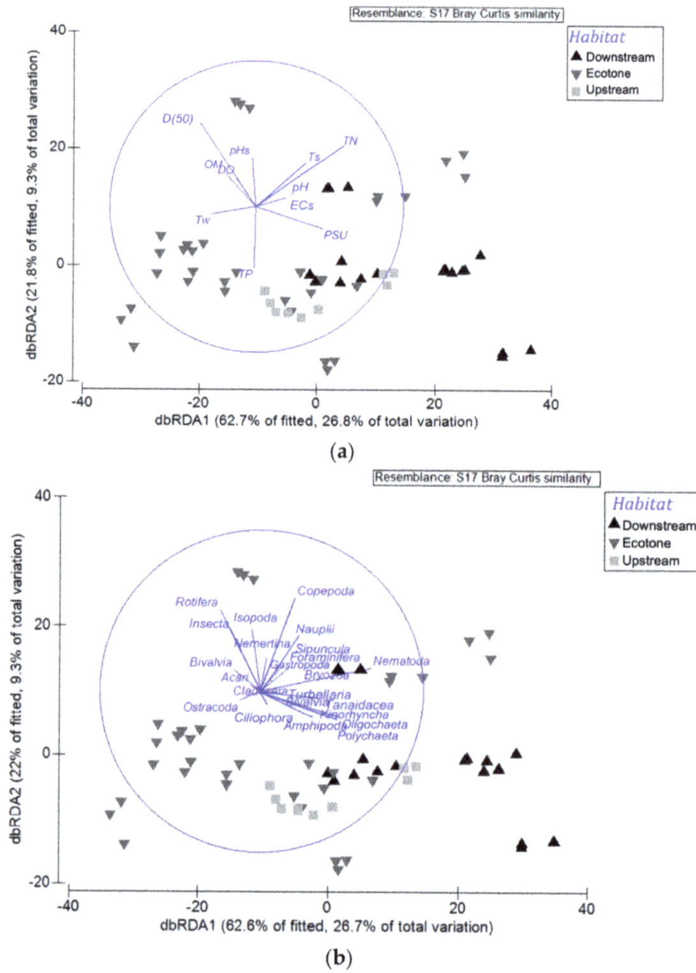

Figure 7. Distance–based redundancy analysis (dbRDA) of meiofaunal community responses to environmental variations showing (**a**) vector overlays of predictor variables and (**b**) vector overlays of species responses. The graph shows the effect of environmental variables on meiofauna community structure based on Spearman's rank correlations.

4. Discussion

Increasing salinization is a growing problem worldwide as it can heighten the stress faced by freshwater organisms and their mortality, thereby adversely impacting ecosystem functionality as well as the services and benefits to human societies that they provide [47]. Given species assemblages can shift to keep pace with climate change [48], gaining insight into the community structure transitions resulting from saltwater intrusion is crucial for managing and promoting coastal resilience [18]. In our study, the signature of saltwater intrusion was shifting of the meiofaunal community structure toward elevated pore water conductivity and surface water salinity along the Van Uc River.

It is widely accepted that one of the main factors influencing species distribution in estuaries is salinity [49–51]. Our findings are, therefore, consistent with several studies, which also identified salinity as an important independent factor determining meiobenthic communities' structure and describing total meiobenthic density and diversity changes [52–54].

Estuarine meiofauna tend to decrease in abundance and number of species as one moves from the sea to freshwater [55–57]. In our study, the meiofaunal richness indices (species richness and Margalef index) of the Van Uc River followed a linear relationship between species richness and salinity (Figure 6). A total of 23 taxa were identified during the dry season in the Van Uc River, with a higher number of taxa found in the brackish ecosystem, then slightly decreasing upstream. Our results revealed higher values compared to other estuaries elsewhere. Soetaert et al. [52] found 13 meiofaunal taxa in five European estuaries, while Pavlyuk et al. [58] recorded 11 taxonomical groups in the Cua-Luc estuary (North Vietnam). In the Mira and Mondengo estuaries, the meiofauna richness were 17 and 12 taxa, respectively [59,60]. Nevertheless, most of the studies in other estuaries were conducted in the intertidal areas with rather high levels of salinity, meaning the studies presented only the meiofaunal assemblage in the brackish ecosystem and neglected the communities in the brackish/freshwater ecotone as well as the freshwater communities. Exceptionally, in the Mekong Delta, where samples were collected across a wider range of salinity (0–25 PSU), a similar result was obtained with a total of 23 taxa recorded along 19 stations of the five estuaries [61].

Different from the richness indices, the meiofauna density in the Van Uc River was higher in the brackish ecosystem (downstream) in comparison with the freshwater ones (upstream), but the highest densities were observed at VU14 and VU17 (ecotone ecosystem) (Figure 4). Such fluctuation in the ecotone meiofaunal density could be related to the high environmental heterogeneity and/or the large amount of total nitrogen in the transitional area (Table 2). It is worth noting that agricultural activities (practices of tillage, rice crop fertilizing, and manuring for polychaeta farm) dominated along the two riverbanks during the sampling time, specifically from VU7 to VU20 (Table 1). Given that such anthropogenic sources can contribute greatly to river nutrient loading [62], the nutrients from the ecotone to freshwater ecosystems in the Van Uc River were considerably higher in comparison with other estuaries [29]. Interestingly, the same patterns of meiofaunal density were also observed at another five estuaries in the Mekong Delta [61], implying the importance of combined effects of salinity and nutrients on meiofaunal assemblages. Such interplay of several crucial environmental factors determining the discrepancy in the distribution of meiobenthic communities is well recognized at several estuarine benthic habitats worldwide [38,52,57,63,64].

In agreement with the above studies, our multivariate analysis revealed that the estuarine environmental gradients were strongly reflected in the meiofauna community structure, with salinity the greatest driving force, followed by TP, TN, and sediment grain size (Figure 7). Unlike the results of PCA, which categorized all stations into three groups based on environmental conditions alone, the distribution pattern of meiofauna assemblages in the Van Uc River during the dry season represented two distinct groups: the brackish and the merging ecotone–freshwater communities. Since meiofauna responses to different environmental variables are often highly species-specific [31], their structural parameters were found to be valuable indicators for detecting environmental changes, providing more pronounced effects on a taxon rather than on total meiofauna [65]. In our study, Nematoda was the most dominant taxon, which represented 72.6% of the total meiofauna density during the dry season. The same result was observed in a few studies worldwide [37,64]. The second abundant group of meiofauna was Rotifera, representing 19.2% during the dry season. This result is different compared to previous studies in other estuaries, where Copepoda was recorded as the second most abundant group [52,66,67], or Sarcomastigophora [68], Polychaeta [69], Tardigrada [70], or even Turbellaria [71].

As meiofauna respond differently to environmental variations depending on their functional traits and life strategy, the occurrence of nematodes and rotifers changed remarkably when there were salinity variations in the Van Uc River. More nematodes were found in the saltier stations, while rotifers preferred less salty ecosystems (Figure 7). The shifts in meiofaunal community structure occurred at VU14 and VU15, which exhibited the minimum amount of Nematoda and maximum density of Rotifera. The same meiofaunal responses were observed at VU3, the brackish station that received freshwater discharge from the irrigation channel. In these three stations, when a sudden reduction in ECs occurred, a large proportion of Nematoda was replaced mainly by the rotifers and some other meiofaunal groups. Rotifer density then decreased at VU16 and VU17, which could have been in correlation with the elevated EC values at these two stations, implying sensitive responses of meiofauna toward salinity variations. These results concurred with the study of Majdi et al. [72] on meiofauna in the sediment of two headwater streams, Ems and Fulbach, in Germany. Ems, which was lower in EC values, was predominantly demarked by rotifers, with much fewer nematodes, while in Fulbach, with higher conductivity, nematodes took the lead, followed by rotifers and other groups. Such drastic change in meiofauna composition pairing with the fluctuation of pore water conductivity emphasizes the important role of electrical conductivity/salinity in regulating riverine ecosystems.

Interestingly, at the brackish and marine-influenced ecotone stations (VU1–VU13), the composition of Rotifera mainly consisted of marine ploima rotifers, whereas from VU14 toward freshwater-influenced stations, freshwater bdelloid rotifers rapidly increased. This calls for further research on what happened to the nematodes and other organisms. Comparing to organisms such as copepods and cladocerans, rotifers are more opportunistic, mainly due to their high reproductive rate. In many cases, they can quickly respond to environmental stresses, showing high sensitivity to elevated salinity [73,74], which makes them a valuable environmental indicator [75,76]. This result indicated that the leading edge of salinity intrusion was penetrating further landward during the late dry season compared to the estimation of salinity intrusion in the study by Nguyen et al. [11], which was conducted in an early month of the dry season. Our findings, therefore, highlight the potential risk of salinization as the water quality at most stations was classified from medium salinity to very high salinity for irrigation purposes during the dry season.

Along with salinity, nutrient enrichment is recognized as another very important factor influencing the meiobenthic taxa composition and abundance patterns [77,78]. In our study, the low density of Copepoda could have been related to the high nutrient contents (TP, TN) and organic matter in the sediment. Additionally, the overall low dissolved oxygen in the river could also have reduced this group as harpacticoid copepods are the most sensitive meiofauna taxon to low oxygen concentrations [79]. In contrast, the substantial decrease in dissolved oxygen (DO < 4 mgL^{-1}) in the upstream ecosystem (VU18–VU20) seems to have adversely impacted the rotifers and other meiofaunal groups, leaving a high percentage of nematodes, owing to their high anaerobic capacity [80]. The role of dissolved oxygen in structuring the benthic meiofauna was evidenced previously in several studies [81–83]. Erikson et al. [84] obtained the same results, showing strong interactions between nutrient inputs and oxygen depletion in tropical lowland rivers. Our findings suggest that the Van Uc River is not only subjected to salinity intrusion but also highly sensitive to pollution by nutrients and organic matter, with substantial impacts on meiofaunal community composition. However, intensive fertilizer use can increase soil salinity [85], which calls for salinity assessments to be reformed to better evaluate the freshwater salinization in the Van Uc River, induced by either the movement of the seawater or excessive fertilization, or both.

5. Conclusions

The meiofaunal community in the Van Uc River was characterized by high abundance and diversity. A total of 23 taxa were identified, with Nematoda the most dominant taxon and playing an important role in controlling the characteristics of the meiofauna assemblages, followed by Rotifera, Copepoda, nauplii, and other groups. Meiofaunal richness indices were higher in the estuary and slightly decreased upriver. In addition, the estuarine gradients were strongly reflected by the meiofaunal community structure, with salinity the greatest driving force, including salinity in the water column and pore water salinity (represented as electrical conductivity) in the sediment. The dynamics of meiofaunal density and community structure were best determined by the interplays among salinity, nutrients, and dissolved oxygen. Both meiofaunal responses and environmental parameters of the Van Uc River indicated a high risk of salinization coupled with pollution induced by nutrients and organic matter. Further studies on nematode and rotifer interactions are needed as they may serve as a potential indicator for salinity intrusion assessment in long-term studies, especially in the context of climate change, in the future. Finally, it is also recommended that both natural salinity intrusion processes and anthropogenic-induced salinization should be considered, to gain insights into the dynamics of the salt front movement, which will support the development of better mitigation and adaptation strategies.

Author Contributions: H.T.N.: Conceptualization, Funding Acquisition, Methodology, Investigation, Formal Analysis, Writing—Original Draft Preparation; Writing—Review and Editing. L.G.: Investigation, Data Curation, Visualization, Writing—Original Draft Preparation. H.V.B., D.T.D. and H.M.: Methodology, Investigation, Data Curation. H.M.D. and T.V.N.: Data Curation, Writing—Review and Editing. S.O.: Writing—Review and Editing All authors have read and agreed to the published version of the manuscript.

Funding: This research was funded by the Graduate University of Science and Technology (GUST), part of the Vietnam Academy of Science and Technology (VAST), grant number GUST.STS.ĐT2020-ST02.

Data Availability Statement: Data are available from the corresponding author on request.

Acknowledgments: The authors gratefully acknowledge the International Joint Laboratory LOTUS—"Land Ocean aTmosphere regional coUpled System"—and the French National Research Institute for Sustainability Development (IRD) for their support.

Conflicts of Interest: The authors declare no conflict of interest.

References

1. Sabater, S.; Elosegui, A. Balancing conservation needs with uses of river ecosystems. *Acta Biol. Colomb.* **2013**, *19*, 3. [CrossRef]
2. Tickner, D.; Parker, H.; Moncrieff, C.R.; Oates, N.E.M.; Ludi, E.; Acreman, M. Managing Rivers for Multiple Benefits—A Coherent Approach to Research, Policy and Planning. *Front. Environ. Sci.* **2017**, *5*, 4. [CrossRef]
3. Strokal, M.; Bai, Z.; Franssen, W.; Hofstra, N.; Koelmans, A.A.; Ludwig, F.; Ma, L.; Van Puijenbroek, P.; Spanier, J.E.; Vermeulen, L.C.; et al. Urbanization: An Increasing Source of Multiple Pollutants to Rivers in the 21st Century. *NPJ Urban Sustain.* **2021**, *1*, 24. [CrossRef]
4. Siddha, S.; Sahu, P. Impact of Climate Change on the River Ecosystem. In *Ecological Significance of River Ecosystems*; Elsevier: Amsterdam, The Netherlands, 2022; pp. 79–104, ISBN 978-0-323-85045-2.
5. Piton, V.; Ouillon, S.; Vinh, V.D.; Many, G.; Herrmann, M.; Marsaleix, P. Seasonal and Tidal Variability of the Hydrology and Suspended Particulate Matter in the Van Uc Estuary, Red River, Vietnam. *J. Mar. Syst.* **2020**, *211*, 103403. [CrossRef]
6. Vinh, V.D.; Ouillon, S.; Thanh, T.D.; Chu, L.V. Impact of the Hoa Binh Dam (Vietnam) on Water and Sediment Budgets in the Red River Basin and Delta. *Hydrol. Earth Syst. Sci.* **2014**, *18*, 3987–4005. [CrossRef]
7. Ho, H.H.; Swennen, R.; Cappuyns, V.; Vassilieva, E.; Neyens, G.; Rajabali, M.; Van Tran, T. Assessment on Pollution by Heavy Metals and Arsenic Based on Surficial and Core Sediments in the Cam River Mouth, Haiphong Province, Vietnam. *Soil Sediment Contam. Int. J.* **2013**, *22*, 415–432. [CrossRef]
8. Nguyen, M.H.; Ouillon, S.; Vu, D.V. Sea-Level Rise in Hai Phong Coastal Area (Vietnam) and Its Response to Enso—Evidence from Tide Gauge Measurement of 1960–2020. *Vietnam. J. Earth Sci.* **2022**, *44*, 109–126. [CrossRef]
9. Nguyen, Y.T.B.; Kamoshita, A.; Dinh, V.T.H.; Matsuda, H.; Kurokura, H. Salinity Intrusion and Rice Production in Red River Delta under Changing Climate Conditions. *Paddy Water Environ.* **2017**, *15*, 37–48. [CrossRef]
10. Nguyen, T.H.; Roberto, R.; Vu, M.C. Potential Use of Reservoirs for Mitigating Saline Intrusion in the Coastal Areas of Red River Delta. In Proceedings of the 2018 IEEE International Conference on Environment and Electrical Engineering and 2018 IEEE Industrial and Commercial Power Systems Europe (EEEIC/I&CPS Europe), Palermo, Italy, 12–15 June 2018; pp. 1–6.
11. Nguyen, T.H.; Vu, M.C.; Roberto, R. Upstream Effects on Salinity Dynamics in the Red River Delta. In *APAC 2019*; Trung Viet, N., Xiping, D., Thanh Tung, T., Eds.; Springer: Singapore, 2020; pp. 1439–1443, ISBN 978-9-81150-290-3.
12. Pereira, C.S.; Lopes, I.; Abrantes, I.; Sousa, J.P.; Chelinho, S. Salinization Effects on Coastal Ecosystems: A Terrestrial Model Ecosystem Approach. *Phil. Trans. R. Soc. B* **2019**, *374*, 20180251. [CrossRef]
13. Du Laing, G.; De Vos, R.; Vandecasteele, B.; Lesage, E.; Tack, F.M.G.; Verloo, M.G. Effect of Salinity on Heavy Metal Mobility and Availability in Intertidal Sediments of the Scheldt Estuary. *Estuar. Coast. Shelf Sci.* **2008**, *77*, 589–602. [CrossRef]
14. Acosta, J.A.; Jansen, B.; Kalbitz, K.; Faz, A.; Martínez-Martínez, S. Salinity Increases Mobility of Heavy Metals in Soils. *Chemosphere* **2011**, *85*, 1318–1324. [CrossRef]
15. Szöcs, E.; Coring, E.; Bäthe, J.; Schäfer, R.B. Effects of Anthropogenic Salinization on Biological Traits and Community Composition of Stream Macroinvertebrates. *Sci. Total Environ.* **2014**, *468–469*, 943–949. [CrossRef]

16. Sowa, A.; Krodkiewska, M. Impact of Secondary Salinisation on the Structure and Diversity of Oligochaete Communities. *Knowl. Manag. Aquat. Ecosyst.* **2020**, *421*, 6. [CrossRef]
17. Tully, K.; Gedan, K.; Epanchin-Niell, R.; Strong, A.; Bernhardt, E.S.; Bendor, T.; Mitchell, M.; Kominoski, J.; Jordan, T.E.; Neubauer, S.C.; et al. Corrigendum: The Invisible Flood: The Chemistry, Ecology, and Social Implications of Coastal Saltwater Intrusion. *BioScience* **2019**, *69*, 760. [CrossRef]
18. Goss, K.F. Environmental Flows, River Salinity and Biodiversity Conservation: Managing Trade-Offs in the Murray—Darling Basin. *Aust. J. Bot.* **2003**, *51*, 619. [CrossRef]
19. Mooney, H.; Larigauderie, A.; Cesario, M.; Elmquist, T.; Hoegh-Guldberg, O.; Lavorel, S.; Mace, G.M.; Palmer, M.; Scholes, R.; Yahara, T. Biodiversity, Climate Change, and Ecosystem Services. *Curr. Opin. Environ. Sustain.* **2009**, *1*, 46–54. [CrossRef]
20. Venâncio, C.; Ribeiro, R.; Lopes, I. Seawater Intrusion: An Appraisal of Taxa at Most Risk and Safe Salinity Levels. *Biol. Rev.* **2022**, *97*, 361–382. [CrossRef]
21. Nguyen, P.T.B.; Koedsin, W.; McNeil, D.; Van, T.P.D. Remote Sensing Techniques to Predict Salinity Intrusion: Application for a Data-Poor Area of the Coastal Mekong Delta, Vietnam. *Int. J. Remote Sens.* **2018**, *39*, 6676–6691. [CrossRef]
22. Li, L.; Zheng, B.; Liu, L. Biomonitoring and Bioindicators Used for River Ecosystems: Definitions, Approaches and Trends. *Procedia Environ. Sci.* **2010**, *2*, 1510–1524. [CrossRef]
23. Parmar, T.K.; Rawtani, D.; Agrawal, Y.K. Bioindicators: The Natural Indicator of Environmental Pollution. *Front. Life Sci.* **2016**, *9*, 110–118. [CrossRef]
24. Hoppit, G.; Schmidt, D.N. A Regional View of the Response to Climate Change: A Meta-Analysis of European Benthic Organisms' Responses. *Front. Mar. Sci.* **2022**, *9*, 896157. [CrossRef]
25. Oertel, N.; Salánki, J. Biomonitoring and Bioindicators in Aquatic Ecosystems. In *Modern Trends in Applied Aquatic Ecology*; Ambasht, R.S., Ambasht, N.K., Eds.; Springer: Boston, MA, USA, 2003; pp. 219–246. ISBN 978-1-4613-4972-3.
26. Pletterbauer, F.; Melcher, A.; Graf, W. Climate Change Impacts in Riverine Ecosystems. In *Riverine Ecosystem Management*; Schmutz, S., Sendzimir, J., Eds.; Springer International Publishing: Cham, Switzerland, 2018; pp. 203–223. ISBN 978-3-319-73249-7.
27. Ferchichi, H.; Ben Hamouda, M.F.; Farhat, B.; Ben Mammou, A. Assessment of Groundwater Salinity Using GIS and Multivariate Statistics in a Coastal Mediterranean Aquifer. *Int. J. Environ. Sci. Technol.* **2018**, *15*, 2473–2492. [CrossRef]
28. Li, Y.; Huang, D.; Sun, W.; Sun, X.; Yan, G.; Gao, W.; Lin, H. Characterizing Sediment Bacterial Community and Identifying the Biological Indicators in a Seawater-Freshwater Transition Zone during the Wet and Dry Seasons. *Environ. Sci. Pollut. Res.* **2022**, *29*, 41219–41230. [CrossRef] [PubMed]
29. Balsamo, M.; Albertelli, G.; Ceccherelli, V.U.; Coccioni, R.; Colangelo, M.A.; Curini-Galletti, M.; Danovaro, R.; D'Addabbo, R.; De Leonardis, C.; Fabiano, M.; et al. Meiofauna of the Adriatic Sea: Present Knowledge and Future Perspectives. *Chem. Ecol.* **2010**, *26*, 45–63. [CrossRef]
30. Zeppilli, D.; Sarrazin, J.; Leduc, D.; Arbizu, P.M.; Fontaneto, D.; Fontanier, C.; Gooday, A.J.; Kristensen, R.M.; Ivanenko, V.N.; Sørensen, M.V.; et al. Is the Meiofauna a Good Indicator for Climate Change and Anthropogenic Impacts? *Mar. Biodivers.* **2015**, *45*, 505–535. [CrossRef]
31. Ridall, A.; Ingels, J. Suitability of Free-Living Marine Nematodes as Bioindicators: Status and Future Considerations. *Front. Mar. Sci.* **2021**, *8*, 685327. [CrossRef]
32. Pusceddu, A.; Gambi, C.; Corinaldesi, C.; Scopa, M.; Danovaro, R. Relationships between Meiofaunal Biodiversity and Prokaryotic Heterotrophic Production in Different Tropical Habitats and Oceanic Regions. *PLoS ONE* **2014**, *9*, e91056. [CrossRef]
33. Semprucci, F.; Sbrocca, C.; Rocchi, M.; Balsamo, M. Temporal Changes of the Meiofaunal Assemblage as a Tool for the Assessment of the Ecological Quality Status. *J. Mar. Biol. Assoc.* **2015**, *95*, 247–254. [CrossRef]
34. Di Lorenzo, T.; Fiasca, B.; Di Cicco, M.; Cifoni, M.; Galassi, D.M.P. Taxonomic and Functional Trait Variation along a Gradient of Ammonium Contamination in the Hyporheic Zone of a Mediterranean Stream. *Ecol. Indic.* **2021**, *132*, 108268. [CrossRef]
35. Cifoni, M.; Boggero, A.; Rogora, M.; Ciampittiello, M.; Martínez, A.; Galassi, D.M.P.; Fiasca, B.; Di Lorenzo, T. Effects of Human-Induced Water Level Fluctuations on Copepod Assemblages of the Littoral Zone of Lake Maggiore. *Hydrobiologia* **2022**, *849*, 3545–3564. [CrossRef]
36. Broman, E.; Raymond, C.; Sommer, C.; Gunnarsson, J.S.; Creer, S.; Nascimento, F.J.A. Salinity Drives Meiofaunal Community Structure Dynamics across the Baltic Ecosystem. *Mol. Ecol.* **2019**, *28*, 3813–3829. [CrossRef] [PubMed]
37. Semprucci, F.; Gravina, M.F.; Magni, P. Meiofaunal Dynamics and Heterogeneity along Salinity and Trophic Gradients in a Mediterranean Transitional System. *Water* **2019**, *11*, 1488. [CrossRef]
38. Leasi, F.; Sevigny, J.L.; Hassett, B.T. Meiofauna as a Valuable Bioindicator of Climate Change in the Polar Regions. *Ecol. Indic.* **2021**, *121*, 107133. [CrossRef]
39. Udalov, A.; Chikina, M.; Azovsky, A.; Basin, A.; Galkin, S.; Garlitska, L.; Khusid, T.; Kondar, D.; Korsun, S.; Kremenetskiy, V.; et al. Integrity of Benthic Assemblages along the Arctic Estuarine-Coastal System. *Ecol. Indic.* **2021**, *121*, 107115. [CrossRef]
40. Baia, E.; Rollnic, M.; Venekey, V. Seasonality of Pluviosity and Saline Intrusion Drive Meiofauna and Nematodes on an Amazon Freshwater-Oligohaline Beach. *J. Sea Res.* **2021**, *170*, 102022. [CrossRef]
41. Officer, C.B. Physical Dynamics of Estuarine Suspended Sediments. *Marine Geology* **1981**, *40*, 1–14. [CrossRef]
42. Karydis, M.; Tsirtsis, G. Ecological Indices: A Biometric Approach for Assessing Eutrophication Levels in the Marine Environment. *Sci. Total Environ.* **1996**, *186*, 209–219. [CrossRef]

43. Richards, L.A. *Diagnosis and Improvement of Saline Alkali Soils, Agriculture, 160, Handbook 60*; US Department of Agriculture: Washington, DC, USA, 1954.
44. Anderson, M.J.; Gorley, R.N.; Clarke, K.N. *PERMANOVA+ for PRIMER: Guide to Software and Statistical Methods*; PRIMER-E Ltd.: Plymouth, UK, 2008.
45. Clarke, K.R.; Warwick, R.M. *Change in Marine Communities: An Approach to Statistical Analysis and Interpretation*; PRIMER-E Ltd.: Plymouth, UK, 2001.
46. Clarke, K.R.; Gorley, R.N. *User Manual/Tutorial*; PRIMER-E Ltd.: Plymouth, UK, 2006.
47. Cunillera-Montcusí, D.; Beklioğlu, M.; Cañedo-Argüelles, M.; Jeppesen, E.; Ptacnik, R.; Amorim, C.A.; Arnott, S.E.; Berger, S.A.; Brucet, S.; Dugan, H.A.; et al. Freshwater Salinisation: A Research Agenda for a Saltier World. *Trends Ecol. Evol.* **2022**, *37*, 440–453. [CrossRef]
48. Root, T.L.; Hall, K.R.; Herzog, M.P.; Howell, C.A. (Eds.) *Biodiversity in a Changing Climate: Linking Science and Management in Conservation*; University of California Press: Oakland, CA, USA, 2015; ISBN 978-0-520-27885-1.
49. Bulger, A.J.; Hayden, B.P.; Monaco, M.E.; Nelson, D.M.; McCormick-Ray, M.G. Biologically-Based Estuarine Salinity Zones Derived from a Multivariate Analysis. *Estuaries* **1993**, *16*, 311. [CrossRef]
50. Attrill, M.J. A Testable Linear Model for Diversity Trends in Estuaries. *J. Anim. Ecol.* **2002**, *71*, 262–269. [CrossRef]
51. McLusky, D.S.; Elliott, M. *The Estuarine Ecosystem*; Oxford University Press: Oxford, UK, 2004; ISBN 978-0-19-852508-0.
52. Soetaert, K.; Vincx, M.; Wittoeck, J.; Tulkens, M. Meiobenthic Distribution and Nematode Community Structure in Five European Estuaries. *Hydrobiologia* **1995**, *311*, 185–206. [CrossRef]
53. Coull, B.C. Long-Term Variability of Estuarine Meiobenthos: An 11 Year Study. *Mar. Ecol. Prog. Ser.* **1985**, *24*, 205–218. [CrossRef]
54. Vincx, M. Diversity of the Nematode Communities in the Southern Bight of the North Sea. *Neth. J. Sea Res.* **1990**, *25*, 181–188. [CrossRef]
55. Coull, B.C. Role of Meiofauna in Estuarine Soft-Bottom Habitats. *Austral. Ecol.* **1999**, *24*, 327–343. [CrossRef]
56. Bouwman, L.A. A Survey of Nematoda from the Ems Estuary: Species Assemblages and Associations. *Zool. Jahrb. Syst. Okologie Geogr. Tiere* **1983**, *110*, 345–376.
57. Heip, C.; Vincx, M.; Vranken, G. The Ecology of Marine Nematodes. In *Oceanography and Marine Biology: An Annual Review, 23*; Aberdeen University Press: Aberdeen, UK, 1985; pp. 399–489. ISBN 0-08-030397-8.
58. Pavlyuk, O.; Trebukhova, Y.; Thanh, N.V.; Tu, N.D. Meiobenthos in Estuary Part of Ha Long Bay (Gulf of Tonkin, South China Sea, Vietnam). *Ocean Sci. J.* **2008**, *43*, 153–160. [CrossRef]
59. Adão, H. Dynamic of Meiofauna Communities in Association with Zostera Noltii Seegrass Beds in the Mira SW Portugal. Ph.D. Thesis, University of Evora, Evora, Portugal, 2003.
60. Alves, A.S.; Adão, H.; Patrício, J.; Neto, J.M.; Costa, M.J.; Marques, J.C. Spatial Distribution of Subtidal Meiobenthos along Estuarine Gradients in Two Southern European Estuaries (Portugal). *J. Mar. Biol. Assoc.* **2009**, *89*, 1529–1540. [CrossRef]
61. Ngo, X.Q.; Smol, N.; Vanreusel, A. The Meiofauna Distribution in Correlation with Environmental Characteristics in 5 Mekong Estuaries, Vietnam. *Cah. Biol. Mar.* **2013**, *54*, 71–83. [CrossRef]
62. Beusen, A.H.W.; Doelman, J.C.; Van Beek, L.P.H.; Van Puijenbroek, P.J.T.M.; Mogollón, J.M.; Van Grinsven, H.J.M.; Stehfest, E.; Van Vuuren, D.P.; Bouwman, A.F. Exploring River Nitrogen and Phosphorus Loading and Export to Global Coastal Waters in the Shared Socio-Economic Pathways. *Glob. Environ. Change* **2022**, *72*, 102426. [CrossRef]
63. Moens, T.; Vincx, M. Temperature, Salinity and Food Thresholds in Two Brackish-Water Bacterivorous Nematode Species: Assessing Niches from Food Absorption and Respiration Experiments. *J. Exp. Mar. Biol. Ecol.* **2000**, *243*, 137–154. [CrossRef]
64. Adão, H.; Alves, A.S.; Patrício, J.; Neto, J.M.; Costa, M.J.; Marques, J.C. Spatial Distribution of Subtidal Nematoda Communities along the Salinity Gradient in Southern European Estuaries. *Acta Oecologica.* **2009**, *35*, 287–300. [CrossRef]
65. Haegerbaeumer, A.; Höss, S.; Ristau, K.; Claus, E.; Heininger, P.; Traunspurger, W. The Use of Meiofauna in Freshwater Sediment Assessments: Structural and Functional Responses of Meiobenthic Communities to Metal and Organics Contamination. *Ecol. Indic.* **2017**, *78*, 512–525. [CrossRef]
66. Warwick, R.; Gee, J. Community Structure of Estuarine Meiobenthos. *Mar. Ecol. Prog. Ser.* **1984**, *18*, 97–111. [CrossRef]
67. Smol, N.; Willems, K.A.; Govaere, J.C.R.; Sandee, A.J.J. Composition, Distribution and Biomass of Meiobenthos in the Oosterschelde Estuary (SW Netherlands). *Hydrobiologia* **1994**, *282–283*, 197–217. [CrossRef]
68. Sinh, N.V.; Kim Phuong, N.T.; Quang, N.X. The Distribution of Meiofauna Community Related to Salinity Gradient in the Ham Luong Estuary, Mekong River. *Acad. J. Biol.* **2014**, *35*, 417–423. [CrossRef]
69. Alongi, D.M. The Role of Soft-Bottom Benthic Communities in Tropical Mangrove and Coral Reef Ecosystems. *Rev. Aquat. Sci.* **1989**, *1*, 243–280.
70. Fonseca, G.; Netto, S.A. Shallow Sublittoral Benthic Communities of the Laguna Estuarine System, South Brazil. *Braz. J. Oceanogr.* **2006**, *54*, 41–54. [CrossRef]
71. Alongi, D.M. Intertidal Zonation and Seasonality of Meiobenthos in Tropical Mangrove Estuaries. *Mar. Biol.* **1987**, *95*, 447–458. [CrossRef]
72. Majdi, N.; Threis, I.; Traunspurger, W. It's the Little Things That Count: Meiofaunal Density and Production in the Sediment of Two Headwater Streams: Meiofauna in Streams. *Limnol. Oceanogr.* **2017**, *62*, 151–163. [CrossRef]
73. Sarma, S.S.S.; Nandini, S.; Morales-Ventura, J.; Delgado-Martínez, I.; González-Valverde, L. Effects of NaCl Salinity on the Population Dynamics of Freshwater Zooplankton (Rotifers and Cladocerans). *Aquat. Ecol.* **2006**, *40*, 349–360. [CrossRef]

74. Medeiros, A.; Barbosa, J.; Medeiros, P.; Rocha, R.; Silva, L. Salinity and Freshwater Discharge Determine Rotifer Distribution at the Mossoró River Estuary (Semiarid Region of Brazil). *Braz. J. Biol.* **2010**, *70*, 551–557. [CrossRef] [PubMed]
75. Gutkowska, A.; Paturej, E.; Kowalska, E. Rotifer trophic state indices as ecosystem indicators in brackish coastal waters. *Oceanologia* **2013**, *55*, 887–899. [CrossRef]
76. Arcifa, M.S.; De Souza, B.B.; De Morais-Junior, C.S.; Bruno, C.G.C. Functional Groups of Rotifers and an Exotic Species in a Tropical Shallow Lake. *Sci. Rep.* **2020**, *10*, 14698. [CrossRef] [PubMed]
77. Widbom, B.; Elmgren, R. Response of Benthic Meiofauna to Nutrient Enrichment of Experimental Marine Ecosystems. *Mar. Ecol. Prog. Ser.* **1988**, *42*, 257–268. [CrossRef]
78. Ristau, K.; Faupel, M.; Traunspurger, W. The Effects of Nutrient Enrichment on a Freshwater Meiofaunal Assemblage: Meiofaunal Response to Nutrient Enrichment. *Freshw. Biol.* **2012**, *57*, 824–834. [CrossRef]
79. De Troch, M.; Roelofs, M.; Riedel, B.; Grego, M. Structural and Functional Responses of Harpacticoid Copepods to Anoxia in the Northern Adriatic: An Experimental Approach. *Biogeosciences* **2013**, *10*, 4259–4272. [CrossRef]
80. Veit-Köhler, G.; Gerdes, D.; Quiroga, E.; Hebbeln, D.; Sellanes, J. Metazoan Meiofauna within the Oxygen-Minimum Zone off Chile: Results of the 2001-PUCK Expedition. *Deep Sea Res. Part II Top. Stud. Oceanogr.* **2009**, *56*, 1105–1111. [CrossRef]
81. Neira, C.; Ingels, J.; Mendoza, G.; Hernandez-Lopez, E.; Levin, L.A. Distribution of Meiofauna in Bathyal Sediments Influenced by the Oxygen Minimum Zone Off Costa Rica. *Front. Mar. Sci.* **2018**, *5*, 448. [CrossRef]
82. Cook, A.A.; John, D.; Lambshead, P.; Hawkins, L.E.; Mitchell, N.; Levin, L.A. Nematode Abundance at the Oxygen Minimum Zone in the Arabian Sea. *Deep Sea Res. Part II Top. Stud. Oceanogr.* **2000**, *47*, 75–85. [CrossRef]
83. Singh, R.; Ingole, B.S. Structure and Function of Nematode Communities across the Indian Western Continental Margin and Its Oxygen Minimum Zone. *Biogeosciences* **2016**, *13*, 191–209. [CrossRef]
84. Eriksen, T.E.; Jacobsen, D.; Demars, B.O.L.; Brittain, J.E.; Søli, G.; Friberg, N. Effects of Pollution-Induced Changes in Oxygen Conditions Scaling up from Individuals to Ecosystems in a Tropical River Network. *Sci. Total Environ.* **2022**, *814*, 151958. [CrossRef] [PubMed]
85. Han, Y.-G.; Kwon, O.; Cho, Y. A Study of Bioindicator Selection for Long-Term Ecological Monitoring. *J. Ecol. Environ.* **2015**, *38*, 119–122. [CrossRef]

water

MDPI

Article

Analyzing the Akinete Protein of the Harmful Freshwater Cyanobacterium, *Dolichospermum circinale*

Keonhee Kim [1], Chae-Hong Park [1] and Soon-Jin Hwang [1,2,*]

[1] Human and Eco-Care Center, Department of Environmental Health Science, Konkuk University, Seoul 05029, Republic of Korea; skyopera@konkuk.ac.kr (K.K.); qkrcoghd@konkuk.ac.kr (C.-H.P.)
[2] Department of Environmental Health Science, Konkuk University, Seoul 05029, Republic of Korea
* Correspondence: sjhwang@konkuk.ac.kr; Tel.: +82-2-450-3748

Abstract: Akinete is a survival structure in cyanobacteria that has overcome unfavorable environmental conditions and influences their perennial blooms in the freshwater system. However, the akinete cellular and biochemical properties are insufficiently explored. We analyzed the akinete structure, as well as akinete-specific proteins and their amino acid sequence. Akinetes of *Dolichospermum circinale* were produced from their vegetative cells isolated from the North Han River, Korea. The akinete protein was obtained using electrophoresis, and utilizing its amino acid sequences, its antibody-binding reaction potential (ig-score) was quantified. Akinete protein masses were 17 kDa–180 kDa, and the akinete protein mass was 110 kDa. The ig score was high (average 5.0121 points) in the first half of the amino acid sequence, indicating a β-turn form. The amino acid sequence, having over 50% homology with the *D. circinale* akinete protein, was not present in GenBank. The homology of the *D. circinale* akinete-specific protein was very low (9.8%) compared to that of *Anabaena variabilis*, indicating that its composition was substantially different, even among phylogenetically close taxa. To the best of our knowledge, this is the first report on the *D. circinale* akinete protein and its amino acid sequence, with preliminary information for their practical application for detecting akinetes in freshwater systems.

Keywords: cyanobacteria; akinete; protein; amino acid; antibody; *Dolichospermum circinale*; North Han River

Citation: Kim, K.; Park, C.-H.; Hwang, S.-J. Analyzing the Akinete Protein of the Harmful Freshwater Cyanobacterium, *Dolichospermum circinale*. *Water* **2023**, *15*, 2746. https://doi.org/10.3390/w15152746

Academic Editors: Eva Papastergiadou and Kostas Stefanidis

Received: 6 July 2023
Revised: 24 July 2023
Accepted: 27 July 2023
Published: 29 July 2023

1. Introduction

Cyanobacteria from the orders of Nostocales and Stigonematales form akinetes as a survival structure [1]. Akinetes located in the sediment can survive in adverse environmental conditions for a long period; when conditions become favorable, they germinate to develop vegetative cells that could proliferate in the water layers [2–4]. Some filamentous cyanobacteria forming akinete, such as *Dolichospermum*, *Aphanizomenon*, and *Cylindrospermopsis*, are known to be harmful because they cause blooms and produce odorous materials and toxins, thereby causing negative effects on water use and ecosystem health [4–7]. Therefore, information on the spatiotemporal distribution of akinetes in the sediment and factors affecting akinete germination provides important evidence for early warnings in water management, since they can serve as one of the precursors to cyanobacterial blooms in the area [8,9].

Akinete detection in situ has some constraints because it is located in the sediment. Analyzing ambient distribution and the abundance of akinetes requires their separation directly from the sediment, followed by their microscopic observation [10–12]. In the conventional method dealing with akinetes from sediment samples, cells are separated individually using micropipettes after ultra-sonication and serial size fractionation [13]. However, it is difficult to completely remove soil particles and organic materials attached to akinetes, preventing the complete separation of the akinete from the sediment. The efficiency of the various separation methods varies; therefore, the need for an effective detection method has been increasingly recognized [14–16].

Molecular-level detection techniques, which are based on DNA and mRNA, could overcome the limitations of the physical separation methods [17–19], because they utilize the whole sediment without an akinete separation process. Several studies have been conducted on cyanobacteria that possess genes producing harmful material, using molecular detection to overcome the limitations of microscopic analysis [20–22]. Similarly, in the field of akinete monitoring, molecular techniques would be more useful to detect akinete cells in the sediment [23]. However, there are limitations to molecular detection on the gene level. With DNA-based methods, it is difficult to either selectively amplify or stain the DNA of akinete cells as demonstrated in the PCR and CARD-FISH (catalyzed reporter deposition–fluorescence in situ hybridization) methods; therefore, it is difficult to verify actual akinete production. Furthermore, there is a time limit of the mRNA-based method for detecting akinete in situ, because mRNA can only be detected for a short period when the akinetes are formed.

The akinete protein persists for a prolonged period in the sediment environment, because the akinete possesses a thick cell wall. In addition, the protein degrades more slowly than mRNA after synthesis. Therefore, if the akinete protein is applied, the limitation of the short detection period that occurs at the mRNA level can be overcome. To date, however, there are limited studies on akinete proteins and the amino acid sequences. Information on the akinete protein and its amino acid sequence for only *Anabaena variabilis* and *Anabaena cylindrica* are available in the global amino acid database (NCBI: National Center for Biotechnology Information) [24–26]. Currently, there is no information regarding the akinete protein of *Dolichospermum circinale*, which frequently causes blooms and produces toxins and geosmin in eutrophic freshwater systems [27].

The purpose of this study was to elucidate the akinete-specific protein and the amino acid sequence of the harmful cyanobacterium, *D. circinale*. The result will be an important addition to akinete protein research and provide preliminary information for developing akinete detection methods.

2. Materials and Methods

2.1. *D. circinale* Isolation and Culture

Vegetative cells of *D. circinale* were collected from the downstream area of the North Han River (37°35′16.72″, 127°20′25.23″) in August 2015, during its bloom. They were stored in a cool icebox immediately after collection and transported to the laboratory within 5 h. The filaments of *D. circinale* were separated into a single strand using the capillary method [28], and the single strand was inoculated on a 96-well plate (Falcon®, Newark, NY, USA) containing the BG-11 medium (Merck Co., Darmstadt, Germany) [29]. Thereafter, the cells were cultured for 30 d in an incubator (VS-8480, Vision, Daejeon, Republic of Korea) at 25 °C and a light intensity of 130 $\mu E/m^2/s$ (14:10 = L:D). A healthy strain showing good growth in the well plate was selected and transferred into a flask with 100 mL BG-11 medium (Merck Co.), and the culture was grown until it reached the log phase.

Akinetes from natural samples to be used for analyzing morphological characteristics were collected from a surface layer of the sediment. The sediment sample was collected using a core sampler (Uwitec, Mondsee, Austria) at the same place as the sampled vegetative cells. The collected sediment sample was then transferred to the laboratory, and 1 g (w/w) was suspended in sterilized water. The suspension was treated twice with ultrasound for 20 s using an ultrasonic device (JAC 4020 type, 60 Hz, 620 w, Ultrasonic, Hwasung-si, Republic of Korea). The pulverized suspension was sequentially filtered through 100, 60, and 10 μm nylon mesh to isolate akinetes. Following the panning method [13], the filtrate was placed on a petri dish (12 cm in diameter), and from the top layer, the floating particles were isolated and mixed with filtered (0.2 μm) sterile water to a final volume of 10 mL. The final solution was stored in a dark brown glass bottle at approximately 4 °C.

2.2. Akinete Preparation

A laboratory designed culture chamber was used to induce akinete formation from *D. circinale* vegetative cells. The chamber could be operated such that the upper and lower parts were separate to simulate the water layer and the sediment, respectively [30]. *D. circinale* vegetative cells were inoculated at a density of 5×10^4 cells/mL in the upper cylinder filled with BG-11 medium. Artificial sediment [31] formulated in the laboratory was placed in the lower cylinder. The chamber, equipped with both the upper and lower cylinders, was placed in an incubator at 20 °C and a light intensity of 30 $\mu E/m^2/s$ (14:10 = L:D) for 5 d to induce akinetes [30]. Akinetes were collected from the sediment of the lower cylinder in the chamber, concentrated using a 10 µm sieve in 10 mL of distilled water, and stored in a dark brown glass bottle under refrigeration at 4 °C.

Subsequently, 1 mL of the refrigerated sample was placed in a Sedgwick–Rafter chamber and the morphological features [32,33] of the akinetes were observed at 400× magnification on an inverted microscope (Axiovert A1, Zeiss, Oberkochen, Germany) (Figure S1). The akinetes were separated individually using a microcapillary [34]. The external morphology of the separated akinetes was examined using an optical microscope (Axio Scope A1, Zeiss) and scanning electron microscope (JSM-7500, Jeol, Akishima, Japan).

2.3. Protein Extraction from Akinete and Vegetative Cell

D. circinale akinetes and vegetative cells were transferred to different microtubes and centrifuged at 20,000 rpm (6708× *g*) for 30 min. The supernatant of each sample was removed and lysis buffer (50 mM Tris-HCl pH 8.0, 150 mM NaCl, 5 mM EDTA,1% NP-40, and protease inhibitor cocktail) was added and left to react for 5 min on ice to facilitate cell lysis. The intracellular substances of the reacted samples were eluted with ultrasound homogenizers (VCX500, Sonics & Materials, Inc., Newtown, CT, USA). The soluble protein present in the supernatant was separated using centrifugation at 4 °C and 13,000 rpm (2834× *g*) for 10 min. Separated proteins were analyzed using the Bradford assay [35]. A 4 × SDS (sodium dodecyl sulfate) sample buffer (#161-0747, Bio-Rad laboratories, Portland, OR, USA) was added to the protein sample. It was then heated to 100 °C for 5 min to obtain a primary protein structure with a negative charge (i.e., a linear shape). Markers were loaded on 12% acrylamide SDS-Poly Acrylamide Gel Electrophoresis (PAGE) gels. Electrophoresis on SDS-PAGE gel was performed at 120 V for 2 h to separate the proteins based on their mass. SDS-PAGE gels were stained with Coomassie Brilliant Blue R-250 solution (Sigma, St. Louis, MO, USA) for more than 24 h and then a destaining solution (10% acetic acid and 45% methanol in 1 L distilled water (DW)) was applied for 1.5 h to remove the background color to obtain protein bands. Protein bands exclusively present in the akinete samples were selected for further analysis.

2.4. Sample Preparation for Mass Spectrometry

Akinete-specific protein bands were cut out from SDS-PAGE gel using a scalpel and transferred to sterilized microtubes. Then, the peptide bonds between amino acids were degraded using an in-gel digestion reaction (in-gel trypsin degradation) [36].

Excised gel spots were destained with 100 µL of destain solution (50% Acetonitrile (ACN)/DW) with shaking for 5 min. After the removal of the solution, gel spots were incubated with 200 mM ammonium bicarbonate (NH_4HCO_3) for 20 min. The gel pieces were dehydrated with 100 µL of ACN and dried in a vacuum centrifuge (VS-30000i, GSI, Republic of Korea). Into the dried gel piece, 50 µL dithiothreitol (10 mM) mixed with 0.1 M ammonium bicarbonate was added and this was incubated at 56 °C for 30 min with shaking. The sample was spun down and the supernatant was removed. Subsequently, 100 µL ACN was added for gel shrinkage. Fifty microliters of iodoacetamide (C_2H_4INO) (55 mM) mixed with 0.1 M ammonium bicarbonate was added and incubated at room temperature for 20 min in the dark. The supernatant was removed and 200 µL ACN was added. Again, it was spun down to remove supernatant, and 200 µL ammonium bicarbonate (0.1 M) was added. The procedure was repeated thrice on the dried gel piece. The dried gel pieces were rehydrated with 20 μL of 50 mM NH_4HCO_3 containing 0.2 g modified trypsin (Promega Co., Madison, WS, USA) for 45 min on ice. After the removal of the solution, 30 μL of 50 mM NH_4HCO_3 was added. The digestion was performed overnight (6–8 h) at 37 °C. The peptide solution was desalted using C18 column (Ziptip, Millipore Co., Burlington, MA, USA). All processed protein samples were analyzed using MALDI-TOF (matrix-assisted laser desorption ionization time of flight) [37] and a protein sequencer (ABI494, Thermo Fisher Scientific, Waltham, MA, USA) (Table S1).

2.5. Analysis of Amino Acid Sequence

The sequenced amino acid was identified by protein BLAST analysis in the Genbank database. The BLAST searching was performed based on the Protein Data Bank proteins (PBD) in Genbank standard database, using the Protein–Protein BLAST (BLASTP) algorithm. We excluded the uncultured/environmental sample sequence, that is, unidentified sequence information, in the search process [38–40]. The hydrophobicity of the amino acid sequence (Kyte–Doolittle hydrophathy) and the immunogenicity score (ig-score) were analyzed based on predicted amino acid secondary structure [41]. The structure was predicted based on Chou–Fasman secondary structure model.

2.6. Statistical Analyses

In the akinete-specific protein amino acid sequence, the correlation between the ig-score and Kyte–Doolittle hydropathy of each section was analyzed to determine the relationship between protein hydrophobicity and antigen–antibody reaction potential. Pearson's correlation analysis was performed using SPSS version 18 (IBM, Armonk, NY, USA). Statistical significance was set at $p < 0.05$.

3. Results and Discussion

3.1. Protein Profile of *D. circinale*

The masses of proteins extracted from akinete ranged from 17 to 180 kDa. Proteins with masses in the range of 40–75 kDa were found in both the akinete and vegetative cells and comprised the largest portions of proteins present (Figure 1). They are speculated to comprise amino acids present in the plasma membranes, carboxysomes, polyphosphates, and cyanophycin granules; these organelles are present both in akinetes and in vegetative cells [42–44]. Small (30–48 kDa) and large proteins (180 kDa) were almost exclusively present in the vegetative cells. Physiologically active enzymes that are formed by various cellular organelles are normally present in active vegetative cells. These active enzymes are known to be very small (<48 kDa) [45–47]. Contrastingly, a miniscule amount of 30 kDa protein was present in akinetes. The 110 kDa protein was solely present in the akinetes (Figure 1). We did not identify whether this protein exclusively originated from the akinete cell wall or within the cell. However, because only the position of the 110 kDa protein showed no band in the extracts of vegetative cells, we considered it a candidate akinete-specific protein [48].

The akinete-specific proteins appear to differ even among similar taxa. The mass of the akinete-specific protein of *D. circinale* was 110 kDa, whereas that of the AvaK protein found in *Anabaena variabilis* akinete was 43 kDa [48]. A kind of akinete-specific protein, AcaK43, which is the AvaK homolog protein, was also found in *A. cylindrica* [49]. Therefore, we suspect that *D. circinale* has an akinete-specific protein, which is different from the AvaK protein in *A. variabilis* and the AcaK43 protein in *A. cylindrica*.

Consequently, the akinete-specific protein mass and peptide may affect the formation of akinetes, resulting in morphological and functional differences in the akinetes of the order Nostocales [50,51]. Morphologically, Nostocale taxa show different akinete shapes. The akinete of *D. circinale* is longer and wider than that of *A. variabilis* by approximately 10 μm (5–10 μm in length and 10–15 μm in width), and the length is similar in *A. cylindrica* akinete, but this is broader in width (length 15–20 μm and width 7–10 μm) (Figure S1). The projections present at both sides of the akinete anodes were present solely in *D. circinale*, and not in *A. variabilis* or *A. cylindrica* [48,52]. Although photosynthesis is functionally not activated in akinetes, *A. cylindrica* showed photosynthetic proteins in both akinete and vegetative cells. However, in akinetes of *A. variabilis*, chlorophyll and phycocyanin disappeared during akinete cell maturation [53].

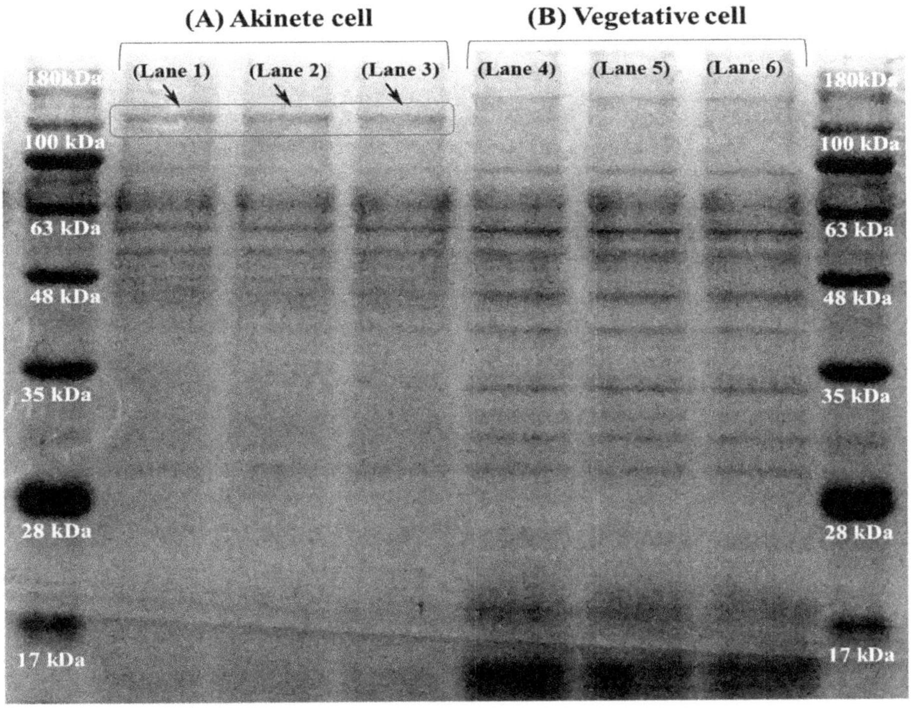

Figure 1. Bands of *Dolichospermum circinale* akinete and vegetative cell proteins displayed on SDS-PAGE gels (12%) stained using Coomassie Brilliant Blue R-250. The sample was loaded in triplicate. Vertically arrayed numbers in both the first and the last lane are the masses of standards in KDa. The arrows and square box indicate potential akinete-specific proteins. (**A**) Total cell protein extracted from akinetes (lane 1, 2, 3). (**B**) Total cell protein extracted from vegetative cells (lane 4, 5, 6).

3.2. Amino Acid Sequence of D. circinale Akinete-Specific Protein

We identified a peptide consisting of 600 amino acid sequences using the peptide mass fingerprint by MALDI-TOF, with the 110 kDa protein solely present in the akinete of *D. circinale*, as inferred from from SDS-PAGE bands (Table 1). The Chou–Fasman secondary structure and Kyte–Doolittle hydropathy index were determined using the peptide amino acid structure. Consequently, the ig-score, which indicates the antibody-binding reaction with the akinete-specific protein, was established (Figure 2, Table S3). The higher the ig-score, the higher the probability of the antibody initiating an immune response with the antigen [15]. The ig-score was higher in the No. 100–200 amino acid group and lower in the No. 470–600 amino acid group (Figure 2). The average ig-score of the whole amino acid sequence was 4.7745 (range: 2.3157–8.1835), whereas its average was 5.4254 between 105 aa and 210 aa. Specifically, the average score was the highest (avg. 6.4564) in the 110 aa–150 aa range. Moreover, the ig-score was high in the 273–345 aa (avg. 5.0602) and 405–475 aa ranges (avg. 5.0121). These results indicate that the amino acid sequence of the 105–349 aa range, particularly the 110–150 aa range, is better for detecting the akinete of *D. circinale*. Contrastingly, the ig-scores of the 5–112 aa, 220–270 aa, and 479–600 aa ranges were lower than the average score. This suggests that the biosensing function (antibody) of the 220–270 aa amino acid sequence range might be relatively low.

Table 1. Amino acid sequence of the *D. circinale* akinete-specific protein analyzed from the 110 kDa sodium dodecyl sulfate poly acrylamide gel electrophoresis (SDS-PAGE) band.

No.	Amino Acid Sequence	No.
1	MKWVTFISLL LLFSSAYSRG VFRRDTHKSE IAHRFKDLGE EHFKGLVLIA	50
51	FSQYLQQCPF DEHVKLVNEL TEFAKTCVAD ESHAGCEKSL HTLFGDELCK	100
101	VASLRETYGD MADCCEKQEP ERNECFLSHK DDSPDLPKLK PDPNTLCDEF	150
151	KADEKKFWGK YLYEIARRHP YFYAPELLYY ANKYNGVFQE CCQAEDKGAC	200
201	LLPKIETMRE KVLASSARQR LRCASIQKFG ERALKAWSVA RLSQKFPKAE	250
251	FVEVTKLVTD LTKVHKECCH GDLLECADDR ADLAKYICDN QDTISSKLKE	300
301	CCDKPLLEKS HCIAEVEKDA IPENLPPLTA DFAEDKDVCK NYQEAKDAFL	350
351	GSFLYEYSRR HPEYAVSVLL RLAKEYEATL EECCAKDDPH ACYSTVFDKL	400
401	KHLVDEPQNL IKQNCDQFEK LGEYGFQNAL IVRYTRKVPQ VSTPTLVEVS	450
451	RSLGKVGTRC CTKPESERMP CTEDYLSLIL NRLCVLHEKT PVSEKVTKCC	500
501	TESLVNRRPC FSALTPDETY VPKAFDEKLF TFHADICTLP DTEKQIKKQT	550
551	ALVELLKHKP KATEEQLKTV MENFVAFVDK CCAADDKEAC FAVEGPKLVV	600

Note: A—alanine, G—glycine, I—isoleucine, L—leucine, P—proline, V—valine, F—phenylalanine, W—tryptophan, Y—tyrosine, D—aspartic acid, E—glutamic acid, R—arginine, H—histidine, K—lysine, S—serine, T—threonine, C—cysteine, M—methionine, N—asparagine, Q—glutamine.

The hydropathy index (Kyte–Doolittle index) of the amino acid sequence, indicating the degree of hydrophobicity [41], shows a significant negative correlation with the ig-score ($r = -0.914$, $p < 0.01$, $n = 593$) (Table S2). This result suggests that antibody binding occurs more easily in the hydrophilic region outside the cell membrane than in the hydrophobic region inside the cell membrane. Therefore, biosensing for akinete detection could be more effective in the hydrophilic condition than in the hydrophobic condition.

While verifying the protein ID of the akinete-specific protein analyzed in this study, we found that the NCBI BLASTP Database (GenBank CDS) did not contain data consistent with the protein amino acid sequence of *D. circinale* akinete. However, when compared to the akinete-specific protein of *A. variabilis*, which is the only one known thus far [48], homology was very low (9.8%). Prior studies were conducted on the endospore of the *Bacillus* species, such as *Bacillus subtilis* and *Bacillus cereus* [54–58]. The protein information of these endogenous spores is composed of short fragments, smaller than 200 aa, and this endospore protein has very low homology (<1%) with the akinete-specific protein found in this study. Both the cyanobacterial akinete and bacterial endogenous spores are formed to overcome harsh conditions and are functionally the same. However, recent proteomic data indicate that, on the contrary to the results of conventional research, akinetes may also play an active role during filamentous growth. Specifically, the akinete was reported to play a role in the fixation of nitrogen and as a carbon storage transfer unit in filaments of

A. cylindrica [49]. Qiu et al. (2020) demonstrated that fixed carbon entered into akinetes from vegetative cells and was converted to glycogen by glycogen synthase, or into trehalose for osmoprotectant during the survival stage [49]. Although both the cyanobacterial akinete and bacterial endogenous spores are functionally the same, they exhibit different degrees of high heat and dry resistance, and different structures at the anatomical level of the cell, such as the cortex components surrounding the spores [58].

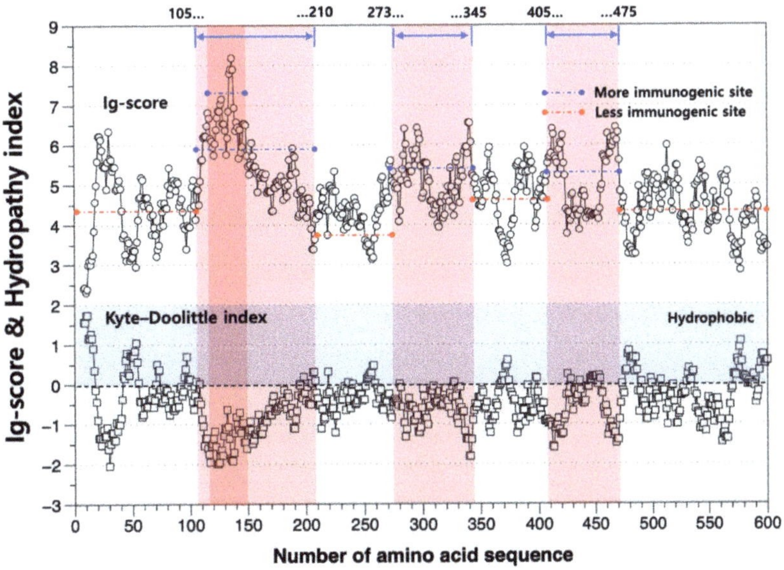

Figure 2. Immunogenicity score (ig-score) and hydropathy index of amino acid sequence of akinete-specific protein. Blue and red dashed lines indicate higher and lower score than the average score, respectively. The more immunogenic site was colored pink. In the hydropathy index graph, the parts that consist of positive values refer to the hydrophobic site.

3.3. *Perspective on Developing the Akinete Screening Technique*

The molecular-level approach has advanced harmful cyanobacterial detection in the field by using specific genes related with harmful material production [20–22]. However, molecular-level applications to identify akinetes exhibit some technical limitations because the akinete cell is surrounded by protein capsule. For this reason, proteomics research could help us to develop a method for the efficient screening of akinetes, such as an antibody biosensor or akinete-specific staining. However, there are some challenging technical hurdles from the practical perspective.

Firstly, it is difficult to verify actual akinete protein binding using the antibody based on the amino acid sequence analyzed in this study, because the best method for making the akinete-specific protein antibody is yet to be established [59]. Therefore, a further study to make an antibody against the akinete protein for screening cyanobacterial akinetes using the antigen–antibody reaction in field samples has high priority.

Secondly, it is necessary to develop a method for removing various non-specific protein polymers or organic particles remaining in the sediment to increase the sensitivity of the reaction with the akinete protein. In environmental proteins, which relate to similar concepts to environmental DNA, bulk proteins could include many non-specific protein polymers and organic particles [60]. They can disturb the activation of akinete-specific antibodies. For this reason, we must focus on establishing a sample purifying method for precise activation of akinete-specific antibodies.

Thirdly, due to the currently very limited information on the akinete protein and peptide, it is necessary to store akinete-specific protein peptide information from the Nostocales taxa [61,62]. Most cyanobacterial proteomics research has been focused on protein interaction networks that govern the lifecycle of cyanobacteria, such as post-translational mechanisms [63,64]. However, since 2002, there has been limited research on akinete-specific proteins. Further studies should include work on akinete peptides using various species from the order of Nostocales to expand our understanding of the species-specific akinete proteins and their role in physiology, and expedite the progress of cyanobacterial proteomics research.

4. Conclusions

D. circinale is the harmful cyanobacterium commonly found in eutrophic freshwater ecosystems. It forms akinetes as a survival structure. In this study, we analyzed the *D. circinale* akinete-specific protein and its amino acid sequence. To the best of our knowledge, the akinete-specific amino acid sequence of the order Nostocales determined in this study is only the third report on this topic, after those on *A. variabilis* and *A. cylindrica*. Although these species have very close phylogenic relationships, their akinete proteins differ in mass and have very low homology, suggesting that the akinete proteins are taxon-specific. Further studies are necessary to accumulate information on akinete proteins and their amino acid sequences among akinete-forming cyanobacteria, in order to establish a biosensing method based on antigen–antibody reactions. However, several technical hurdles need to be overcome related to the protein-level screening of akinetes for application in the field. Therefore, future research on akinetes in cyanobacteria must go beyond simple physiological studies and involve in-depth investigations at the metaomics level, which address the technical limitations at the protein level.

Supplementary Materials: The following supporting information can be downloaded at: https://www.mdpi.com/article/10.3390/w15152746/s1, Figure S1: Images of *D. circinale* akinete, Table S1: Analytical conditions of MALDI-TOF. Table S2: Result of the correlation analysis between the ig score and Kyte–Doolittle hydropathy index value. Table S3: Chou–Fasman secondary structure prediction for calculating ig score of *D. circinale*'s akinete-specific protein.

Author Contributions: Conceptualization, S.-J.H. and K.K.; methodology, K.K. and C.-H.P.; software, K.K.; validation, S.-J.H. and K.K.; formal analysis, K.K. and C.-H.P.; data curation, S.-J.H., K.K. and C.-H.P.; writing—original draft preparation, K.K.; writing—review and editing, S.-J.H.; supervision, S.-J.H. All authors have read and agreed to the published version of the manuscript.

Funding: This research was funded by Konkuk university in 2020 (2020-A019-0197).

Informed Consent Statement: Not applicable.

Data Availability Statement: The data used and analyzed during this study are available from the corresponding author on reasonable request.

Acknowledgments: The authors are very grateful to late Tae-Hyun Youm who helped with the protein extraction experiment.

Conflicts of Interest: The authors declare no conflict of interest.

References

1. Suikkanen, S.; Kaartokallio, H.; Hällfors, S.; Huttunen, M.; Laamanen, M. Life cycle strategies of bloom-forming, filamentous cyanobacteria in the Baltic Sea. *Deep Sea Res. Part II Top* **2010**, *57*, 199–209. [CrossRef]
2. Anderson, D.M.; Wall, D. Potential importance of benthic cysts of Gonyaulax tamarensis and G. excavata in initiating toxic dinoflagellate blooms 1, 2, 3. *J. Phycol.* **1978**, *14*, 224–234. [CrossRef]
3. Andersen, D.M.; Keafer, B.A. An endogenous annual clock in the toxic marine dinoflagellate Gonyaulax tamarensis. *Nature* **1987**, *325*, 616. [CrossRef]
4. Baker, P.D.; Bellifemine, D. Environmental influences on akinete germination of *Anabaena circinalis* and implications for management of cyanobacterial blooms. *Hydrobiologia* **2000**, *427*, 65–73. [CrossRef]

5. Huber, A.L. Factors affecting the germination of akinetes of *Nodularia spumigena* (Cyanobacteriaceae). *Appl. Environ. Microbiol.* **1985**, *49*, 73–78. [CrossRef]
6. Ståhl-Delbanco, A.; Hansson, L.-A. Effects of bioturbation on recruitment of algal cells from the "seed bank" of lake sediments. *Limnol. Oceanogr.* **2002**, *47*, 1836–1843. [CrossRef]
7. van Dok, W.; Hart, B.T. Akinete germination in *Anabaena circinalis* (cyanophta). *J. Phycol.* **1997**, *33*, 12–17.
8. Anderson, D.M.; Glibert, P.M.; Burkholder, J.M. Harmful algal blooms and eutrophication: Nutrient sources, composition, and consequences. *Estuaries* **2002**, *25*, 704–726. [CrossRef]
9. Norris, G.; Mcandrews, J.H. Dinoflagellate cysts from post-glacial lake muds, Minnesota (USA). *Rev. Paleobot. Palynol.* **1970**, *10*, 131–156. [CrossRef]
10. Thiel, T.; Wolk, C.P. Metabolic activities of isolated akinetes of the cyanobacterium Nostoc spongiaeforme. *J. Bacteriol. Res.* **1983**, *156*, 369–374. [CrossRef]
11. Moore, D.; O'Donohue, M.; Garnett, C.; Critchley, C.; Shaw, G. Factors affecting akinete differentiation in *Cylindrospermopsis raciborskii* (Nostocales, Cyanobacteria). *Freshw. Biol.* **2005**, *50*, 345–352. [CrossRef]
12. Sukenik, A.; Kaplan-Levi, R.N.; Viner-Mozzini, Y.; Lupu, A.; Sela, D. Induction, Isolation and Counting of Akinetes in *Aphanizomenon ovalisporum*. *Bio-protoc* **2016**, *6*, e1808. [CrossRef]
13. Kwon, D.; Kim, K.; Jo, H.; Lee, S.D.; Yun, S.M.; Park, C. Environmental factors affecting akinete germination and resting cell awakening of two cyanobacteria. *Appl. Microsc.* **2023**, *53*, 1–13. [CrossRef]
14. Legrand, B.; Miras, Y.; Beauger, A.; Dussauze, M.; Latour, D. Akinetes and ancient DNA reveal toxic cyanobacterial recurrences and their potential for resurrection in a 6700-year-old core from a eutrophic lake. *Sci. Total Environ.* **2019**, *687*, 1369–1380. [CrossRef]
15. Kang, Y.-J.; Moon, C.-H.; Cho, H.-J. Comparison of panning and sodium polytungstate methods for separating dinoflagellate cysts. *J. Fish Aquat. Sci.* **2008**, *41*, 228–231. [CrossRef]
16. Kang, Y.J. *Distribution of Dinoflagellate Cyst and Germination Using SPT Solution in Eutrophic Region of Southern Coastal Waters of Korea*; Pukyong National University: Busan, Republic of Korea, 2007.
17. D'agostino, P.M.; Woodhouse, J.N.; Makower, A.K.; Yeung, A.C.; Ongley, S.E.; Micallef, M.L.; Moffitt, M.C.; Neilan, B.A. Advances in genomics, transcriptomics and proteomics of toxin-producing cyanobacteria. *Environ. Microbiol. Rep.* **2016**, *8*, 3–13. [CrossRef]
18. Liu, Y.; Ren, Z.; Qu, X.; Zhang, M.; Yu, Y.; Zhang, Y.; Peng, W. Microbial community structure and functional properties in permanently and seasonally flooded areas in Poyang Lake. *Sci. Rep.* **2020**, *10*, 1–11.
19. Rasmussen, J.P.; Monis, P.T.; Saint, C.P. *Early Detection of Cyanobacterial Toxins Using Genetic Methods*; American Water Works Association Research Foundation: Denver, CO, USA, 2008.
20. Tsao, H.-W.; Michinaka, A.; Yen, H.-K.; Giglio, S.; Hobson, P.; Monis, P.; Lin, T.-F. Monitoring of geosmin producing *Anabaena circinalis* using quantitative PCR. *Water Res.* **2014**, *49*, 416–425. [CrossRef]
21. Ueno, Y.; Nagata, S.; Tsutsumi, T.; Hasegawa, A.; Watanabe, M.F.; Park, H.-D.; Chen, G.-C.; Chen, G.; Yu, S.-Z. Detection of microcystins, a blue-green algal hepatotoxin, in drinking water sampled in Haimen and Fusui, endemic areas of primary liver cancer in China, by highly sensitive immunoassay. *Carcinogs* **1996**, *17*, 1317–1321. [CrossRef]
22. Vaitomaa, J.; Rantala, A.; Halinen, K.; Rouhiainen, L.; Tallberg, P.; Mokelke, L.; Sivonen, K. Quantitative real-time PCR for determination of microcystin synthetase E copy numbers for *Microcystis* and *Anabaena* in lakes. *Appl. Environ. Microbiol.* **2003**, *69*, 7289–7297. [CrossRef]
23. Ramm, J.; Lupu, A.; Hadas, O.; Ballot, A.; Rücker, J.; Wiedner, C.; Sukenik, A. A CARD-FISH protocol for the identification and enumeration of cyanobacterial akinetes in lake sediments. *FEMS Microbiol. Ecol.* **2012**, *82*, 23–36. [CrossRef]
24. Wootton, J.C.; Federhen, S. Statistics of local complexity in amino acid sequences and sequence databases. *Comput. Chem.* **1993**, *17*, 149–163. [CrossRef]
25. Kodama, Y.; Shumway, M.; Leinonen, R. The Sequence Read Archive: Explosive growth of sequencing data. *Nucleic Acids Res.* **2011**, *40*, D54–D56. [CrossRef]
26. Park, Y.M.; Squizzato, S.; Buso, N.; Gur, T.; Lopez, R. The EBI search engine: EBI search as a service—Making biological data accessible for all. *Nucleic Acids Res.* **2017**, *45*, W545–W549. [CrossRef]
27. Li, X.; Dreher, T.W.; Li, R. An overview of diversity, occurrence, genetics and toxin production of bloom-forming *Dolichospermum* (*Anabaena*) species. *Harmful Algae* **2016**, *54*, 54–68. [CrossRef]
28. Stein, J.R. *Handbook of Phycological Methods: Culture Methods and Growth Measurements*; CUP archive; Cambridge University Press: Cambridge, UK, 1979; Volume 1.
29. Stanier, R.Y.; Kunisawa, R.; Mandel, M.; Cohen-Bazire, G. Purification and properties of unicellular blue-green algae (order Chroococcales). *Bacteriol. Rev.* **1971**, *35*, 171. [CrossRef]
30. Park, C.-H.; Park, M.-H.; Kim, K.H.; Park, J.-H.; Kwon, D.-R.; Kim, N.Y.; Lim, B.-J.; Hwang, S.-J. Akinete germination chamber: An experimental device for cyanobacterial akinete germination and plankton emergence. *Harmful Algae* **2018**, *72*, 74–81. [CrossRef]
31. Egeler, P.; Römbke, J.; Meller, M.; Knacker, T.; Nagel, R. Bioaccumulation test with Tubificid Sludgeworms in artificial media–development of a standardisable method. *Hydrobiologia* **1999**, *406*, 271–280. [CrossRef]
32. Li, R.; Watanabe, M.; Watanabe, M.M. Taxonomic studies of planktic species of *Anabaena* based on morphological characteristics in cultured strains. *Hydrobiologia* **2000**, *438*, 117–138. [CrossRef]

33. Park, C.H. *Study on the Akinete Life Cycle and Vegetative Cell Dynamics in a Harmful Cyanobacterium, Dolichospermum circinale (Nostocales)*; Konkuk University: Seoul, Republic of Korea, 2018.
34. Guillard, R.R. Culture of phytoplankton for feeding marine invertebrates. In *Culture of Marine Invertebrate Animals*; Springer: Berlin/Heidelberg, Germany, 1975; pp. 29–60.
35. Kruger, N.J. The Bradford method for protein quantitation. In *The Protein Protocols Handbook*; Springer: Berlin/Heidelberg, Germany, 2009; pp. 17–24.
36. Shevchenko, A.; Tomas, H.; Havli, J.; Olsen, J.V.; Mann, M. In-gel digestion for mass spectrometric characterization of proteins and proteomes. *Nat. Protoc.* **2006**, *1*, 2856. [CrossRef]
37. Maier, T.; Kostrzewa, M. Fast and reliable MALDI-TOF MS-based microorganism identification. *Chim. Oggi* **2007**, *25*, 68. [CrossRef]
38. O'Leary, N.A.; Wright, M.W.; Brister, J.R.; Ciufo, S.; Haddad, D.; McVeigh, R.; Rajput, B.; Robbertse, B.; Smith-White, B.; Ako-Adjei, D. Reference sequence (RefSeq) database at NCBI: Current status, taxonomic expansion, and functional annotation. *Nucleic Acids Res.* **2015**, *44*, D733–D745. [CrossRef]
39. Tatusova, T.; DiCuccio, M.; Badretdin, A.; Chetvernin, V.; Nawrocki, E.P.; Zaslavsky, L.; Lomsadze, A.; Pruitt, K.D.; Borodovsky, M.; Ostell, J. NCBI prokaryotic genome annotation pipeline. *Nucleic Acids Res.* **2016**, *44*, 6614–6624. [CrossRef]
40. Brister, J.R.; Ako-Adjei, D.; Bao, Y.; Blinkova, O. NCBI viral genomes resource. *Nucleic Acids Res.* **2014**, *43*, D571–D577. [CrossRef]
41. Pisitkun, T.; Hoffert, J.D.; Saeed, F.; Knepper, M.A. NHLBI-AbDesigner: An online tool for design of peptide-directed antibodies. *Am. J. Physiol. Cell Physiol.* **2011**, *302*, C154–C164. [CrossRef]
42. Cardemil, L.; Wolk, C.P. The polysaccharides from heterocyst and spore envelopes of a blue-green alga. Methylation analysis and structure of the backbones. *Biol. Chem.* **1976**, *251*, 2967–2975. [CrossRef]
43. Cardemil, L.; Wolk, C.P. Polysaccharides from the envelopes of heterocysts and spores of the blue-green algae *Anabaena variabilis* and *Cylindrospermum licheniforme*. *J. Phycol.* **1981**, *17*, 234–240. [CrossRef]
44. Graham, L.E.; Wilcox, L.W. *Algae*; Prentice Hall: Upper Saddler River, NJ, USA, 2000; Volume 25, p. 29.
45. Maheshwari, R.; Bharadwaj, G.; Bhat, M.K. Thermophilic fungi: Their physiology and enzymes. *Microbiol. Mol. Biol. Rev.* **2000**, *64*, 461–488. [CrossRef]
46. Raugei, G.; Ramponi, G.; Chiarugi, P. Low molecular weight protein tyrosine phosphatases: Small, but smart. *Cell Mol. Life Sci.* **2002**, *59*, 941–949. [CrossRef]
47. da Silva Aires, R.; Steindorff, A.S.; Ramada, M.H.S.; de Siqueira, S.J.L.; Ulhoa, C.J. Biochemical characterization of a 27 kDa 1, 3-β-D-glucanase from *Trichoderma asperellum* induced by cell wall of Rhizoctonia solani. *Carbohydr. Polym.* **2012**, *87*, 1219–1223. [CrossRef]
48. Zhou, R.; Wolk, C.P. Identification of an akinete marker gene in *Anabaena variabilis*. *J. Bacteriol. Res.* **2002**, *184*, 2529–2532. [CrossRef]
49. Qiu, Y.; Gu, L.; Brözel, V.; Whitten, D.; Hildreth, M.; Zhou, R. Unique proteomes implicate functional specialization across heterocysts, akinetes, and vegetative cells in *Anabaena cylindrica*. *bioRxiv* **2020**. [CrossRef]
50. Bryant, D.A. *The Molecular Biology of Cyanobacteria*; Springer Science & Business Media: Berlin/Heidelberg, Germany, 2006; Volume 1.
51. Flores, F.G. *The Cyanobacteria: Molecular Biology, Genomics, and Evolution*; Horizon Scientific Press: Poole, UK, 2008.
52. Perez, R.; Forchhammer, K.; Salerno, G.; Maldener, I. Clear differences in metabolic and morphological adaptations of akinetes of two Nostocales living in different habitats. *Microbiology* **2016**, *162*, 214–223. [CrossRef]
53. Singh, H.; Sunita, K. A biochemical study of spore germination in the blue-green alga *Anabaena doliolum*. *J. Exp. Bot.* **1974**, *25*, 837–845. [CrossRef]
54. Zheng, L.; Donovan, W.P.; Fitz-James, P.C.; Losick, R. Gene encoding a morphogenic protein required in the assembly of the outer coat of the Bacillus subtilis endospore. *Genes Dev.* **1988**, *2*, 1047–1054. [CrossRef]
55. Abhyankar, W.; de Koning, L.J.; Brul, S.; de Koster, C.G. Spore proteomics: The past, present and the future. *FEMS Microbiol. Lett.* **2014**, *358*, 137–144. [CrossRef]
56. Setlow, B.; Hand, A.; Setlow, P. Synthesis of a Bacillus subtilis small, acid-soluble spore protein in Escherichia coli causes cell DNA to assume some characteristics of spore DNA. *J. Bacteriol.* **1991**, *173*, 1642–1653. [CrossRef]
57. Zhang, G.; An, Y.; Zabed, H.M.; Guo, Q.; Yang, M.; Yuan, J.; Li, W.; Sun, W.; Qi, X. Bacillus subtilis Spore Surface Display Technology: A Review of Its Development and Applications. *J. Microbiol. Biotechnol.* **2019**, *29*, 179–190.
58. Abel-Santos, E. Endospores, sporulation and germination. In *Molecular Medical Microbiology*; Elsevier: Amsterdam, The Netherlands, 2015; pp. 163–178.
59. Ow, S.Y.; Cardona, T.; Taton, A.; Magnuson, A.; Lindblad, P.; Stensjö, K.; Wright, P.C. Quantitative shotgun proteomics of enriched heterocysts from Nostoc sp. PCC 7120 using 8-plex isobaric peptide tags. *J. Proteome Res.* **2008**, *7*, 1615–1628. [CrossRef]
60. Brandão-Dias, P.F.; Rosi, E.J.; Shogren, A.J.; Tank, J.L.; Fischer, D.T.; Egan, S.P. Fate of environmental proteins (eProteins) from genetically Engineered crops in Streams is controlled by Water pH and ecosystem metabolism. *Environ. Sci.Technol.* **2021**, *55*, 4688–4697. [CrossRef]
61. Pandey, S.; Rai, R.; Rai, L.C. Proteomics combines morphological, physiological and biochemical attributes to unravel the survival strategy of *Anabaena* sp. PCC7120 under arsenic stress. *J. Proteom.* **2012**, *75*, 921–937. [CrossRef]

62. Panda, B.; Basu, B.; Rajaram, H.; Kumar Apte, S. Methyl viologen responsive proteome dynamics of *Anabaena* sp. strain PCC7120. *Proteomics* **2014**, *14*, 1895–1904. [CrossRef]
63. Battchikova, N.; Muth-Pawlak, D.; Aro, E.-M. Proteomics of cyanobacteria: Current horizons. *Curr. Opin. Biotechnol.* **2018**, *54*, 65–71. [CrossRef]
64. Sound, J.K.; Bellamy-Carter, J.; Leney, A.C. The increasing role of structural proteomics in cyanobacteria. *Essays Biochem.* **2023**, *67*, 269–282.

water

MDPI

Article

Effect of Land Use on Stream Water Quality and Biological Conditions in Multi-Scale Watersheds

Jong-Won Lee, Se-Rin Park and Sang-Woo Lee *

Department of Forestry and Landscape Architecture, Konkuk University, Gwangjin-gu, Seoul 05029, Republic of Korea; jwlee8901@naver.com (J.-W.L.); serin87@konkuk.ac.kr (S.-R.P.)
* Correspondence: swl7311@konkuk.ac.kr; Tel.: +82-2-450-4120

Abstract: Understanding the relation between watershed land use and stream conditions is critical for watershed planning and management. This study investigated the effects of land use on stream water quality and biological conditions in sub-watersheds and micro-watersheds across the Han River watershed in South Korea. We developed random forest models for each water quality and biological indicator using the proportions of urban, agricultural, and forested areas. Our results indicate that water quality and biological indicators were significantly affected by forest area at both scales, and the sub-watershed models performed better than the micro-watershed models. Accumulated local effects were used to interpret the effect of each explanatory variable on the response variable. The plots for water quality and biological indicators with proportions of watershed land use demonstrated similar patterns at both scales, although the relation between land use and stream conditions was slightly more sensitive in micro-watersheds than in sub-watersheds. Urban and agricultural areas showed a lower proportion of water quality and biological condition variability in the micro-watersheds than in the sub-watersheds, while forests showed the opposite results. The findings of this study suggest that different spatial scales should be considered when developing effective watershed management strategies to maintain stream ecosystems.

Keywords: land use/cover; water quality; biological indicator; watershed scale; stream condition

Citation: Lee, J.-W.; Park, S.-R.; Lee, S.-W. Effect of Land Use on Stream Water Quality and Biological Conditions in Multi-Scale Watersheds. *Water* **2023**, *15*, 4210. https://doi.org/10.3390/w15244210

Academic Editors: Eva Papastergiadou and Kostas Stefanidis

Received: 30 October 2023
Revised: 4 December 2023
Accepted: 5 December 2023
Published: 6 December 2023

1. Introduction

River and stream ecosystems are under pressure from various anthropogenic activities over multiple spatial scales [1]. In particular, anthropogenic land-use changes in watersheds are major factors hindering the integrity of river and stream ecosystems, including water quality and aquatic organism health. The negative effects and complex processes associated with urban and agricultural areas in watersheds cause changes to river and stream ecosystems at various spatial scales. To respond to these negative changes in stream ecosystems, it is necessary to understand the impact of various land uses in watersheds on stream ecosystems. Therefore, the impacts of land use in watersheds on water quality and aquatic organisms have been studied extensively.

In general, urban areas are the source of many anthropogenic pollutants that negatively affect streams [2,3]. Chemical pollutants, such as excess nutrients, heavy metals, and organic compounds, flow into stream ecosystems from urban areas due to increases in impermeable surfaces, such as pavements and rooftops [4,5]. Runoff with high concentrations of pollutants from industrial facilities or sewage treatment plants also acts as a stressor for aquatic organisms. The influx of these artificial pollutants destabilizes the physical and chemical runoff processes and disrupts the stream ecosystem [6–8]. For example, although nutrients can benefit aquatic organisms, heavy metals can have negative effects, and the combined effects of these stressors can disrupt ecosystem mechanisms [9]. Agricultural areas are known to promote eutrophication and harmful algal blooms in streams owing to excessive fertilizer runoff, sediment influx, and agricultural landscapes [10,11]. Conversely, forest areas and riparian vegetation in the watershed improve the stability of stream channels, regulate water temperature by preventing light from penetrating the canopy, and provide habitat and shelter for aquatic organisms. They also reduce nutrient runoff from the watershed, thereby mitigating erosion and eutrophication and improving biodiversity and ecosystem function by improving riparian vegetation ecosystem conditions and habitat quality [12,13].

The relationship between land use and river ecosystems is also related to the watershed size. Several studies have reported that the impact of land use on stream water quality and aquatic organisms varies across different spatial scales [14–17]. For example, the parameters on the riparian or reach scale may predict impacts on water quality and aquatic organisms better than those on the catchment scale. Dala-Corte et al. [18] and Pan et al. [19] reported that the land use adjacent to streams had a more important effect on chemical water quality and biotic communities compared to that observed from land use on the catchment scale across grasslands in southern Brazil and the Willamette Valley Ecoregion in Oregon, USA. Assessing land-use impacts on streams within these watersheds can allow us to more accurately estimate causal pathways in smaller watersheds, which are potentially vital for biodiversity and stream health [20]. However, studies have reported that watershed-scale land-use parameters better account for biotic diversity and water quality, including aquatic community patterns and species composition [7,17]. Tudesque et al. [21] explained that in the Adour–Garonne River in southwest France, the watershed scale is an important determinant of biological community structure. Ding et al. [14] and Zhang et al. [22] reported that the impact of land use on water quality in the Dongjiang River Basin and Three Gorges Reservoir area in China is better explained on the catchment scale than on the riparian, reach, or buffer scales. In addition, large-scale watersheds are more spatially dependent on land use due to the direct impact of the physical and chemical water quality and the cumulative anthropogenic influence on stream ecosystems related to water quality and aquatic organisms [15,21,23].

Recently, ecosystem research has used machine learning models to replace traditional statistical models that were previously built as description-driven limited models [24]. Machine learning, a branch of artificial intelligence, effectively overcomes the limitations of data-dependent bivariate and multivariate statistical methods, enabling prediction-driven models to estimate highly predictive models [8,25]. Unlike the general linear model, which traditionally predicts variables, tree-based random forest (RF) models make no assumptions about data distribution. They also integrate missing values with numeric or categorical predictors, tolerate general changes, such as the size of the data, remain resilient to predictor outliers [26], and perform well against complex and nonlinear interactions between variables [26]. The biggest advantage is that the prediction accuracy generally surpasses that of conventional methods. Another advantage of these RF models is the possibility of revealing complex nonlinear relations between land cover characteristics and stream water quality. Park et al. [8] found that these models accurately depict the complex nonlinear relation between landscape characteristics and stream water quality. Further benefits include the accuracy of machine learning models tested on new datasets and their application in predicting characteristics of aquatic ecosystems, such as the stream water quality. Moreover, such models can support future land-use planning scenarios to aid in policy decisions. Previous studies effectively employed the RF model to analyze the impact of land use on water quality and biological integrity. The relationships among stream water quality, aquatic organisms, and land use often exhibit disproportionate and nonlinear patterns, indicating abrupt points of change (thresholds). For example, significant changes in fish community composition and species abundance of 10.9% and 17.5%, respectively, were observed in a southeastern Queensland watershed [27]. These studies identified the point at which changes occur in the impact of land use, revealing an abrupt nonlinear ecological threshold [27–29]. Thresholds are needed to protect or restore water quality and aquatic organisms amid continuous land-use changes [21,29,30]. When ecological thresholds are surpassed, such as by pollutants flowing into streams, stream ecosystems can be damaged. Identifying these thresholds can help determine the impact of land use on streams and facilitate land-use planning [27].

This study aimed to quantify the relative importance and effects of the spatial scale of land use on the water quality and health of aquatic organisms in the Han River Basin. RF models were developed to evaluate the relations between land use and water quality and biological indicators. Most related studies have focused on forecasting changes and determining explanatory power according to spatial scale to identify the relation between land use and stream water quality and biological indicators [11,30]. In particular, the accumulated local effect (ALE) is used to compensate for the shortcomings of existing partial dependency plots (PDPs) and to display the results of the RF model more clearly. This can be a powerful tool for demonstrating the relation between highly correlated water quality and biological indicators. Our study advances the existing approach by analyzing differences in the impact of land use on water quality and aquatic organisms at multiple spatial scales and by comparing and analyzing the thresholds of land use according to different watershed scales. The results of this study can reduce uncertainty and provide insights for effective decision making in developing contaminated watershed management strategies and restoration plans. Different spatial scales should be considered when developing effective watershed management strategies to sustain stream ecosystems. This work is expected to help establish criteria for estimating land-use proportions to meet the health goals for stream ecosystems in watershed management.

2. Materials and Methods

2.1. Study Area

The Han River, located at 126°–129° E and 36°–38° N, originates from Taebaek Mountain and flows into the Yellow Sea. It is the second largest river in Korea, spanning a length of 494 km, and has a basin area of approximately 35,770 km^2. Spring (March–May) and autumn (October–November) are generally sunny and dry owing to the influence of mobile cyclones. Summer (June–September) accounts for two thirds of the total annual precipitation, with high temperatures due to the influence of the North Pacific high pressure. Winter (December–February) is cold and dry owing to the influence of temperate cyclones. Overall, the Han River basin comprises highlands over 1000 m above sea level on the east and lowlands on the west [31], with an average annual precipitation of 1208.3 mm. The main type of land cover comprises forests, occupying approximately 68% of the land, while urban and arable land account for approximately 17% and 7%, respectively. In this study, the Han River basin was divided into large- and small-scale watersheds (Figure 1). The large-scale watershed unit (3rd–4th-order streams) was determined by the Ministry of Environment for water management, encompassing an area ranging from 38.96 km^2 to 447.87 km^2. The small-scale watershed unit (1st–2nd order streams) was defined by local governments based on stream management goals and ranges from 0.11 km^2 to 12.97 km^2.

2.2. Data Source and Preprocessing

2.2.1. Water Quality and Biological Indicators

In Korea, the National Aquatic Ecology Monitoring Program (NAEMP) has developed evaluation standards and sampling protocols for monitoring rivers and streams. This program evaluates the overall ecological health status of streams, including the chemical water quality, biological condition, habitat quality, riparian vegetation, and land use around the stream [32]. For stream evaluation, the NAEMP investigates the ecological health and water quality of the five major streams and tributaries in Korea twice a year (spring: April–May; autumn: October–November) using aquatic organisms. The ecological health of streams was evaluated using the Trophic Diatom Index (TDI), Benthic Macroinvertebrate Index (BMI), and Fish Assessment Index (FAI) based on biomonitoring at 3035 sites nationwide. In addition, the Riparian Vegetation Index and Habitat and Riparian Index were included to evaluate the riparian vegetation and habitat. Table 1 shows the equations used to estimate the biological indicators used in the NAEMP. The NAEMP also measures various water quality parameters at the biomonitoring sites. Biochemical oxygen demand (BOD5), ammonia/ammonium (NH_3-N), nitrate (NO_3-N), total nitrogen (TN), total phosphorus (TP), phosphate (PO_4-P), and chlorophyll-*a* (Chl-*a*) were measured. Water quality was analyzed using an official test method conducted by the Korean Ministry of Environment. In this study, the biological indicators TDI, BMI, and FAI were used.

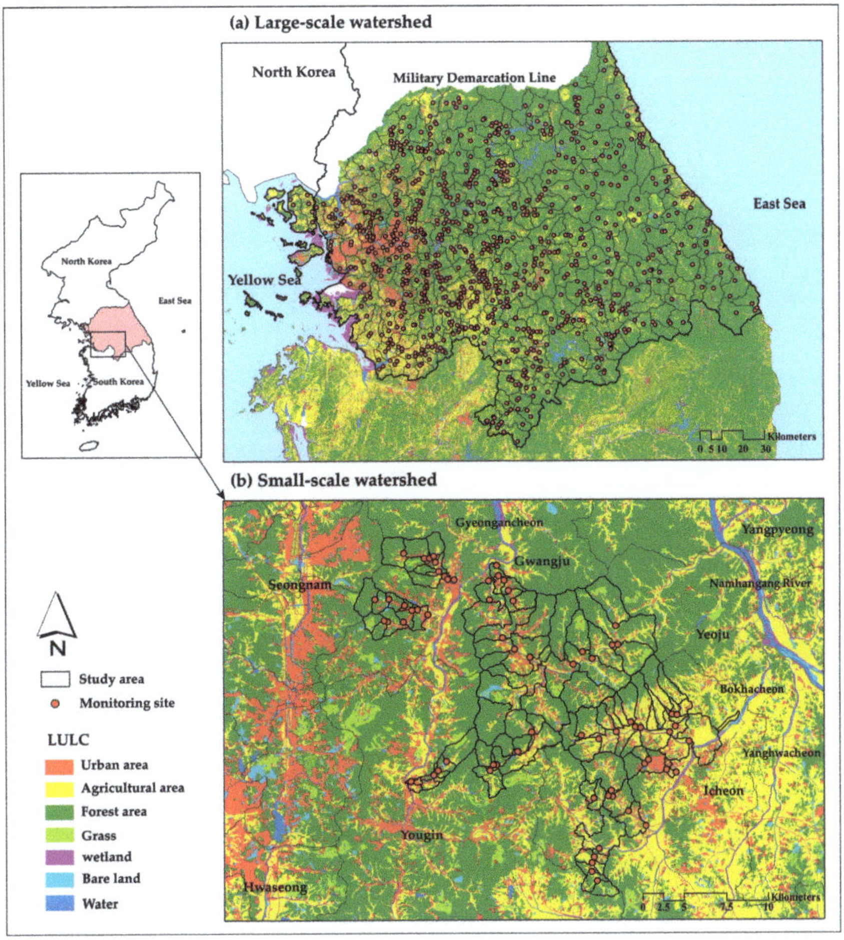

Figure 1. Monitoring sites and land use/land cover in the Han River basin. (**a**) Large-scale watersheds showing 3rd–4th-order stream basins and (**b**) small-scale watersheds showing 1st–2nd-order stream watersheds.

In this study, biological indicators and water quality were measured in the same manner across both watershed scales. The data used were obtained from basic environmental survey projects conducted by the Han River Watershed Environmental Management Office between 2018 and 2022 to diagnose the health status of streams. Table 1 shows the equations used to calculate the values for biological indicators.

Table 1. Equations for biological indicators used in the National Aquatic Ecology Monitoring Program.

Biological Indicators	Equations
TDI (Trophic Diatom Index)	$TDI = 100 - \{(WMS \times 25) - 25\}$ WMS: weighted mean sensitivity $WMS = \sum Aj \cdot Sj \cdot Vj / \sum Aj \cdot Vj$ Aj: proportion (relative abundance) of species in sample, % Sj: pollution sensitivity of species, $1 \leq S \leq 5$ Vj: indicator value of species, $1 \leq V \leq 3$
BMI (Benthic Macroinvertebrate Index)	$BMI = \left(4 - \frac{\sum_{i=1}^{n} s_i \cdot h_i \cdot g_i}{\sum_{i=1}^{n} h_i \cdot g_i}\right) \times 25$ i: number assigned to the species n: number of species s_i: unit saprobic value of the species i h_i: frequency of the species i g_i: indicator weight value of the species i
FAI (Fish Assessment Index)	FAI = sum of 8 metrics Metric 1 (M1): number of Korean native species Metric 2 (M2): number of rifle benthic species Metric 3 (M3): number of sensitive species Metric 4 (M4): percentage of tolerant species Metric 5 (M5): percentage of omnivores Metric 6 (M6): percentage of insectivores Metric 7 (M7): amount of native species Metric 8 (M8): percentage of fish abnormalities

2.2.2. Land Use/Land Cover

The land-use proportion of the stream watershed was determined using a land cover map provided by the Environmental Geographic Information Service. This land cover map had a spatial resolution of 5 m, and a 1:25,000 topographic map was used for the analysis in ArcGIS 10.6.1. The same land cover maps were applied to both the large- and small-scale watersheds. These maps were used to estimate the effects of land use on water quality and biological indicators by extracting urban, agricultural, and forest areas, which are the major land uses in the watershed, from the seven classified categories. Urban areas included residential, industrial, commercial, cultural, sports, leisure, transportation, and public facilities. Agricultural areas included cultivation facilities, orchards, and other croplands, while forested areas included broadleaf trees, conifers, and mixed forests. The impact of land use was assumed to remain constant throughout the study period.

2.3. *Statistical Analysis*

2.3.1. Independent Two-Sample *T*-Test and Pearson's Correlation

In this study, three analytical methods were used to estimate the impact on stream ecosystems on the watershed scale. Initially, an independent two-sample *t*-test was performed to verify the statistical differences between large- and small-scale watersheds. Actual differences in the mean values for water quality (BOD, TN, and TP), biological indicators (TDI, BMI, and FAI), and proportion of land use (urban, agricultural, and forest areas) across scales were identified. Furthermore, a correlation analysis was performed before applying the RF model to determine the impact of land use on water quality and biological indicators in both watersheds. In particular, PDPs, which are often used in RF model analysis, were affected by correlations between factors. If the correlation between variables is strong, then a bias may occur in the interpretation of the PDP [33]. Therefore, a correlation analysis was performed before visualizing the relations between the variables based on the RF model. Correlations were calculated for each water quality variable, biological indicator, and land-use proportion based on Pearson correlation analysis. The correlation coefficient ranges from −1 to +1. A value of 0 indicates that there is no statistical association between the variables, a value closer to +1 indicates a strong positive association, and a value closer to −1 indicates a strong negative association [34]. The parameters used in this study exhibited a normal distribution, except for the urban proportion. However, since this study investigates threshold values of water quality and biological indicators according to the actual urban proportion, this proportion was used without conversion.

2.3.2. RF Regression Algorithm

The RF regression algorithm was used to predict and evaluate the relation between predictor variables (land-use proportion) and response variables (water quality and biological indicators) using non-parametric ensemble machine learning. The goal of RF regression is to create a real tree-like ensemble from which multiple regression trees are built to produce a regression [35–37].

The RF model derives the final prediction result using the average of the prediction results from each tree model through multiple decision trees. In the model training process, which is based on bagging (bootstrap aggregation), several decision tree models are gathered to form a forest, and the model is trained [35,36]. Bagging applies bootstrapping to extract the training dataset by allowing duplication in the entire dataset so that each tree model uses a different training dataset. It is a prediction method that summarizes the results of each model by making the size as large as the number of original data [37]. A prerequisite for improving the ensemble model's performance is to secure diversity through bagging and assuring randomness by dividing it into random subspaces. It considers only these variables by randomly selecting fewer variables than the number of original variables. Random predictor selection reduces the correlation between trees and variance and, in particular, depends only on the number of predictors selected by the model user [38,39]. The RF model considers the mean and variance of the out-of-bag error difference between the out-of-bag error of the original dataset and a random mixture of the values of a particular variable. This identifies the importance and choice of variables to be used in the model. The variable importance in RF determines the extent to which each variable contributes to the accuracy and node impurity improvement. In this study, when the accuracy of the tree constructed via changing the order of specific variables was reconstructed, it was assessed as the mean squared error, which is the average of the differences in the reduced accuracy. To optimize the RF model, the numbers of regression trees and input variables per node were adjusted [39,40]. The number of regression trees was 200, and 1/3 of the total number of variables were optimized via setting them as random sampling variables. In this study, the data were analyzed via division into training and test data—70% and 30%, respectively. The model was trained on the training data and then the root-mean-square error and the mean absolute error were used to assess the accuracy of the model performance.

RFs do not yield single trees that can be graphed; however, the results can be expressed as PDPs or ALEs [29]. PDPs are commonly used to visualize the effects of predictors in black-box supervised learning models. This parameter shows the marginal effect on the prediction result of the machine learning model, and the relationship between the target and a feature can be determined via plotting this effect. In addition, it can be used to graphically characterize the relation between the probability of existence of one predictor after averaging the effects of other predictors in the model [41]. Therefore, the PDP provides a graphical depiction of the marginal effect of a variable on the regression [42]. Moreover, PDPs show the average effect of the features on the predictions and can be clearly interpreted if there is no correlation. They are also easy to implement in a plot and can analyze the relationships between features and predictors [33,43]. Thus, many studies using RF models have employed PDPs [8,44,45]. However, the most vulnerable PDP assumes that there is no correlation between a feature and other features. If there is a correlation among the features, then new data will be generated in the distribution regions where the actual probability is low, and the result will differ from the real values. Therefore, the ALEs can be selected as an unbiased choice for the PDP when the features are correlated. In addition, the ALEs visualize the main effects of individual predictors in the black-box supervised learning model and explain how each characteristic value averages and affects the prediction of the model [46]. Moreover, they operate conditionally instead of constraining the distribution, thereby solving the problem of independence. Because a correlation was observed among water quality, organisms, and land use in this study (Table 2), the ALE was used [43,46]. The RF regression model and ALEs were constructed using the RF ALE Plot packages in R Studio [46,47].

Table 2. Descriptive statistics of the water quality parameters, biological indicators, and land-use proportion of small watersheds in the Han River basin.

Variables	Min.	Max.	Mean	S.D.
TDI (0–100)	13.8	87.2	59.9	17.2
BMI (0–100)	23.7	94.0	65.3	17.9
FAI (0–100)	6.3	100	56.3	17.8
BOD (mg/L)	0.6	7.8	2.2	1.1
TN (mg/L)	0.97	10.28	3.37	1.42
TP (mg/L)	0.006	0.390	0.085	0.074
Urban (%)	0.0	68.6	10.4	11.0
Agricultural (%)	0.2	86.1	28.5	18.0
Forest (%)	0.0	96.7	54.3	22.0

Notes: n = 157. S.D., standard deviation; Min., minimum; Max., maximum; BOD, biochemical oxygen demand; TN, total nitrogen; TP, total phosphorus; TDI, Trophic Diatom Index; BMI, Benthic Macroinvertebrate Index; FAI, Fish Assessment Index.

3. Results

3.1. Descriptive Statistics and Independent Sample T-Test Analysis

Tables 2 and 3 describe the water quality variables, biological indicators, and land-use proportions used in this study. The average BOD, TN, and TP concentrations were lower in the large watersheds compared to those in the small watersheds. However, the maximum BOD and TN concentrations were higher in the small watersheds compared to those in the large watersheds. In both large and small watersheds, urban areas had a relatively low average proportion compared to that in farmland and forests, with forests showing a relatively high proportion. As for biological indicators, the TDI, BMI, and FAI values were better in large watersheds compared to those in small watersheds. According to the Ministry of Environment's aquatic ecosystem health evaluation, the average TDI, BMI, and FAI had grades of C, B, and B for large watersheds, respectively, and C, B, and C for small watersheds, respectively.

Table 3. Descriptive statistics of water quality parameters, biological indicators, and land-use proportion for large watersheds in the Han River basin.

Variables	Min.	Max.	Mean	S.D.
TDI (0–100)	11.4	98.7	65.1	18.0
BMI (0–100)	27.0	94.6	72.2	16.8
FAI (0–100)	3.2	100	66.8	21.1
BOD (mg/L)	0.5	5.9	1.6	0.8
TN (mg/L)	1.04	8.16	2.88	1.20
TP (mg/L)	0.007	0.368	0.051	0.049
Urban (%)	0.0	77.8	7.7	14.5
Agricultural (%)	0.0	100	17.0	13.8
Forest (%)	0.0	100	69.3	21.6

Notes: n = 177. S.D., standard deviation; Min., minimum; Max., maximum; BOD, biochemical oxygen demand; TN, total nitrogen; TP, total phosphorus; TDI, Trophic Diatom Index; BMI, Benthic Macroin-vertebrate Index; FAI, Fish Assessment Index.

Regarding aquatic ecosystem health, grades A and B suggest the presence of relatively abundant trophic diatoms sensitive to nutrients, many benthic macroinvertebrate species vulnerable to organic pollution, and a higher abundance of fish species sensitive to environmental changes. Grades D and E indicate the opposite, with the Ministry of Environment classifying streams with these grades as impaired. Grade C signifies a normal stream that is neither healthy nor damaged.

The independent sample *t*-test results for the differences in water quality, biological indicators, and land-use proportion according to the watershed scale are shown in Table 4. Water quality (BOD, TN, TP), biological indicators (TDI, BMI, FAI), and land use (agricultural, forest) presented statistically significant differences according to the watershed scale.

Table 4. *T*-test for water quality, biological indicators, and land-use proportion between large and small watersheds.

Variables	Levene		*t*-Value	*p*-Value
	F	Sig.		
TDI (0–100)	0.001	0.981	2.669	0.008
BMI (0–100)	2.106	0.148	3.626	0.000
FAI (0–100)	9.441	0.002	4.931	0.000
BOD (mg/L)	14.043	0.000	−5.327	0.000
TN (mg/L)	1.10	0.29	−3.41	0.00
TP (mg/L)	24.057	0.000	−4.819	0.000
Urban (%)	1.237	0.267	−1.874	0.062
Agricultural (%)	20.991	0.000	−6.455	0.000
Forest (%)	0.126	0.723	6.257	0.000

Notes: BOD, biochemical oxygen demand; TN, total nitrogen; TP, total phosphorus; TDI, Trophic Diatom Index; BMI, Benthic Macroinvertebrate Index; FAI, Fish Assessment Index.

3.2. Correlation Analysis

Water quality and biological indicators were correlated with land-use properties in multiple watersheds (Figure 2). In the large- and small-scale watersheds, the relations between all variables were found to be significant. Correlations between land-use proportion, water quality, and biological indicators in the large-scale watersheds were greater than those in the small-scale watersheds. In urban and agricultural areas, a positive correlation was observed with the water quality parameters BOD, TN, and TP, while a negative correlation was observed with the biological indicators TDI, BMI, and FAI. In particular, in the large- and small-scale watersheds, forest areas showed a stronger correlation between water quality and biological indicators than that for other land uses/land covers.

3.3. RF Models for Water Quality and Biological Indicators in Large- and Small-Scale Watersheds

RF models were created for each watershed scale for the water quality and biological indicators, and the performance of each model was compared (Table 5). Overall, better root-mean-square error and mean absolute error values were observed for the large watersheds, and better TN values were only observed in small watersheds.

Table 5. Water quality and biological indicator prediction performance of random forest models.

Scale	Evaluate	BOD	TN	TP	TDI	BMI	FAI
Large-scale	RMSE	0.40	0.80	0.024	12.71	9.52	11.26
	MAE	0.30	0.59	0.017	9.90	7.00	7.91
Small-scale	RMSE	0.69	0.63	0.044	15.45	12.29	16.06
	MAE	0.55	0.53	0.032	12.31	9.75	12.81

Notes: RMSE, root-mean-square Error; MAE, mean absolute error, BOD, biochemical oxygen demand; TN, total nitrogen; TP, total phosphorous; TDI, Trophic Diatom Index; BMI, Benthic Macroinvertebrate Index; FAI, Fish Assessment Index.

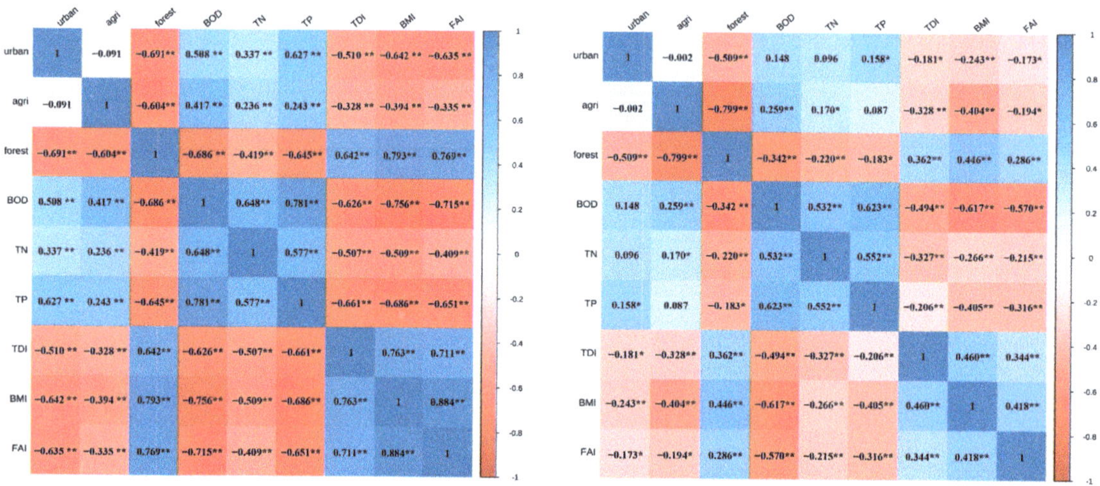

Figure 2. Relationships among biological indicators, water quality parameters, and land use in the large-scale watersheds (n = 177) (**left figure**) and small-scale watersheds (n = 157) (**right figure**) expressed as Pearson correlation coefficients (** $p < 0.01$, * $p < 0.05$).

3.4. Analysis of the ALE Plots

The ALE plot shows the effect of the land-use proportion on the water quality prediction probability (Figure 3). In large-scale watersheds, the water quality concentration increased sharply when the urban proportion was approximately 10% or higher. In agricultural areas, the BOD and TN concentrations were likely to increase when the agricultural proportion was approximately 5–10% or higher. However, the TP concentration was likely to increase when the agricultural proportion was approximately 25% or higher. In small-scale watersheds, the water quality concentration increased when the urban proportion was in the range of 5–10% or higher, and the agricultural proportion was in the range of 20–30% or higher (Figure 4). This means that the threshold proportion for water quality may vary depending on the watershed scale. The likelihood of water quality deterioration gradually decreased as the proportion of forests increased on both scales.

Figures 5 and 6 illustrate the impact of the land-use proportion on the predicted probabilities of TDI, BMI, and FAI. In large-scale watersheds, all indicator values decreased when the urban proportion was approximately 5% or higher. In agricultural areas, the TDI decreased by more than 25% and the BMI and FAI populations decreased by approximately 5%. In small-scale watersheds, the TDI was likely to decrease when the urban proportion was greater than 3%, and BMI and FAI were more likely to increase when the proportion was greater than 10%. In agricultural areas, the TDI increased nonlinearly up to approximately 30% and then decreased. The BMI decreased when the agricultural proportion was over 10%, and the FAI increased and decreased nonlinearly until the agricultural proportion reached approximately 30% and then decreased. In large-scale watersheds, the forest proportion gradually increased over the 25–35% range, and in small-scale watersheds, the forest proportion increased over 40–50%.

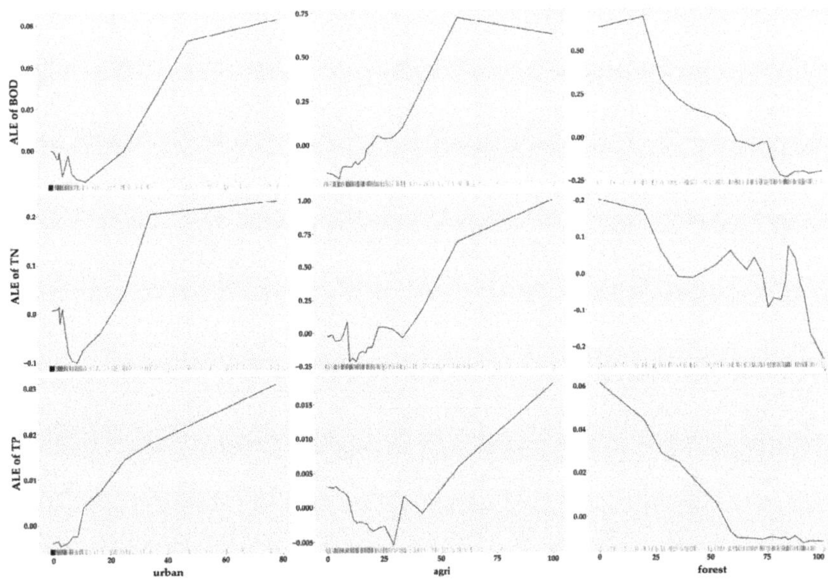

Figure 3. Accumulated local effect plots for random forest regressions of water quality parameters and land-use variables in large-scale watersheds. ALE, accumulated local effect; BOD, biochemical oxygen demand; TN, total nitrogen; TP, total phosphorous.

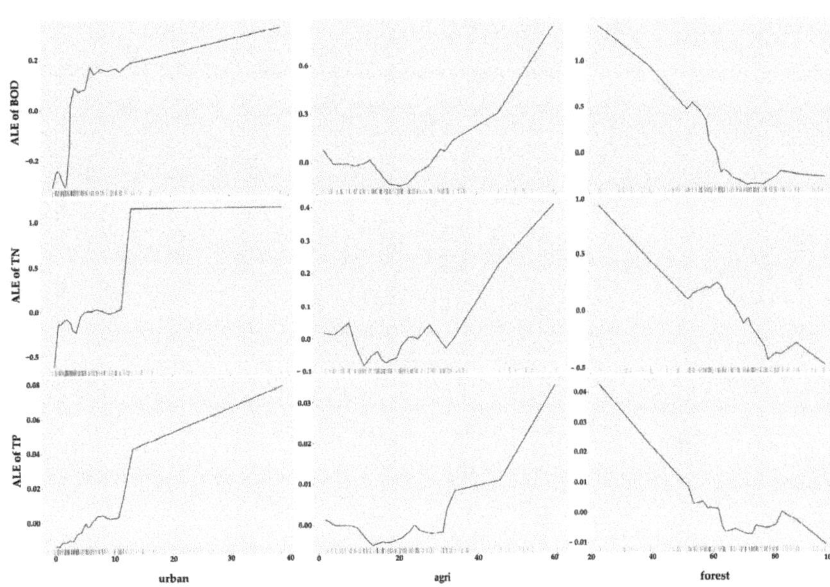

Figure 4. Accumulated local effect plots for random forest regressions of water quality parameters and land-use variables in the small-scale watersheds. ALE, accumulated local effect; BOD, biochemical oxygen demand; TN, total nitrogen; TP, total phosphorous.

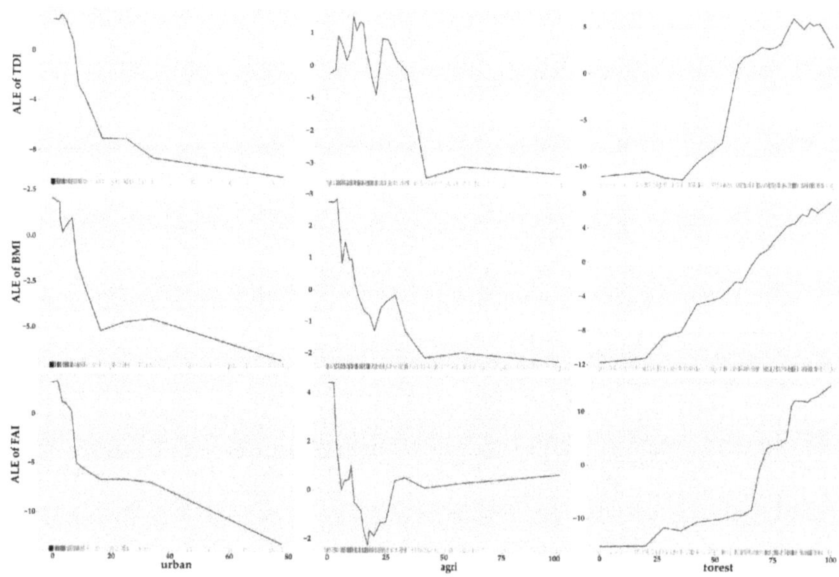

Figure 5. Accumulated local effect plots for random forest regressions for biological indicators and land-use variables in the large-scale watersheds. ALE, accumulated local effect; FAI, Fish Assessment Index; TN, total nitrogen; TP, total phosphorous.

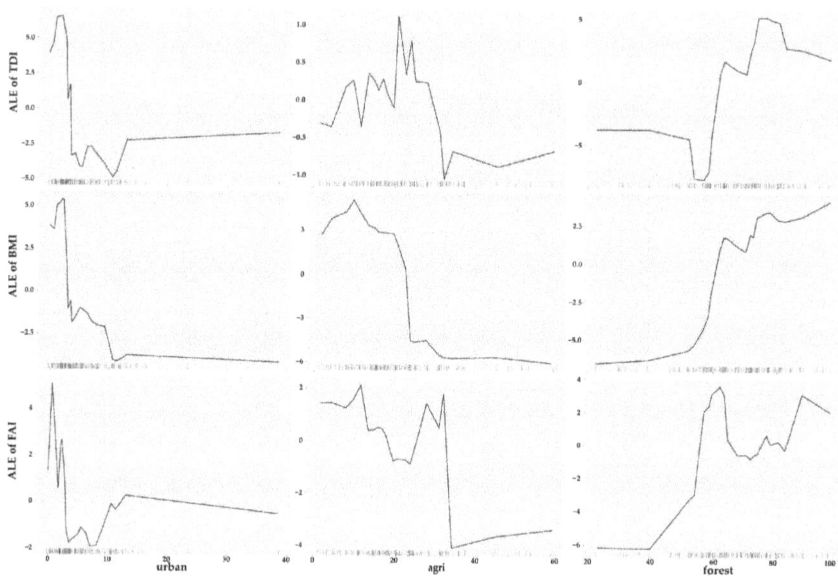

Figure 6. Accumulated local effect plots for random forest regressions for biological indicators and land-use variables in the small-scale watersheds. ALE, accumulated local effect; FAI, Fish Assessment Index; TN, total nitrogen; TP, total phosphorous.

4. Discussion

4.1. Land-Use Thresholds for Water Quality and Biological Indicators

The stream water quality and biological indicators responded nonlinearly as the proportion of land use increased, indicating critical values. Identifying the critical points at which water quality and biological indicators respond to changes in the proportion of land use in a watershed is important for land-use planning. Therefore, many studies have identified associated ecological thresholds [48,49].

This study visualized the nonlinear relationship of land-use proportions with water quality and biological indicators using ALE main effect plots (Figures 3–6). The ALE main effect plot showed that urban and agricultural areas have a negative effect on water quality and biological indicators on both watershed scales, while forest areas have a positive effect on water quality and biological indicators. In addition, the plot showed that the critical point of the impact of land-use proportion on water quality and biological indicators was different for different watershed scales.

This study confirmed that the range of urban proportions over which BOD, TN, and TP concentrations began to increase varied depending on the watershed scale. In large watersheds, the water quality concentration increased in 35–45% of urban areas, and in small watersheds, the water quality concentration increased in 5–10% of urban areas [1]. This is similar to previous studies in which water quality concentrations increased when the urban fraction was 10–50%. Additionally, in this study, the BOD, TN, and TP concentrations were highest when the urban proportion was 35–45% and 5–10% in the large and small watersheds, respectively. Tromboni and Dodds [50] showed that in Brazil, maximum nutrient concentrations are reached at 10–46% urban area, and water quality concentrations may further increase as urbanization accelerates. In the case of biological indicators in urban areas, a decrease in biological indicator values was observed when the urban area proportion was over 5% and 3% in large and small watersheds, respectively, with a rapid decrease observed for large watersheds. This result is similar to that of previous studies. For example, studies have reported a sharp decrease in benthic macroinvertebrate total taxa richness, Ephemeroptera, Plecoptera, and Trichoptera taxa richness, and the Shannon–Wiener diversity index in approximately 3–15% of the urban areas of relatively small watersheds (1–70 km^2) [51]. Additionally, in Korea's riparian area, the fish community composition is low when the urban proportion is 2–19% [27]. This suggests that to preserve the biological status and minimize the impact on the development of large and small watershed urban areas, the proportion of development sites should be calculated according to the watershed scale.

In agricultural areas of large-scale watersheds, the values of water quality and biological indicators decreased at approximately 5–25% of the percentage of agricultural land use. Generally, the anthropogenic input of pollutants from agricultural areas is a major cause of increased nitrogen and phosphorus concentrations in streams [52]. For example, non-point pollutants entering streams from agricultural areas include fertilizers, sediments, nitrogen-fixing crops, and animal waste [53,54]. However, the threshold was found to be low compared to that in other studies. For example, Roth et al. [17] observed no change in the fish community in Wisconsin, USA, until the cropland proportion of the watershed reached 50%; Grimstead et al. [55] found that the abundance of benthic macroinvertebrate communities decreases when croplands exceed approximately 70% of the catchment area; and D'Amario et al. [56] found that the concentrations of nitrogen and phosphorus also increased with an increase in the agricultural area of the watershed and found that the excess concentration occurred at approximately 34–43%. An increase in agricultural areas leads to an increase in stream nutrients, which also affects algae and benthic macroinvertebrates [56]. However, previous studies [17,55] were conducted in intensive agricultural areas in large-scale watersheds. Intensive agricultural areas generally have negative impacts on water quality and aquatic organisms. However, riparian vegetation in these regions can effectively buffer the negative impacts of agricultural activities. Although this study did not consider the impact of riparian vegetation on stream water quality and aquatic organisms, it shows that large-scale watersheds have different impacts on streams than small-scale watersheds, even with a small percentage of disturbed land use/land cover within the watersheds. These characteristics may be related to the land-use/land-cover composition and structure of the watersheds, including riparian vegetation [57]. High landscape diversity in agricultural areas and forests in riparian areas can improve water quality by reducing nutrient and organic material losses [58–61]. It has been reported that agricultural areas in small watersheds have a negative impact on stream water quality by increasing land-use diversity and fragmenting the natural environment [17]. Land-use proportions alone cannot capture the diverse and complex responses of watersheds to spatial patterns and scales [62]. Contaminant sources in agricultural areas can be negative for aquatic organisms, but the threshold value can change due to the influence of several other factors, such as nutrients flowing into streams; habitats, including sediment delivery systems; and government policies [63].

In the case of forests, water quality and biological indicators increased as the forest proportion increased. These results are consistent with those of previous studies showing that forested areas play an important role in maintaining stream biological integrity and water quality [64]. Brogna et al. [65] found that watershed forests were highly positively correlated with stream and biological water quality. Forests have various benefits for streams, including acting as filters to remove pollutants carried by surface runoff and reducing nutrient concentrations. In this study, the ALEs showed a decrease in water quality concentration in approximately 20–25% of the large- and small-scale watersheds, and an increase in biological indicators in 40–60% of the watersheds. Clément et al. [29] found that having at least 50% forested areas could promote good water quality in Canadian rivers. A previous study suggested an environmental conservation goal of approximately 60% forest cover [8]. In conclusion, securing a forest cover proportion of 25–60% is the minimum requirement for maintaining stream water quality and biological integrity.

4.2. Watershed Scale

The association between land use and water quality and biological indicators assessed using the RF model showed that land use had a significant impact on water quality and biological indicators. In the RF model, large-scale watersheds showed better predictions than small-scale watersheds. This is consistent with other studies showing that the cumulative effect of land use throughout the watershed influences stream water quality and biological conditions [1,66]. In general, catchment scale is reported to have an indirect effect on aquatic organisms by integrating all environmental factors that appear in small watersheds, such as the riparian and reach scales [67,68]. These results may be due to the contribution of all land use in the watershed to the stream ecosystem due to geographical factors, such as the slope and elevation of the watershed, gravity, and various patterns of land use [66]. However, studies have also reported a stronger relation between water quality and biological indicators on smaller scales [69]. Larger watersheds are difficult to estimate because the paths of pollutants moving into streams are more complex and diverse than those in smaller watersheds [14,15,70]. Such differences in the scale of the study results may be due to the effects of water quality parameters and spatial resolution differences in the research design [3,7,71,72]. It is clear that the structure and function of stream ecosystems, including stream water quality and aquatic organisms, depend on the spatial scale [73,74]. This is an important criterion in watershed management and restoration. Current watershed management aims to synthesize the characteristics of watersheds and manage individual streams. However, goals or management strategies based on the watershed may ignore the watershed characteristics of individual streams. This implies that a watershed plan must be set differently for each scale. Land use at various spatial scales affects stream ecosystems through a series of processes. Therefore, understanding the relationship between land use, stream water quality, and aquatic organisms within a watershed is an important part of watershed planning, management, and restoration [24,63]. Understanding the impact of land use on the water quality and biological properties of streams and effectively managing watersheds require insight into their spatial scales. To understand the impact of land use on stream water quality and aquatic organisms and to manage watersheds effectively, it is necessary to consider the spatial scale.

5. Conclusions

This study found that the relation between land use and water quality and biological indicators in the Han River Basin is nonlinear and that critical points appear differently depending on the scale. The results showed similar patterns of relationships between land use and water quality and biological indicators in both the large and small watersheds. As shown in the correlation analysis results, urban areas and agricultural areas had a negative impact on water quality and biological indicators, while forests had a positive impact. The ALE plot results showed that water quality and biological indicators fluctuated in urban areas at lower land-use ratios compared to those of agricultural areas. Both water quality and biological indicators showed more sensitivity to the land-use ratio in small watersheds than in large watersheds. In forested areas, the water quality and biological indicators fluctuated within a certain land-use ratio range, regardless of the watershed scale. In particular, in large watersheds, when the forest proportion was over approximately 25%, the biological indicators increased rapidly. Water quality showed a decreasing trend as the forest rate increased, and when the forest rate was over 25%, it decreased sharply. Therefore, we conclude that achieving at least 25% forest area is necessary to maintain water quality and biological health. The results of this study suggest that to develop effective watershed management strategies to maintain stream ecosystems, various spatial scales must be considered, and a minimum forest area of 25% or higher must be maintained.

Author Contributions: J.-W.L. designed the research, performed the data analysis, and wrote the manuscript, interpreted the results of the analysis. S.-R.P. edited the manuscript; S.-W.L. and S.-R.P. reviewed and finalized the manuscript. All authors have read and agreed to the published version of the manuscript.

Funding: This work was supported by the Korea Environment Industry and Technology Institute through The Decision Support System Development Project for Environmental Impact Assessment, funded by the Korea Ministry of Environment (No. 2020002990009). This paper and the research were supported by the Korea Forest Service (Korea Forestry Promotion Institute) (FTIS 2021331A00-2223-AA01).

Data Availability Statement: The data supporting the findings of this investigation can be obtained from [Water Environment Information System]. Readers should contact the corresponding author for details.

Conflicts of Interest: The authors declare no conflict of interest.

References

1. Allan, J.D. Landscapes and riverscapes: The influence of land use on stream ecosystems. *Annu. Rev. Ecol. Evol. Syst.* **2004**, *35*, 257–284. [CrossRef]
2. Diem, J.E.; Hill, T.C.; Milligan, R.A. Diverse multi-decadal changes in streamflow within a rapidly urbanizing region. *J. Hydrol.* **2018**, *556*, 61–71. [CrossRef]
3. Schiff, R.; Benoit, G. Effects of impervious Cover at multiple spatial scales on coastal watershed streams 1. *J. Am. Water Resour. Assoc.* **2007**, *43*, 712–730. [CrossRef]
4. Wicke, D.; Matzinger, A.; Sonnenberg, H.; Caradot, N.; Schubert, R.-L.; Dick, R.; Heinzmann, B.; Dünnbier, U.; von Seggern, D.; Rouault, P. Micropollutants in urban stormwater runoff of different land uses. *Water* **2021**, *13*, 1312. [CrossRef]
5. Hamid, A.; Bhat, S.U.; Jehangir, A. Local determinants influencing stream water Quality. *Appl. Water Sci.* **2020**, *10*, 24. [CrossRef]
6. Teurlincx, S.; Kuiper, J.J.; Hoevenaar, E.C.; Lurling, M.; Brederveld, R.J.; Veraart, A.J.; Janssen, A.B.; Mooij, W.M.; de Senerpont Domis, L.N. Towards restoring urban waters: Understanding the main pressures. *Curr. Opin. Environ. Sustain.* **2019**, *36*, 49–58. [CrossRef]
7. Shi, P.; Zhang, Y.; Li, Z.; Li, P.; Xu, G. Influence of land use and land cover patterns on seasonal water Quality at multi-spatial scales. *CATENA* **2017**, *151*, 182–190. [CrossRef]
8. Park, S.R.; Kim, S.; Lee, S.W. Evaluating the relationships between riparian land cover characteristics and biological integrity of streams using random forest algorithms. *Int. J. Environ. Res. Public Health* **2021**, *18*, 3182. [CrossRef]
9. Johnson, R.C.; Jin, H.; Carreiro, M.M.; Jack, J.D. Macroinvertebrate community structure, secondary production and trophic-level dynamics in urban streams affected by non-point-source pollution. *Freshw. Biol.* **2013**, *58*, 843–857. [CrossRef]
10. Liu, B.; Chen, S.; Liu, H.; Guan, Y. Changes in the ratio of benthic to planktonic diatoms to eutrophication status of muskegon lake through time: Implications for a valuable indicator on water Quality. *Ecol. Indic.* **2020**, *114*, 106284. [CrossRef]

11. Smucker, N.J.; Vis, M.L. Using diatoms to assess human impacts on streams benefits from multiple-habitat sampling. *Hydrobiologia* **2010**, *654*, 93–109. [CrossRef]
12. Sweeney, B.W.; Newbold, J.D. Streamside forest buffer width needed to protect stream water Quality, habitat, and organisms: A literature review. *J. Am. Water Resour. Assoc.* **2014**, *50*, 560–584. [CrossRef]
13. Turunen, J.; Elbrecht, V.; Steinke, D.; Aroviita, J. Riparian forests can mitigate warming and ecological degradation of agricultural headwater streams. *Freshw. Biol.* **2021**, *66*, 785–798. [CrossRef]
14. Ding, J.; Jiang, Y.; Liu, Q.; Hou, Z.; Liao, J.; Fu, L.; Peng, Q. Influences of the land use pattern on water Quality in low-order streams of the Dongjiang River Basin, China: A multi-scale analysis. *Sci. Total Environ.* **2016**, *551–552*, 205–216. [CrossRef]
15. Buck, O.; Niyogi, D.K.; Townsend, C.R. Scale-dependence of land use effects on water Quality of streams in agricultural catchments. *Environ. Pollut.* **2004**, *130*, 287–299. [CrossRef]
16. Lammert, M.; Allan, J.D. Assessing biotic integrity of streams: Effects of scale in measuring the influence of land use/Cover and habitat structure on fish and macroinvertebrates. *Environ. Manag.* **1999**, *23*, 257–270. [CrossRef]
17. Roth, N.E.; Allan, J.D.; Erickson, D.L. Landscape influences on stream biotic integrity assessed at multiple spatial scales. *Lands. Ecol.* **1996**, *11*, 141–156. [CrossRef]
18. Dala-Corte, R.B.; Giam, X.; Olden, J.D.; Becker, F.G.; Guimarães, T.d.F.; Melo, A.S. Revealing the pathways by which agricultural land-use affects stream fish communities in South Brazilian grasslands. *Freshw. Biol.* **2016**, *61*, 1921–1934. [CrossRef]
19. Pan, Y.; Herlihy, A.; Kaufmann, P.; Wigington, J.; Van Sickle, J.; Moser, T. Linkages among land-use, water Quality, physical habitat conditions and lotic diatom assemblages: A multi-spatial scale assessment. *Hydrobiologia* **2004**, *515*, 59–73. [CrossRef]
20. Oeding, S.; Taffs, K.H.; Cox, B.; Reichelt-Brushett, A.; Sullivan, C. The influence of land use in a highly modified catchment: Investigating the importance of scale in riverine health assessment. *J. Environ. Manag.* **2018**, *206*, 1007–1019. [CrossRef]
21. Tudesque, L.; Tisseuil, C.; Lek, S. Scale-dependent effects of land Cover on Water physico-chemistry and diatom-based metrics in a major river system, the Adour-Garonne Basin (south Western France). *Sci. Total Environ.* **2014**, *466–467*, 47–55. [CrossRef]
22. Zhang, J.; Li, S.; Jiang, C. Effects of Land Use on water Quality in a River Basin (Daning) of the Three Gorges Reservoir Area, China: Watershed versus riparian Zone. *Ecol. Indic.* **2020**, *113*, 106226. [CrossRef]
23. Young-Kyu, S. Comparison of water Quality between forested and agricultural subcatchments in Daegwallyong Area. *Korean Geogr. Soc.* **2004**, *39*, 544–561.
24. Lee, J.W.; Lee, S.W.; An, K.J.; Hwang, S.J.; Kim, N.Y. An estimated structural equation model to assess the effects of land use on water Quality and benthic macroinvertebrates in streams of the Nam-Han River system, South Korea. *Int. J. Environ. Res. Public Health* **2020**, *17*, 2116. [CrossRef]
25. Wang, F.; Wang, Y.; Zhang, K.; Hu, M.; Weng, Q.; Zhang, H. Spatial heterogeneity modeling of water quality based on random forest regression and model interpretation. *Environ. Res.* **2021**, *202*, 111660. [CrossRef]
26. Hastie, T.; Tibshirani, R.; Friedman, J.H.; Friedman, J.H. *The Elements of Statistical Learning: Data Mining, Inference, and Prediction*; Springer: Berlin/Heidelberg, Germany, 2009; p. 2.
27. Chen, K.; Olden, J.D. Threshold responses of riverine fish communities to land use conversion across regions of the world. *Glob. Chang. Biol.* **2020**, *26*, 4952–4965. [CrossRef]
28. Munsch, S.H.; Andrews, K.S.; Crozier, L.G.; Fonner, R.; Gosselin, J.L.; Greene, C.M.; Harvey, C.J.; Lundin, J.I.; Pess, G.R.; Samhouri, J.F.; et al. Potential for ecological nonlinearities and thresholds to inform pacific salmon management. *Ecosphere* **2020**, *11*, e03302. [CrossRef]
29. Clément, F.; Ruiz, J.; Rodríguez, M.A.; Blais, D.; Campeau, S. Landscape diversity and forest edge density regulate stream water Quality in agricultural catchments. *Ecol. Indic.* **2017**, *72*, 627–639. [CrossRef]
30. Foudi, S.; Spadaro, J.V.; Chiabai, A.; Polanco-Martínez, J.M.; Neumann, M.B. The climatic dependencies of urban ecosystem services from green roofs: Threshold effects and non-linearity. *Ecosyst. Serv.* **2017**, *24*, 223–233. [CrossRef]
31. Chang, H. Spatial analysis of water quality trends in the Han River Basin, South Korea. *Water Res.* **2008**, *42*, 3285–3304. [CrossRef]
32. Lee, S.-W.; Hwang, S.-J.; Lee, J.-K.; Jung, D.-I.; Park, Y.-J.; Kim, J.-T. Overview and application of the National Aquatic Ecological Monitoring Program (NAEMP) in Korea. *Ann. Limnol. Int. J. Lim.* **2011**, *47*, S3–S14. [CrossRef]
33. Greenwell, B.M.; Boehmke, B.C.; McCarthy, A.J. A simple and effective model-based variable importance measure. *arXiv* **2018**, arXiv:1805.04755.
34. Baak, M.; Koopman, R.; Snoek, H.; Klous, S. A new correlation coefficient between categorical, ordinal and interval variables with Pearson characteristics. *Comput. Stat. Data Anal.* **2020**, *152*, 107043. [CrossRef]
35. Breiman, L. Random forests. *Mach. Learn.* **2001**, *45*, 5–32. [CrossRef]
36. Ratolojanahary, R.; Houé Ngouna, R.H.; Medjaher, K.; Junca-Bourié, J.; Dauriac, F.; Sebilo, M. Model selection to improve multiple imputation for handling high rate missingness in a water quality dataset. *Expert Syst. Appl.* **2019**, *131*, 299–307. [CrossRef]
37. Grömping, U. Variable Importance Assessment in Regression: Linear Regression versus Random Forest. *Am. Stat.* **2009**, *63*, 308–319. [CrossRef]
38. Prasad, A.M.; Iverson, L.R.; Liaw, A. Newer classification and regression tree techniques: Bagging and random forests for ecological prediction. *Ecosystems* **2006**, *9*, 181–199. [CrossRef]
39. Probst, P.; Wright, M.N.; Boulesteix, A. Hyperparameters and tuning strategies for random Forest. *WIREs Data Min. Knowl.* **2019**, *9*, e1301. [CrossRef]

40. Wang, L.; Zhou, X.; Zhu, X.; Dong, Z.; Guo, W. Estimation of biomass in wheat using random forest regression algorithm and remote sensing data. *Crop J.* **2016**, *4*, 212–219. [CrossRef]
41. Cutler, D.R.; Edwards, T.C., Jr.; Beard, K.H.; Cutler, A.; Hess, K.T.; Gibson, J.; Lawler, J.J. Random forests for classification in ecology. *Ecology* **2007**, *88*, 2783–2792. [CrossRef]
42. RColorBrewer, S.; Liaw, M.A. *Package 'Randomforest.'*; University of California: Berkeley, CA, USA, 2018.
43. Molnar, C. *Interpretable Machine Learning*; Lulu.com: Morrisville, NC, USA, 2020; ISBN 0244768528.
44. Yu, Q.; Ji, W.; Prihodko, L.; Ross, C.W.; Anchang, J.Y.; Hanan, N.P. Study becomes insight: Ecological learning from machine learning. *Methods Ecol. Evol.* **2021**, *12*, 2117–2128. [CrossRef]
45. Stritih, A.; Senf, C.; Seidl, R.; Grêt-Regamey, A.; Bebi, P. The impact of land-use legacies and recent management on natural disturbance susceptibility in mountain forests. *For. Ecol. Manag.* **2021**, *484*, 118950. [CrossRef]
46. Apley, D.W.; Zhu, J. Visualizing the effects of predictor variables in black box supervised learning models. *J. R. Stat. Soc. B* **2020**, *82*, 1059–1086. [CrossRef]
47. Liaw, A.; Wiener, M. Classification and regression by RandomForest. *R News* **2002**, *2*, 18–22.
48. Barnosky, A.D.; Hadly, E.A.; Bascompte, J.; Berlow, E.L.; Brown, J.H.; Fortelius, M.; Getz, W.M.; Harte, J.; Hastings, A.; Marquet, P.A.; et al. Approaching a state shift in earth's biosphere. *Nature* **2012**, *486*, 52–58. [CrossRef]
49. Groffman, P.M.; Baron, J.S.; Blett, T.; Gold, A.J.; Goodman, I.; Gunderson, L.H.; Levinson, B.M.; Palmer, M.A.; Paerl, H.W.; Peterson, G.D.; et al. Ecological thresholds: The key to successful environmental management or an important concept with no practical application? *Ecosystems* **2006**, *9*, 1–13. [CrossRef]
50. Tromboni, F.; Dodds, W.K. Relationships between land use and stream nutrient concentrations in a highly urbanized tropical region of brazil: Thresholds and riparian zones. *Environ. Manage.* **2017**, *60*, 30–40. [CrossRef]
51. Wang, B.; Liu, D.; Liu, S.; Zhang, Y.; Lu, D.; Wang, L. Impacts of urbanization on stream habitats and macroinvertebrate communities in the tributaries of Qiangtang River, China. *Hydrobiologia* **2012**, *680*, 39–51. [CrossRef]
52. Ni, X.; Parajuli, P.B.; Ouyang, Y.; Dash, P.; Siegert, C. Assessing land use change impact on stream discharge and stream water Quality in an agricultural watershed. *CATENA* **2021**, *198*, 105055. [CrossRef]
53. Savci, S. An agricultural pollutant: Chemical fertilizer. *Int. J. Environ. Sci. Dev.* **2012**, *3*, 73–80. [CrossRef]
54. Mangadze, T.; Wasserman, R.J.; Froneman, P.W.; Dalu, T. Macroinvertebrate functional feeding group alterations in response to habitat degradation of headwater austral streams. *Sci. Total Environ.* **2019**, *695*, 133910. [CrossRef]
55. Grimstead, J.P.; Krynak, E.M.; Yates, A.G.; Land Cover, S.-S. Thresholds for conservation of stream invertebrate communities in agricultural landscapes. *Landsc. Ecol.* **2018**, *33*, 2239–2252. [CrossRef]
56. D'Amario, S.C.; Rearick, D.C.; Fasching, C.; Kembel, S.W.; Porter-Goff, E.; Spooner, D.E.; Williams, C.J.; Wilson, H.F.; Xenopoulos, M.A. The prevalence of nonlinearity and detection of ecological breakpoints across a land use gradient in streams. *Sci. Rep.* **2019**, *9*, 3878. [CrossRef]
57. Marzin, A.; Verdonschot, P.F.M.; Pont, D. The relative influence of catchment, riparian corridor, and reach-scale anthropogenic pressures on fish and macroinvertebrate assemblages in French Rivers. *Hydrobiologia* **2013**, *704*, 375–388. [CrossRef]
58. Uuemaa, E.; Roosaare, J.; Mander, Ü. Scale dependence of landscape metrics and their indicatory value for nutrient and organic matter losses from catchments. *Ecol. Indic.* **2005**, *5*, 350–369. [CrossRef]
59. Jones, K.B.; Neale, A.C.; Nash, M.S.; Van Remortel, R.D.; Wickham, J.D.; Riitters, K.H.; O'Neill, R.V. Predicting nutrient and sediment loadings to streams from landscape metrics: A multiple watershed study from the United States mid-Atlantic region. *Landsc. Ecol.* **2001**, *16*, 301–312. [CrossRef]
60. Gergel, S.E.; Turner, M.G.; Miller, J.R.; Melack, J.M.; Stanley, E.H. Landscape indicators of human impacts to riverine systems. *Aquat. Sci.* **2002**, *64*, 118–128. [CrossRef]
61. Fitzpatrick, F.A.; Scudder, B.C.; Lenz, B.N.; Sullivan, D.J. Effects of multi-scale environmental characteristics on agricultural stream biota in Eastern Wisconsin 1. *J. Am. Water Resour. Assoc.* **2001**, *37*, 1489–1507. [CrossRef]
62. Hawkins, C.P.; Norris, R.H.; Gerritsen, J.; Hughes, R.M.; Jackson, S.K.; Johnson, R.K.; Stevenson, R.J. Evaluation of the use of landscape classifications for the prediction of freshwater biota: Synthesis and recommendations. *J. N. Am. Benthol. Soc.* **2000**, *19*, 541–556. [CrossRef]
63. Meador, M.R.; Goldstein, R.M. Assessing water Quality at large geographic scales: Relations among land use, Water physicochemistry, riparian condition, and fish community structure. *Environ. Manag.* **2003**, *31*, 504–517. [CrossRef]
64. De Mello, K.; Valente, R.A.; Randhir, T.O.; dos Santos, A.C.A.; Vettorazzi, C.A. Effects of Land Use and Land Cover on water Quality of Low-Order Streams in Southeastern Brazil: Watershed versus riparian Zone. *CATENA* **2018**, *167*, 130–138. [CrossRef]
65. Brogna, D.; Dufrêne, M.; Michez, A.; Latli, A.; Jacobs, S.; Vincke, C.; Dendoncker, N. Forest cover correlates with good biological water Quality. Insights from a regional study (Wallonia, Belgium). *J. Environ. Manag.* **2018**, *211*, 9–21. [CrossRef]
66. Hunsaker, C.T.; Levine, D.A. Hierarchical approaches to the study of water Quality in rivers. *BioScience* **1995**, *45*, 193–203. [CrossRef]
67. Villeneuve, B.; Piffady, J.; Valette, L.; Souchon, Y.; Usseglio-Polatera, P. Direct and indirect effects of multiple stressors on stream invertebrates across watershed, reach and site scales: A structural equation modelling better informing on hydromorphological impacts. *Sci. Total Environ.* **2018**, *612*, 660–671. [CrossRef]

68. Forio, M.A.E.; Burdon, F.J.; De Troyer, N.; Lock, K.; Witing, F.; Baert, L.; De Saeyer, N.; Rîșnoveanu, G.; Popescu, C.; Kupilas, B.; et al. A bayesian belief network learning tool integrates multi-scale effects of riparian buffers on stream invertebrates. *Sci. Total Environ.* **2022**, *810*, 152146. [CrossRef]
69. Zhou, T.; Wu, J.; Peng, S. Assessing the effects of landscape pattern on river water Quality at multiple scales: A case study of the Dongjiang River watershed, China. *Ecol. Indic.* **2012**, *23*, 166–175. [CrossRef]
70. Nash, M.S.; Heggem, D.T.; Ebert, D.; Wade, T.G.; Hall, R.K. Multi-scale landscape factors influencing stream water Quality in the State of Oregon. *Environ. Monit. Assess.* **2009**, *156*, 343–360. [CrossRef]
71. Pratt, B.; Chang, H. Effects of land Cover, topography, and built structure on seasonal water Quality at multiple spatial scales. *J. Hazard. Mater.* **2012**, *209–210*, 48–58. [CrossRef]
72. Allan, J.D.; Johnson, L. Catchment-scale analysis of aquatic ecosystems. *Freshw. Biol.* **1997**, *37*, 107–111. [CrossRef]
73. Meyer, J.L.; Strayer, D.L.; Wallace, J.B.; Eggert, S.L.; Helfman, G.S.; Leonard, N.E. The contribution of headwater streams to biodiversity in river networks 1. *J. Am. Water Resour. Assoc.* **2007**, *43*, 86–103. [CrossRef]
74. Wipfli, M.S.; Richardson, J.S.; Naiman, R.J. Ecological linkages between headwaters and downstream ecosystems: Transport of organic matter, invertebrates, and wood down headwater channels 1. *J. Am. Water Resour. Assoc.* **2007**, *43*, 72–85. [CrossRef]

water

MDPI

Article

Water and Sediment Chemistry as Drivers of Macroinvertebrates and Fish Assemblages in Littoral Zones of Subtropical Reservoirs

Linton F. Munyai [1,*], Thendo Liphadzi [2], Thendo Mutshekwa [3,4], Mulalo I. Mutoti [5], Lubabalo Mofu [6] and Florence M. Murungweni [2]

1 School of Biology and Environmental Sciences, University of Mpumalanga, Nelspruit 1200, South Africa
2 Department of Geography and Environmental Sciences, University of Venda, Thohoyandou 0950, South Africa
3 Department of Freshwater Invertebrates, Albany Museum, Makhanda 6139, South Africa
4 Department of Zoology and Entomology, Rhodes University, Makhanda 6139, South Africa
5 Department of Environmental, Water, and Earth Sciences, Tshwane University of Technology, Pretoria 0001, South Africa
6 South African Institute for Aquatic Biodiversity (SAIAB), Makhanda 6140, South Africa
* Correspondence: munyailinton@gmail.com

Abstract: Reservoirs are human-made ecosystems with diverse purposes that benefit humans both directly and indirectly. They however cause changes in geomorphological processes such as sediment cycling and influence the composition and structure of aquatic biota. This study aimed to identify water and sediment quality parameters as drivers of macroinvertebrates and fish communities during the cool-dry and hot-wet seasons in the littoral zones of three subtropical reservoirs (Albasini, Thathe and Nandoni). Macroinvertebrates and fish were collected from three sites (n = 3 from each site) in each reservoir. A total of 501 and 359 macroinvertebrates and fish individuals were collected throughout the sampling period, respectively. The present study employed a two-way ANOVA in conjunction with redundancy analysis (RDA) to assess the relationships that exist between water and sediment variables, macroinvertebrates diversity and species abundances across seasons. Based on the two-way ANOVA model, significant differences were observed across reservoirs for evenness, Simpson's diversity, and total abundance, while seasonal differences were observed for most metrics, with exception for evenness. The RDA results identified four water variables (i.e., water temperature, oxidation–reduction potential, pH and conductivity) and one sediment metal (Mg) as the most important parameters in driving the fish community structure. Field observations and metal results attest that the Nandoni reservoir shows high concentrations of metals in sediments as compared to other reservoirs, suggesting that anthropogenic activities such as car washing, brick making, recreation, fishing, wastewater treatment work and landfill site may be the major contributor of metals to the Nandoni reservoir, which accumulate in the littoral zones. Findings of this study highlight the need to analyze reservoir ecological conditions at several scales. The study of macroinvertebrates and fish, water, and sediment chemistry in the littoral zone laid the groundwork for proposing measures for conserving aquatic ecosystems.

Keywords: aquatic ecosystems; biomonitoring; community structure; littoral macroinvertebrates; macrozoobenthos; reservoir ecology

Citation: Munyai, L.F.; Liphadzi, T.; Mutshekwa, T.; Mutoti, M.I.; Mofu, L.; Murungweni, F.M. Water and Sediment Chemistry as Drivers of Macroinvertebrates and Fish Assemblages in Littoral Zones of Subtropical Reservoirs. *Water* **2024**, *16*, 42. https://doi.org/10.3390/w16010042

Academic Editors: Eva Papastergiadou and Kostas Stefanidis

Received: 10 November 2023
Revised: 15 December 2023
Accepted: 19 December 2023
Published: 21 December 2023

1. Introduction

Freshwater is among vital natural resources that play an essential role in supporting and maintaining life on Earth [1]. Freshwater constitutes approximately 2.5 percent of the overall water volume on planet Earth, but in turn, it serves as habitat and the foundation of aquatic biodiversity [2]. Aquatic ecosystems such as lakes, ponds, streams, rivers, and wetlands provide important services for human life such as water security, food security, and economic productivity [3–5]. Water is not only a life-sustaining resource for all biotic components, but also a major factor contributing to all developmental activities associated with environmental and cultural processes [6]. However, as tolerating as freshwater ecosystems are, they are vulnerable to the increasing pressure and threats posed by global anthropogenic pollution from activities such as agriculture, development of urban areas, and mining [7–9].

According to Dalu et al. [10], water pollution is a major management concern in developing countries, with many rivers and streams classified as endangered, as they are subjected to untreated solid waste, wastewater discharge, higher population, stormwater, and agricultural runoff. A study by Dalu and Chauke [11] indicated that anthropogenic activities (for example, cattle grazing and cultivation) contribute to increased nutrient concentration in freshwater ecosystems such as wetlands, which have a significant negative effect on macroinvertebrate communities. Enough evidence from various studies shows that activities within agricultural sectors tend to have a noticeable effect on water systems in comparison to activities such as forestry and urban areas [12–14].

Among the aquatic biota, macroinvertebrates and fish are generally highly sensitive organisms and their dynamics can seriously be affected by changes in water quality [15,16]. Due to anthropogenic activities, fish and macroinvertebrate communities are threatened, leading to an unbalanced food-web structure in freshwater ecosystems [17,18]. These organisms possess various attributes such as a high production rate, a faster growth rate, and the ability to provide a meaningful ecological state over time in a particular water body [19,20]; thus, they can be used for bioassessment of the ecological integrity of freshwater ecosystems. Although these organisms usually respond to changes in the quality of the water, their sensitivity may differ significantly [21].

Macroinvertebrates and fish communities are comprised of a variety of taxa with varying degrees of pollution tolerance [22,23]. While this has been well studied in different aquatic ecosystems, changes in community structure of macroinvertebrates and fish have been less explored in subtropical standing water bodies [11,17]. The physicochemical conditions of reservoirs water and sediments are intrinsically intertwined; it is therefore difficult to disregard the dynamics of sediment flora and fauna if one wants to completely comprehend these aquatic ecosystems [24,25]. For instance, physical and chemical variable concentrations are important in structuring the diversity of macroinvertebrates and fish, with implications for biodiversity conservation in these water bodies [26,27]. These dynamics are important to understand because catchment activities can be sources of various metals and nutrients that are washed to the reservoirs through runoff and can enter the reservoirs, thus altering water quality and aquatic food-web functioning [28,29]. Sediment may also contain nutrients, heavy metals, and organochlorine pesticides that are released into the water column and then into the food chain [30]. As a result, they contribute significantly to a variety of biotic and abiotic processes in reservoirs [31].

Given the importance of freshwater in promoting healthy aquatic biodiversity, rapid assessments are required to detect management and monitoring strategies to ensure the safety of aquatic ecosystems and to aid in the documentation of policies to support these measures. This study assessed the seasonal patterns in macroinvertebrates and fish communities, and their diversity matrices in response to seasonal variations in the littoral zones of subtropical reservoirs. These reservoirs are important ecological features because they are habitats for many fauna and flora, foraging ground for crocodiles and a source of water supply for many local communities around the Vhembe district. This study hypothesized that (i) water and sediment chemistry would cause changes in macroinvertebrates and fish communities due to various anthropogenic activities practiced in the catchment as communities respond differently to changes in anthropogenic pressures and (ii) the hot-wet season will have more abundant and diverse macroinvertebrates and fish than the cool-dry season because different seasons changes aquatic dynamics.

2. Methods and Materials

2.1. Ethical Clearance

Macroinvertebrates and fish samples were collected following ethical clearance approved by the research committee of the University of Venda (Ethical clearance No. SES/20/ERM/14/1611).

2.2. Study Area

This research was conducted during the cool-dry and hot-wet seasons of the year 2022 in three selected water supply reservoirs, namely Nandoni (22°59′20″ S, 30°36′27″ E), Thathe (22°56′45″ S 30°20′7″ E) and Albasini (23°6′30″ S 30°7′48.2″ E), within the Vhembe district, Limpopo Province, South Africa (Figure 1). These reservoirs are located within the Luvuvhu River Catchment (LRC), which connects the three reservoirs and many tributaries [32], and forms part of the larger Limpopo River system. The selection of these reservoirs for sampling is based on: (1) the river connection that exists within the catchment with the reservoirs, (2) the biodiversity they inhabit, (3) the activities that are being practiced adjacent to these dams which are associated with land degradation, and (4) the role that they play in local communities (water provision). The Luvuvhu River rises as a steep mountain stream in the south-easterly slopes of the Soutpansburg mountains and runs through a variety of landscapes for approximately 200 km before joining the Limpopo River near to Pafuri in the Kruger National Park [33,34]. The Nandoni reservoir is located in the middle section of the LRC, whereas the Albasin and Thathe reservoirs are located in the upper LRC, with Thathe captured along the tributary of Mutshundudi River, which is steep, narrow, and dominated by cobble riffles [33]. The mean annual rainfall in the LRC is 608 mm and the mean annual runoff is 520×10^6 m^3. The topography varies from 200 m to 1500 m, which influences rainfall and runoff distribution in the catchment [35].

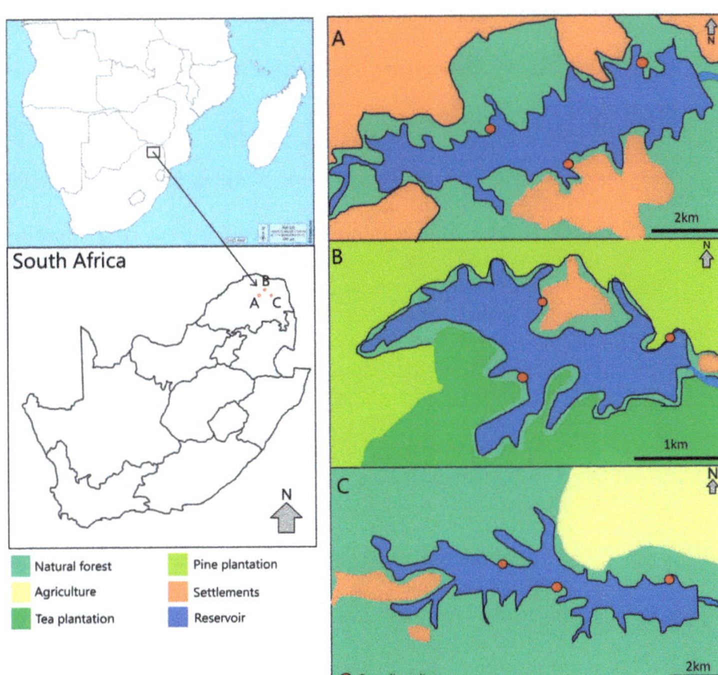

Figure 1. Map showing the location of the sample collection sites and land use activities across three reservoirs, (**A**) the Nandoni reservoir, (**B**) the Thathe reservoir, and (**C**) the Albasini reservoir, within the Luvuvhu River Catchment, Limpopo prince, South Africa.

2.3. Determining Environmental Variables

2.3.1. Physicochemical Parameters of Water

During sampling, pH, conductivity (µS cm^{-1}), total dissolved solids (mg L^{-1}), resistivity (ohm), salinity (ppm), oxygen redox potential (mV), and water temperature (°C) were measured in situ at each site ($n = 3$; ~3 m apart) per season using a portable multiparameter probe PCTestr 35 (Eutech/Oakton Instruments, Singapore). All physicochemical parameters measurements were performed before macroinvertebrates and fish sampling. These measurements were performed 3 m away from where macroinvertebrates and fish were collected to avoid habitat disturbance [8].

2.3.2. Sediment Chemistry Variables

Sediment samples ($n = 2$) were collected from each site per season (2 m away from where macroinvertebrates and fish were sampled) using acid-washed wooden splints, and in order to prevent cross-contamination, each integrated sample was placed in a fresh plastic Ziplock bag. The composite sediment samples were promptly sealed and packaged in an ice-filled cooler box before being transported to the University of Venda Pollution Laboratory for analysis within 24 h. Upon arrival to the laboratory, the samples were placed in an oven and dried at 60 °C for 72 h. Samples were then taken out and disaggregated in a porcelain mortar. After using a riffle splitter to homogenize the dried sediment samples, a sediment subsample weighing 0.5 kg was separated and transported to BEMLAB in Cape Town for further analysis. The concentrations of the following elements were analyzed for each site and season: pH, phosphorus (P), potassium (K^+), calcium (Ca^{2+}), magnesium (Mg^{2+}), sodium (Na^+), copper (Cu), zinc (Zn^{2+}), manganese (Mn), boron (B), iron (Fe), carbon (C), sulfur (S), ammonium ($NH_4{}^+$), and soil organic matter (SOM). Briefly, for each sediment sample, ammonia and phosphorus were analyzed using a SEAL Auto-Analyzer 3 coupled with high resolution and Bray-2 extract as described by Bray and Kurtz [36] and following Allen et al. [37], an inductively coupled plasma atomic emission spectroscopy (ICPMS) instrument was used to measure elements such as K, Zn, Na Ca, Cu, Mg, Mn, B, Fe, and S (see Dalu et al. [38,39] for detailed methods).

2.4. Littoral Macroinvertebrates Sampling

Macroinvertebrates were collected at each site (water depth 0.1–0.6 m) and season using a handheld kick net (frame 30 cm × 30 cm, mesh size 500 µm) by submerging the sample kick net for five minutes while kicking the benthic substrate to loosen any attached taxa on sand and rocks, then sweeping and pulling the net through macrophytes along the 10 m transect in the littoral zones. This sampling procedure allowed a good and enough representation of macroinvertebrates individual per unit time (see Dalu et al. [23] for methodology). The net was then taken out of the water column and macroinvertebrates were carefully separated from the net, and then preserved in 70% ethanol in labelled 500 mL polyethylene containers for further processing in the laboratory. In the laboratory, macroinvertebrates were sorted under a dissecting Olympus microscope using forceps and identified to the family level following the guide by Fry [40]. The results of macroinvertebrates were presented as relative abundances.

2.5. Littoral Fish Sampling

A seine net was used to collect fish taxa in the littoral zones of the reservoirs (<1.5 m deep). A seine net (5 m long) was pulled into the reservoir in a straight-line motion and dragged across a 20 × 20 m transect within the reservoir by swiping all fish species found in the covered section ($n = 2$). All captured fish were removed from the seine net and put into different 20 L buckets containing water. Fish were identified to the species level on site, and the data were recorded; then fish were released back to the reservoir immediately after identifying each of them. The results of fish were presented as relative abundances.

2.6. Data Analyses

Macroinvertebrate diversity metrices (evenness, Margalef's diversity taxa richness, Shannon–Wiener diversity, and Simpson's diversity) were calculated using a macroinvertebrate community dataset in PAST version 4.03. The effects of different reservoirs (three levels: Albasini, Nandoni, and Thathe), and seasons (two seasons; hot-wet, cool dry), and their interaction, on environmental variables (i.e., water and sediments) and macroinvertebrates diversity metrics, i.e., evenness, Margalef's diversity index, taxa richness, Shannon–Wiener diversity, Simpson's diversity index and total abundance were examined using a factorial two-way ANOVA with Tukey's post hoc after testing for homogeneity of variances (Levene's test, $p > 0.05$) and normality of distribution (Shapiro–Wilk test, $p > 0.05$) SPSS v16.0 (SPSS Inc., Chicago, IL, USA, 2007) were determined. In all analyses, significance was inferred at $p < 0.05$.

Furthermore, distance-based Permutational Analysis of Variance (PERMANOVA) based on Bray–Curtis dissimilarities were employed for biological data and 9999 permutations along with Monte Carlo tests were utilized to analyze differences in fish and macroinvertebrate communities using PERMANOVA + PRIMER version 6 [41].

For all community analyses, all macroinvertebrate and fish abundance data were square-root transformed in order to reduce skewness, while all environmental variable data except for pH were log $(x + 1)$ transformed. We used detrended canonical correspondence analysis (DCCA) to determine whether unimodal or linear methods would be appropriate for the ordination analysis. The length of the gradient was examined and since the longest gradient was <3.0, a linear constrained method, i.e., redundancy analysis (RDA), was employed. An RDA was then performed on the transformed macroinvertebrate and fish abundance data to mainly examine the links between species composition and the selected environmental variables. All variables that were significant ($p < 0.05$) were subjected to RDA forward selection based on environmental variables and sediment chemistry variables to identify a minimal subset of environmental variables that were important to drive the macroinvertebrate and fish community structure (Monte Carlo test with 9999 permutations). The software Canoco version 5.1 was used for the analysis [42].

3. Results

3.1. Environmental Variables (Water and Sediment Chemisty)

Differences in water chemistry variables over the sampling reservoirs and different seasons are highlighted in Table 1. Across reservoirs and seasons, temperature (mean range, 18.7–28.4 °C), pH (mean range, 7.5–8.4), conductivity (mean range, 78.3–421.2 μS cm^{-1}), TDS (mean range, 58.1–142.0), salinity (mean range, 462.6–2246.1 ppm), resistivity (mean range, 720.3–1665.0 °C ohm) and ORP (mean range, 41.8–54.6 mV) differed substantially (Table 1). Significant differences were observed for temperature (ANOVA: F = 145.270; $p < 0.001$), pH (ANOVA: F= 9.631; $p = 0.003$), conductivity (ANOVA: F = 148.528; $p < 0.001$), TDS (ANOVA: F = 127.415; $p < 0.001$), and ORP (ANOVA: F = 48.958; $p < 0.001$), while seasonal differences were observed for temperature (ANOVA: F = 1862.535; $p < 0.001$), pH (ANOVA: F = 21.715; $p = 0.001$), conductivity (ANOVA: F = 1732.485; $p < 0.001$), TDS (ANOVA: F = 16.652; $p = 0.002$), and salinity (ANOVA: F = 4.825; $p = 0.048$) and reservoirs × seasons interactions were observed for temperature (ANOVA: F = 22.449; $p < 0.001$), pH (ANOVA: F = 3.915; $p = 0.049$), conductivity (ANOVA: F = 119.397; $p < 0.001$), TDS (ANOVA: F = 199.144; $p < 0.001$), and ORP (ANOVA: F = 4.543; $p = 0.034$) (Table 2). In general, significant differences were detected when comparing reservoirs.

According to post hoc comparison, across reservoirs, significant differences were observed for temperature, i.e., Albasini vs. Thathe ($p < 0.001$) and Nandoni vs. Thathe ($p < 0.001$), pH, i.e., Albasini vs. Thathe ($p = 0.018$), Nandoni vs. Thathe ($p = 0.003$), conductivity, i.e., Albasini vs. Nandoni ($p < 0.001$), Albasini vs. Thathe ($p < 0.001$), Nandoni vs. Thathe ($p < 0.001$), TDS, i.e., Albasini vs. Thathe ($p < 0.001$), Nandoni vs. Thathe ($p < 0.001$) and ORP, i.e., Albasini vs. Nandoni ($p < 0.001$), Nandoni vs. Thathe ($p < 0.001$).

Table 1. Mean (±standard deviation) water chemistry variables measured across sampling reservoirs (i.e., Albasini, Nandoni and Thathe) and different seasons (hot-wet and cool-dry). Abbreviations: TDS—total dissolved solids; ORP—oxidation–reduction potential.

Variables	Unit	Seasons	Reservoirs		
			Albasini	Nandoni	Thathe
Temperature	°C	Cool-dry	22.4 ± 10.3	20.8 ± 9.5	18.7 ± 8.5
		Hot-wet	27.6 ± 12.9	28.4 ± 13.3	25.5 ± 11.9
pH		Cool-dry	7.5 ± 2.9	8.3 ± 3.3	7.5 ± 2.9
		Hot-wet	8.3 ± 3.3	8.4 ± 3.3	8.1 ± 3.2
Conductivity	µS cm^{-1}	Cool-dry	128.1 ± 148.2	421.2 ± 209.8	316.0 ± 157.1
		Hot-wet	205.7 ± 102.3	169.4 ± 85.0	78.3 ± 39.6
TDS	mg L^{-1}	Cool-dry	106.8 ± 52.5	132.5 ± 65.4	113.3 ± 55.8
		Hot-wet	142.0 ± 70.2	123.2 ± 60.8	58.1 ± 29.7
Salinity	ppm	Cool-dry	481.5 ± 331.0	1902.4 ± 1417.6	2246.1 ± 1749.6
		Hot-wet	462.6 ± 232.6	686.7 ± 501.6	688.3 ± 504.9
Resistivity	ohm	Cool-dry	1541.5 ± 950.2	778.0 ± 743.4	720.3 ± 787.9
		Hot-wet	1665.0 ± 842.3	1397.0 ± 835.5	1304.6 ± 810.1
ORP	mV	Cool-dry	43.3 ± 20.8	50.1 ± 24.3	43.5 ± 20.9
		Hot-wet	41.8 ± 20.1	54.6 ± 26.5	46.5 ± 22.4

Table 2. Two-way analysis of variance (ANOVA) based on water and sediment variables and macroinvertebrate and fish diversity metrics for sampled reservoirs (i.e., Albasini, Nandoni and Thathe: *df* = 2) and seasons (hot-wet and cool-dry: *df* = 1). F-values are discerned with type III sums of squares via. Satterthwaite's method. Significant *p*-values are in bold.

Variables	Reservoirs		Seasons		Reservoirs × Seasons	
	F	*p*	F	*p*	F	*p*
Water chemistry variables						
Temperature	**145.270**	**<0.001**	**1862.535**	**<0.001**	22.449	**<0.001**
pH	**9.631**	**0.003**	**21.715**	**0.001**	**3.915**	**0.049**
Conductivity	**148.528**	**<0.001**	**1732.485**	**<0.001**	**119.397**	**<0.001**
TDS	**127.415**	**<0.001**	**16.652**	**0.002**	**199.144**	**<0.001**
Salinity	2.100	0.165	**4.825**	**0.048**	1.212	0.332
Resistivity	1.491	0.264	2.115	0.172	0.276	0.764
ORP	**48.958**	**<0.001**	5.620	0.035	**4.543**	**0.034**
Sediment chemistry variables						
pH	0.76	0.927	0.757	0.401	**4.882**	**0.028**
P	2.655	0.111	0.207	0.657	2.038	0.176
NH_4^+	0.681	0.525	0.531	0.480	0.567	0.582
K	1.690	0.226	1.625	0.227	0.482	0.629
Ca	0.312	0.738	0.285	0.603	0.303	0.744
Mg	1.242	0.323	0.571	0.465	1.841	0.201
Cu	0.403	0.677	0.001	1.000	0.244	0.787
Zn	0.131	0.878	0.001	0.971	3.154	0.079
Mn	0.582	0.574	0.096	0.763	2.639	0.112
B	1.165	0.317	1.559	0.236	0.324	0.730
Fe	0.087	0.917	0.048	0.830	1.024	0.388
C	0.956	0.412	3.119	0.103	0.719	0.507
S	1.165	0.345	0.235	0.637	**5.459**	**0.021**
Macroinvertebrate diversity metrics						
Evenness	**49.408**	**<0.001**	1.353	0.267	**4.228**	**0.041**
Margalef's diversity	**40.745**	**<0.001**	0.267	0.615	**18.737**	**<0.001**
Taxa richness	**52.125**	**<0.001**	0.042	0.842	**7.292**	**0.008**
Shannon–Wiener diversity	**15.887**	**<0.001**	3.650	0.080	3.399	0.068
Simpson's diversity	**7.549**	**0.008**	**5.985**	**0.031**	2.046	0.172
Total abundance	**51.128**	**<0.001**	0.139	0.716	**4.794**	**0.029**

Table 2. *Cont.*

Variables	Reservoirs		Seasons		Reservoirs × Seasons	
	F	p	F	p	F	p
Fish diversity metrics						
Evenness	29.031	**<0.001**	0.008	0.928	**10.045**	**0.003**
Margalef's diversity	0.309	0.740	**8.112**	**0.015**	**4.409**	**0.037**
Taxa richness	3.040	0.085	**16.000**	**0.002**	2.560	0.119
Shannon–Wiener diversity	1.487	0.265	**20.071**	**0.001**	**9.163**	**0.004**
Simpson's diversity	**9.824**	**0.003**	**17.173**	**0.001**	**15.023**	**0.001**
Total abundance	**15.031**	**0.001**	**6.870**	**0.022**	0.046	0.956

Generally, in the Albasini reservoir, high mean sediment chemistry variable concentrations were observed for pH (mean, 7.2) and Fe (565.3 mg kg^{-1}). In the Nandoni reservoir, high mean sediment chemistry variable concentrations were observed for P (mean, 249 mg kg^{-1}) and Ca (mean, 21.3 mg kg^{-1}). Lastly, in the Thathe reservoir, high mean sediment chemistry variable concentrations were observed for pH (mean, 7.2), NH_4^+ (mean, 123.5), K (mean, 666.7 mg kg^{-1}), Mg (mean, 20.2 mg kg^{-1}), Cu (mean, 16.7 mg kg^{-1}), Zn (mean, 2.2 mg kg^{-1}), Mn (mean, 398.1 mg kg^{-1}), B (mean, 1.2 mg kg^{-1}), C (mean, 2.9%) and S (mean, 155.9 mg kg^{-1}). With regard to seasons across reservoirs, high mean sediments chemistry was observed for pH, P, Ca, Cu, Zn, Mn, C during the hot-wet seasons and pH, NH_4^+, K, Mg, B, Fe and S during the cool-dry season. According to two-way ANOVA, all the sediment variables were found to be similar between the study reservoirs and seasons ($p > 0.05$; Table 2).

3.2. *Macroinvertebrate Communities*

A total of 501 macroinvertebrate individuals belonging to 21 families and 7 orders were identified within three selected water bodies (Table 3). In terms of orders, Hemiptera (28.6%) was the dominating order followed by Odonata (23.8%) and Coleoptera (14.3%). Overall, bladder snails Physidae were the most dominant family group, accounting for 31.7% of the total individuals across sampling reservoirs and seasons, with freshwater shrimp Atyidae being the second most abundant family group (12.0%). Non-biting midges Chironomidae were the third most abundant accounting for 8.0%. Based on the two-way ANOVA model, significant differences were observed across sampling reservoirs for all five metrics ($p < 0.05$; Table 2), while seasonal differences were observed for Simpson's diversity (ANOVA: F = 5.989; $p = 0.031$). Reservoirs and season interaction indicated significant differences for most of the metrics ($p < 0.05$) (Table 2), with the exception for Shannon–Wiener diversity and Simpson's diversity ($p > 0.005$).

Table 3. Mean relative abundances (%) of the dominant macroinvertebrate species and diversity metrics observed across two seasons for the study site categories: Albasini, Nandoni and Thathe reservoirs.

Order	Taxa	Cool-Dry			Hot-Wet		
		Albasini	Nandoni	Thathe	Albasini	Nandoni	Thathe
Mollusca	Thiaridae		26.9	15.7			5.7
Mollusca	Lymnaeidae	3.6			2.3		
Mollusca	Physidae	57.1		15.7	37.9	5.6	
Crustacea	Atyidae	3.6		5.9	23.5	21.3	2.9
Crustacea	Potamonautidae	3.6	19.2	5.9			8.6
Odonata	Aeshnidae	2.4		5.9	3.0	6.7	
Odonata	Lestidae			3.9		4.5	
Odonata	Libellulidae	3.0		7.8	0.8	10.1	14.3
Odonata	Coenagrionidae	1.8	15.4	3.9	3.8		8.6
Odonata	Gomphidae	2.4	15.4	2.0	7.6	1.1	

Table 3. *Cont.*

Order	Taxa	Cool-Dry			Hot-Wet		
		Albasini	Nandoni	Thathe	Albasini	Nandoni	Thathe
Coleoptera	Dytiscidae	4.8	19.2	7.8		5.6	
Coleoptera	Gyrinidae	6.0			0.8		
Coleoptera	Hydroptilidae	3.0					
Hemiptera	Nepidae	3.6		5.9	6.1	10.1	
Hemiptera	Notonectidae			9.8		9.0	
Hemiptera	Gerridae				3.0	7.9	
Hemiptera	Belostomatidae	4.2		2.0			8.6
Hemiptera	Aphelocheiridae			3.9	2.3		
Hemiptera	Corixidae		3.9	3.9			
Ephemeroptera	Baetidae	1.2					17.1
Diptera	Chironomidae				9.1	18.0	34.3
Diversity metrics							
Evenness		0.4	0.9	0.9	0.6	0.9	0.9
Margalef's diversity		2.8	3.2	1.1	2.4	2.5	2.1
Taxa richness		14	6	15	12	11	8
Shannon–Wiener diversity		1.7	2.2	1.0	1.8	2.1	1.6
Simpson's diversity		0.7	0.9	0.6	0.8	0.9	0.8

Using PERMANOVA, macroinvertebrate community structure was found to differ significantly among reservoirs (PERMANOVA: Pseudo-F = 9.8429, p(MC) < 0.001) and seasons (PERMANOVA: Pseudo-F = 7.07, p(MC) < 0.001). The interaction between reservoirs and seasons was also found to be different (PERMANOVA: Pseudo-F = 2.7995, p(MC) < 0.001). Pairwise comparisons highlighted significant differences in macroinvertebrate community structure between reservoirs, i.e., Albasini vs. Nandoni (p(MC) = 0.0018), Albasini vs. Thathe p(MC) = 0.0023) and Nandoni vs. Thathe (p(MC) = 0.0015).

3.3. Fish Communities

In total, 359 fish individuals were identified within the three reservoirs over the study period, with 111, 176, and 69 individuals being identified in Albasini, Nandoni, and Thathe, respectively. Overall, Mozambique tilapia *Oreochromis mossambicus* (Peters, 1852) was the most dominant taxa accounting for 33.7% of the total individuals, followed by Banded tilapia *Tilapia Sparrmanii* (Smith, 1840), largemouth Bass *Micropterus salmoides* (Lacépède, 1802), Straightfin Barb *Enteromius paludinosus* (Peters, 1852) and sharptooth catfish *Clarias gariepinus* (Burchell, 1822) accounting for 9.7%, 8.9%, 8.6% and 7.5%, respectively (Table 4). During the hot-wet and cool-dry seasons across all reservoirs, *O. mossambicus* was the most dominant taxa accounting for 29.2% and 40.1%, respectively (Table 4).

Table 4. Mean relative abundances (%) of the dominant fish species and diversity metrics observed across two seasons for the study site categories: Albasini, Nandoni and Thathe reservoirs.

Species	Cool-Dry			Hot-Wet		
	Albasini	Nandoni	Thathe	Albasini	Nandoni	Thathe
Clarias gariepinus (Burchell, 1822)			13.3	12.2	11.9	16.1
Micropterus dolomieu (Lacépède, 1802)	7.9			39.0	18.6	
Micropterus salmoides (Lacépède, 1802)	23.7	5.9			6.8	16.1
Oreochromis mossambicus (Peters, 1852)	15.8	52.9	26.7	7.3	23.7	6.5
Coptodon rendalli (Boulenger, 1897)	13.2	5.9	26.7	7.3	5.1	9.7
Tilapia sparrmanii (Smith. 1840)	18.5	14.7		4.9	5.1	6.5
Enteromius afrohamiltoni (Crass, 1960)	10.5				3.4	
Gambusia affinis (Baird and Girard, 1853)			33.3			6.5
Enteromius paludinosus (Peters, 1852)		14.7		12.2	13.6	16.1
Enteromius unitaeniatus (Günther, 1866)	13.2	5.9			8.5	6.5

Table 4. *Cont.*

Species	Cool-Dry			Hot-Wet		
	Albasini	Nandoni	Thathe	Albasini	Nandoni	Thathe
Labeo rosae (Steindachner, 1894)					3.4	
Mesobola brevianalis (Boulenger, 1908)				7.3		16.1
Glossogobius guiris (Hamilton, 1822)				4.9		
Labeo cylindricus (Peters, 1852)				4.9		
Diversity metrics						
Evenness	0.8	0.5	0.5	0.7	0.7	0.8
Margalef's diversity	1.9	1.3	1.3	2.1	2.4	2.0
Taxa richness	7	6	3	9	10	9
Shannon–Wiener diversity	1.6	1.0	1.0	1.7	1.9	1.7
Simpson's diversity	0.8	0.5	0.5	0.8	0.8	0.8

Based on the two-way ANOVA model, significant differences were observed across sampling reservoirs for evenness, Simpson's diversity and total abundance ($p < 0.05$; Table 2), while seasonal differences were observed for most metrics, with exception for evenness ($p > 0.05$), whereas, reservoirs and season interaction indicated significant differences for most of the metrics ($p < 0.05$) (Table 2), with exception for taxa richness and total abundance ($p > 0.05$). Using PERMANOVA, no significant differences in fish community structure were observed across sites PERMANOVA: Pseudo-F = 1.168, p(MC) = 1.149), seasons (PERMANOVA: Pseudo-F = 0.616, p(MC) = 0.627) and sites and seasons interaction (PERMANOVA: Pseudo-F = 1.352, p(MC) = 0.239).

3.4. The Influence of Environmental Parameters on Macroinvertebrate Communities

The RDA first and second axes of the selected exploratory variables accounted for 37.99% of the total macroinvertebrate community data variance. Of the 19 physicochemical variables, the macroinvertebrate community structure across the three reservoirs and the two sampling seasons was found to be significantly associated with water temperatures, ORP, TDS, and conductivity (Figure 2a). Water temperature and TDS were positively associated with the first axis, while conductivity was negatively associated with the second axis (Figure 2a). The hot-wet season was clearly separated along the first axis from the cool-dry season. According to their correlation with the axes, the environmental variables determining the gradients in the RDA diagrams were temperature, conductivity, ORP and TDS (Table 5). Only A1 (Albasani reservoir) appears to be clearly associated to Aeshnidae, Nepidae and Gerridae which were positively correlated with high water temperature and high TDS (Figure 2a).

3.5. The Influence of Environmental Parameters on Fish Communities

The RDA first and second axes of the selected exploratory variables accounted for 56.38% of the total fish community data variance (Figure 2b). The RDA identified five variables (i.e., water temperature, ORP, water pH, Mg (in sediments), and conductivity (water) as the most important in structuring the fish community structures (Figure 2b). Water temperature, ORP, and pH were positively associated with the first axis while conductivity was negatively associated with the second axis (Figure 2a), while water temperature is positively associated with the first axis, and conductivity is negatively associated with the second axis (Figure 2b). The hot-wet season was separated along the second axis from the cool-dry season. According to their correlation with the axes, the environmental variables determining the gradients in the RDA diagrams were temperature, pH, ORP and conductivity (Table 5). The fish that were associated with the hot-wet seasons were *E. unitaeniatus*, *E. paludinosus*, and *L. rosae* which were associated with high water temperature and high ORP. While the fish species that were in the cool-dry season were *G. affinis*, *C. rendalli*, and *T. sparrmanii* (Figure 2b).

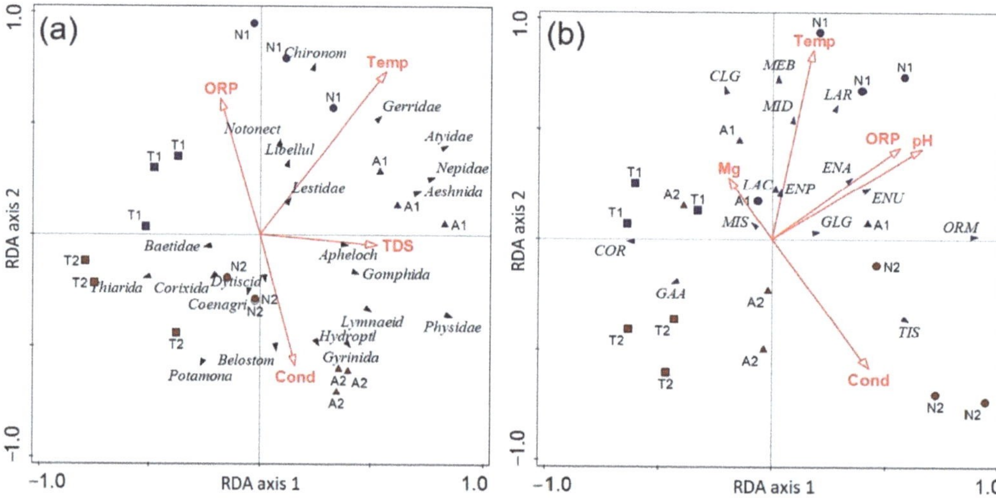

Figure 2. (**a**) Redundancy analysis (RDA) plot highlighting the relationship between measured significant environmental variables with macroinvertebrate communities between the cool-dry and hot-wet seasons across three reservoirs, namely Nandoni, Thathe and Albasini. Abbreviations: Temp: temperature; Cond: conductivity; TDS: total dissolved solids: ORP: oxidation–reduction potential: T: Thathe; A: Albasini; N: Nandoni. The numbers correspond to: 1 = cool-dry; 2 = hot-wet. Abbreviations: Chironom: Chironomidae; Aeshnida: Aeshnidae; Libellul: Libellulidae; Notonect: Notonectidae; Apheloch: Aphelocheiridae; Gomphida: Gomphidae; Lymnaeid: Lymnaeidae; Hydroptl: Hydroptilidae; Gyrinida: Gyrinidae; Thiarida: Thiaridae; Corixida: Corixidae; Dytiscid: Dytiscidae; Coenagri: Coenagrionidae; Belostom: Belostomatidae; Potamon: Potamonautidae. (**b**) Redundancy analysis (RDA) plot highlighting the relationship between measured significant environmental variables with fish communities between the cool-dry and hot-wet seasons across three reservoirs, namely Nandoni, Thathe and Albasini. Abbreviations: Temp: temperature; Cond: conductivity; Mg: magnesium; ORP: oxidation–reduction potential: T: Thathe; A: Albasini; N: Nandoni; CLG: *Clarias gariepinus;* COR: *Coptodon rendalli;* ENA: *Enteromius afrohamiltoni;* ENP: *Enteromius paludinosus;* ENU: *Enteromius unitaeniatus;* GAA: *Gambusia affinis;* GLG: *Glossogobius guiris;* LAC: *Labeo cylindricus;* LAR: *Labeo rosae;* MEB: *Mesobola brevianalis;* MID: *Micropterus dolomieu;* MIS: *Micropterus salmoides;* ORM: *Oreochromis mossambicus;* TIS: *Tilapia sparrmanii.*

Table 5. Correlation coefficients between environmental variables and the first two axes of RDA ordination.

Figure 2a	**Axis 1**	**Axis 2**
Temperature	0.72	0.68
Conductivity	0.13	−0.99
ORP	−0.73	0.68
TDS	0.57	−0.82
Figure 2b		
Temperature	0.26	0.96
pH	0.52	0.85
ORP	0.51	0.77
Mg	−0.94	0.31
Conductivity	0.82	−0.56

4. Discussion

Understanding the dynamics of water quality, sediment chemistry and macroinvertebrate diversity indices is crucial for monitoring and assessing the health of aquatic ecosystems. Benthic macroinvertebrates and various fish species are some of the effective tools for evaluating the ecological state of aquatic environments [43,44]. Macroinvertebrates (21 taxa) and fish species (14 taxa) recorded in the three reservoirs over a period of two seasons demonstrated a relatively similar taxa richness when compared to other large reservoirs studies in other regions such as 32 macroinvertebrates taxa found in small reservoirs in the district [8]. Results of the present study show that macroinvertebrate and fish community diversity tended to change between seasons and associated changes in the physicochemical water quality variables. Littoral macroinvertebrates and fish were mainly driven by water quality variables, with only sediment Mg showing a significant association with fish assemblages. These results are similar to that of a study by Kolpin et al. [45] and Mofu et al. [46], which indicated that water quality variables can strongly structure fish assemblages.

From the present study, the Hemiptera order was dominant, followed by Odonata and Coleoptera, and these are among the most diverse aquatic insect orders that characterize almost all environmental conditions (from extremely pristine to degraded ecological conditions) [47]. One of the most significant and prevalent families of aquatic insects is the family Chironomidae, which belongs to the Diptera order. Its presence in any environment can serve as an indication of contamination [48]. Hence, in this study, high to low diversity of species within studied reservoirs (reservoirs situated in high catchment area to low catchment area) suggested that habitat conditions deteriorate from upstream to downstream. In this study, macroinvertebrate abundance decreases with season (cool-dry), which is not significant compared to those observed by Trottier et al. [49]. Therefore, there is still a lot of unanswered questions regarding the ecological processes involved and the direct and indirect causes of these ecosystem changes.

Throughout the study sites in all three reservoirs, we observed a high concentration in some water and sediment variables, suggesting that there may be various land uses activities within the catchment providing contaminants to these aquatic systems. Through field observation, most land adjacent to Albasini dam is dominated by commercial agriculture, which may be the primary contributors of nutrients to this reservoir. Furthermore, with an increase in land use activities adjacent to this reservoir, we predicted the decline in water and sediment quality at Albasini and Thathe in the near future. Looking at the Nandoni reservoir, water and sediment quality decline is associated with the various activities such as sewage treatment works, landfill site, recreation, fishing, washing in the reservoir, bathing, and ritual activities. The decline in water and sediment quality in aquatic environments can lead to a decline in absolute abundance, species richness and diversity, which can later result in declines in aquatic biodiversity [50]. Similarly, results of the present study agree with others that have demonstrated human activities changes to potentially change environmental variables in aquatic systems [51–53]. Therefore, understanding human activities in the watershed is crucial for the sustainable management of the aquatic environment.

The RDA suggests that some macroinvertebrate taxa such as Chironomidae, Gerridae, Dytiscidae, Gomphidae, Belostomatidae and Libellulidae were mostly affected by water variables (i.e., water temperature, ORP, TDS, and conductivity), with no sediment variable inducing any significant changes. Similarly, results were observed in other studies, where physicochemical parameters such as pH, conductivity and temperature drive macroinvertebrates communities [11,54,55]. Furthermore, fish communities within the three reservoirs in the study sites and across the two seasons were driven by water temperature, ORP, pH, and conductivity with an addition of sediment metal magnesium. According to Sylvain et al. [56], metals such as magnesium have been documented to drive microbiota and fish compositional shifts along widespread hydrochemical gradients. The high magnesium concentration in sediment was recorded in the littoral zones where fish sampling was performed, suggesting a direct influence of Mg on fish structure. However, we observed some resistant fish species (*M. salmoides*, *O. mossambicus*, *C. rendalli* and *T. sparrmanii*) which were found in littoral zones of the Nandoni reservoir, which recorded a high concentration of sediment metals and physicochemical variables. Furthermore, significant changes in fish and macroinvertebrates diversity indices were observed, whereby species richness and Shannon wiener diversity decreased with water quality concentration across sites.

According to Zhang et al. [57], reservoirs with a substantial amount of anthropogenic activity input are likely to have higher sediment metal concentrations, which has major negative impacts on aquatic biota. Similarly, Tamiru [58] highlighted that in the lake Tana (Northwestern Ethiopia), changes in water chemistry had a considerable impact on aquatic biodiversity, leading to significant changes in biomass, productivity, and littoral biogeochemistry. Therefore, since species reacted to changes in environmental factors, these findings may suggest that fish species are the primary determinants of biotic community structure within reservoirs with exception to few variables. The results of the present study agree with the hypothesis that fish and macroinvertebrates community diversity will be high during the hot-wet seasons since most of these reservoirs increase water levels during the hot-wet season, which provides a conducive environment for fish and macroinvertebrates to thrive.

5. Conclusions

The present study indicates that the abundances and diversity of macroinvertebrates and fish taxa are driven by seasonal patterns in the littoral zones of subtropical reservoirs. In addition, selected water and sediment chemistry variables were able to drive the species assemblages within the reservoirs. Among the local environmental factors, sediment metal has little impact on the species communities as compared to water parameters, allowing only a narrow spectrum of sediment and water chemistry to drive the structure of macroinvertebrates and fish. Seasonal changes were shown to contribute highly to the communities based on diversity indices, which significantly changes across seasons. These findings suggest that management strategies aimed at maintaining high levels of aquatic biodiversity in reservoirs should be directed towards reducing the level of anthropogenic pressure (i.e., contaminants from the catchment, including those in water and benthic sediments), while ensuring the continuous assessment of aquatic fauna to determine the ecological status.

Author Contributions: L.F.M., conceptualization, investigation, data curation, formal analysis, supervision, resources, writing—original draft, and review and editing. T.L., conceptualization, investigation, data curation, and writing and editing. T.M., conceptualization, investigation, writing—original draft, and review and editing. L.M., conceptualization, investigation, formal analysis, data curation, and review and editing. M.I.M., writing—original draft and review and editing. F.M.M., conceptualization, supervision, and review and editing. All authors have read and agreed to the published version of the manuscript.

Funding: We are grateful for the financial support of the National Research Foundation (UID: 129098) received by Linton Munyai. We are also grateful for the financial support of the University of Venda Niche Grant (FSEA/21/GGES/02).

Data Availability Statement: All the data collected during this research are presented in the manuscript.

Conflicts of Interest: All co-authors have seen and agree with the contents of the manuscript and there is no financial interest to report. Thus, all authors have declared that no competing interest exist.

References

1. Kushawaha, J.; Borra, S.; Kushawaha, A.K.; Singh, G.; Singh, P. Climate change and its impact on natural resources. In *Water Conservation in the Era of Global Climate Change*; Elsevier Science: Amsterdam, The Netherlands, 2021; pp. 333–346.
2. Dotaniya, M.L.; Meena, V.D.; Saha, J.K.; Dotaniya, C.K.; Mahmoud, A.E.D.; Meena, B.L.; Meena, M.D.; Sanwal, R.C.; Meena, R.S.; Doutaniya, R.K.; et al. Reuse of poor-quality water for sustainable crop production in the changing scenario of climate. *Environ. Dev. Sustain.* **2023**, *25*, 7345–7376. [CrossRef] [PubMed]
3. Wassie, S.B. Natural resource degradation tendencies in Ethiopia: A review. *Environ. Syst. Res.* **2020**, *9*, 1–29. [CrossRef]
4. Bănăduc, D.; Simić, V.; Cianfaglione, K.; Barinova, S.; Afanasyev, S.; Öktener, A.; McCall, G.; Simić, S.; Curtean-Bănăduc, A. Freshwater as a sustainable resource and generator of secondary resources in the 21st century: Stressors, threats, risks, management and protection strategies, and conservation approaches. *Int. J. Environ. Res. Public Health* **2022**, *19*, 16570. [CrossRef] [PubMed]

5. Datta, S.; Sinha, D.; Chaudhary, V.; Kar, S.; Singh, A. Water pollution of wetlands: A global threat to inland, wetland, and aquatic phytodiversity. In *Handbook of Research on Monitoring and Evaluating the Ecological Health of Wetlands*; IGI Global: Hershey, PA, USA, 2022; pp. 27–50.
6. Jamion, N.A.; Lee, K.E.; Mokhtar, M.; Goh, T.L.; Simon, N.; Goh, C.T.; Bhat, I.U.H. The integration of nature values and services in the nature-based solution assessment framework of constructed wetlands for carbon–water nexus in carbon sequestration and water security. *Environ. Geochem. Health* **2023**, *45*, 1201–1230. [CrossRef] [PubMed]
7. Parris, K. Impact of agriculture on water pollution in OECD countries: Recent trends and future prospects. *Int. J. Water Resour. Dev.* **2011**, *27*, 33–52. [CrossRef]
8. Dalu, T.; Dlamini, P.; Wasserman, R.J.; Mokgoebo, M.J.; Mutshekwa, T.; Dondofema, F.; Cuthbert, R.N. Effects of environmental variables on littoral macroinvertebrate community assemblages in subtropical reservoirs. *Chem. Ecol.* **2021**, *37*, 419–436. [CrossRef]
9. Hojjati-Najafabadi, A.; Mansoorianfar, M.; Liang, T.; Shahin, K.; Karimi-Maleh, H. A review on magnetic sensors for monitoring of hazardous pollutants in water resources. *Sci. Total Environ.* **2022**, *824*, 153844. [CrossRef] [PubMed]
10. Dalu, T.; Wasserman, R.J.; Tonkin, J.D.; Mwedzi, T.; Magoro, M.L.; Weyl, O.L.F. Water or sediment? Partitioning the role of water column and sediment chemistry as drivers of macroinvertebrate communities in an austral South African stream. *Sci. Total Environ.* **2017**, *607*, 317–325. [CrossRef]
11. Dalu, T.; Chauke, R. Assessing macroinvertebrate communities in relation to environmental variables: The case of Sambandou wetlands, Vhembe Biosphere Reserve. *Appl. Water Sci.* **2020**, *10*, 1–11. [CrossRef]
12. Massawe, P.I.; Mvena, A.; Nyoki, D.; Chambile, E.L. Effects of anthropogenic activities on availability of clean and safe water: A case of Uluguru forest catchment areas of Morogoro, Tanzania. *South. Asian J. Dev. Res.* **2019**, *1*, 114–123.
13. Polasky, S.; Nelson, E.; Pennington, D.; Johnson, K.A. The impact of land-use change on ecosystem services, biodiversity and returns to landowners: A case study in the state of Minnesota. *Environ. Resour. Econ.* **2011**, *48*, 219–242. [CrossRef]
14. Ayivor, J.S.; Gordon, C. Impact of land use on river systems in Ghana. *West. Afr. J. Appl. Ecol.* **2012**, *20*, 83–95.
15. Hughes, R.M.; Zeigler, M.; Stringer, S.; Linam, G.W.; Flotemersch, J.; Jessup, B.; Joseph, S.; Jacobi, G.; Guevara, L.; Cook, R.; et al. Biological assessment of western USA sandy bottom rivers based on modeling historical and current fish and macroinvertebrate data. *River Res. Appl.* **2022**, *38*, 639–656. [CrossRef] [PubMed]
16. Feio, M.J.; Hughes, R.M.; Serra, S.R.; Nichols, S.J.; Kefford, B.J.; Lintermans, M.; Robinson, W.; Odume, O.N.; Callisto, M.; Macedo, D.R.; et al. Fish and macroinvertebrate assemblages reveal extensive degradation of the world's rivers. *Glob. Change Biol.* **2023**, *29*, 355–374. [CrossRef] [PubMed]
17. Aziz, M.S.B.; Hasan, N.A.; Mondol, M.M.R.; Alam, M.M.; Haque, M.M. Decline in fish species diversity due to climatic and anthropogenic factors in Hakaluki Haor, an ecologically critical wetland in northeast Bangladesh. *Heliyon* **2021**, *7*, e05861. [CrossRef]
18. Sudarso, J.; Suryono, T.; Yoga, G.P.; Samir, O.; Imroatushshoolikhah Ibrahim, A. The impact of anthropogenic activities on benthic macroinvertebrates community in the Ranggeh River. *J. Ecol. Eng.* **2021**, *22*, 179–190. [CrossRef]
19. Gunkel, G.; Michels, U.; Scheideler, M. Water lice and other macroinvertebrates in drinking water pipes: Diversity, abundance and health risk. *Water* **2021**, *13*, 276. [CrossRef]
20. Jacks, F.; Milošević, D.; Watson, V.; Beazley, K.F.; Medeiros, A.S. Bioassessment of the ecological integrity of freshwater ecosystems using aquatic macroinvertebrates: The case of Sable Island National Park Reserve, Canada. *Environ. Monit. Assess.* **2021**, *193*, 1–16. [CrossRef]
21. Rasifudi, L.; Addo-Bediako, A.; Bal, K.; Swemmer, T.M. Distribution of benthic macroinvertebrates in the Selati River of the Olifants River system, South Africa. *Afr. Entomol.* **2018**, *26*, 398–406. [CrossRef]
22. Passy, S.I.; Bode, R.W.; Carlson, D.M.; Novak, M.A. Comparative environmental assessment in the studies of benthic diatom, macroinvertebrate, and fish communities. *Int. Rev. Hydrobiol.* **2004**, *89*, 121–138. [CrossRef]
23. Dalu, T.; Clegg, B.; Nhiwatiwa, T. Macroinvertebrate communities associated with littoral zone habitats and the influence of environmental factors in Malilangwe Reservoir, Zimbabwe. *Knowl. Manag. Aquat. Ecosyst.* **2012**, *406*, 6. [CrossRef]
24. Ghaleno, O.R. Potential ecological risk assessment of heavy metals in sediments of water reservoir case study: Chah Nimeh of Sistan. *Proc. Int. Acad. Ecol. Environ. Sci.* **2015**, *5*, 89.
25. Moura, D.S.; Neto, I.E.L.; Clemente, A.; Oliveira, S.; Pestana, C.J.; de Melo, M.A.; Capelo-Neto, J. Modeling phosphorus exchange between bottom sediment and water in tropical semiarid reservoirs. *Chemosphere* **2020**, *246*, 125686. [CrossRef] [PubMed]
26. Dos Santos, N.C.L.; Garcia-Berthou, E.; Dias, J.D.; Lopes, T.M.; Affonso, I.D.P.; Severi, W.; Gomes, L.C.; Agostinho, A.A. Cumulative ecological effects of a Neotropical reservoir cascade across multiple assemblages. *Hydrobiologia* **2018**, *819*, 77–91. [CrossRef]
27. Liu, S.; Xie, G.; Wang, L.; Cottenie, K.; Liu, D.; Wang, B. Different roles of environmental variables and spatial factors in structuring stream benthic diatom and macroinvertebrate in Yangtze River Delta, China. *Ecol. Indic.* **2016**, *61*, 602–611. [CrossRef]
28. Glibert, P.M.; Fullerton, D.; Burkholder, J.M.; Cornwell, J.C.; Kana, T.M. Ecological stoichiometry, biogeochemical cycling, invasive species, and aquatic food webs: San Francisco Estuary and comparative systems. *Rev. Fish. Sci.* **2011**, *19*, 358–417. [CrossRef]
29. Rettig, J.E.; Smith, G.R. Relative strength of top-down effects of an invasive fish and bottom-up effects of nutrient addition in a simple aquatic food web. *Environ. Sci. Pollut. Res.* **2021**, *28*, 5845–5853. [CrossRef]
30. Kang, M.; Tian, Y.; Peng, S.; Wang, M. Effect of dissolved oxygen and nutrient levels on heavy metal contents and fractions in river surface sediments. *Sci. Total Environ.* **2019**, *648*, 861–870. [CrossRef]

31. Singh, J.; Kalamdhad, A.S. Effects of heavy metals on soil, plants, human health and aquatic life. *Int. J. Res. Chem. Environ.* **2011**, *1*, 15–21.
32. Barnhoorn, I.E.; Van Dyk, J.C.; Pieterse, G.M.; Bornman, M.S. Intersex in feral indigenous freshwater Oreochromis mossambicus, from various parts in the Luvuvhu River, Limpopo Province, South Africa. *Ecotoxicol. Environ. Saf.* **2010**, *73*, 1537–1542. [CrossRef]
33. Odiyo, J.O.; Makungo, R.; Nkuna, T.R. Long-term changes and variability in rainfall and streamflow in Luvuvhu River Catchment, South Africa. *South. Afr. J. Sci.* **2015**, *111*, 1–9. [CrossRef] [PubMed]
34. Ramulifho, P.; Ndou, E.; Thifhulufhelwi, R.; Dalu, T. Challenges to implementing an environmental flow regime in the Luvuvhu River Catchment, South Africa. *Int. J. Environ. Res. Public Health* **2019**, *16*, 3694. [CrossRef] [PubMed]
35. Patience, M.T.; Elumalai, V.; Rajmohan, N.; Li, P. Occurrence and distribution of nutrients and trace metals in groundwater in an intensively irrigated region, Luvuvhu catchment, South Africa. *Environ. Earth Sci.* **2021**, *80*, 1–15. [CrossRef]
36. Bray, R.H.; Kurtz, L.T. Determination of total, organic, and available forms of phosphorus in soils. *Soil Sci.* **1945**, *59*, 39–46. [CrossRef]
37. Allen, K.; Corre, M.D.; Tjoa, A.; Veldkamp, E. Soil nitrogen-cycling responses to conversion of lowland forests to oil palm and rubber plantations in Sumatra, Indonesia. *PLoS ONE* **2015**, *10*, e0133325. [CrossRef]
38. Dalu, T.; Cuthbert, R.N.; Taylor, J.C.; Magoro, M.L.; Weyl, O.L.; Froneman, P.W.; Wasserman, R.J. Benthic diatom-based indices and isotopic biomonitoring of nitrogen pollution in a warm temperate Austral river system. *Sci. Total Environ.* **2020**, *748*, 142452. [CrossRef]
39. Dalu, T.; Murudi, T.; Dondofema, F.; Wasserman, R.J.; Chari, L.D.; Murungweni, F.M.; Cuthbert, R.N. Balloon milkweed Gomphocarpus physocarpus distribution and drivers in an internationally protected wetland. *BioInvasions Rec.* **2020**, *9*, 627–641. [CrossRef]
40. Fry, C. *Field Guide to the Freshwater Macroinvertebrates of South Africa*; Jacana Media: Auckland Park, South Africa, 2021.
41. Anderson, M.J.; Gorley, R.N.; Clarke, K.R. *PERMANOVA+ for PRIMER: Guide to Software and Statistical Methods*; PRIMER-E Ltd.: Plymouth, UK, 2008.
42. Ter Braak, C.J.F.; Šmilauer, P. *CANOCO Reference Manual and Canodraw for Windows User's Guide: Software for Canonical Community Ordination (Version 5)*; Cacono: Ithaca, NY, USA, 2012.
43. Purcell, A.H.; Bressler, D.W.; Paul, M.J.; Barbour, M.T.; Rankin, E.T.; Carter, J.L.; Resh, V.H. Assessment tools for urban catchments: Developing biological indicators based on benthic macroinvertebrates 1. *J. Am. Water Resour. Assoc.* **2009**, *45*, 306–319. [CrossRef]
44. Halliwell, D.B.; Langdon, R.W.; Daniels, R.A.; Kurtenbach, J.P.; Jacobson, R.A. Classification of freshwater fish species of the northeastern United States for use in the development of indices of biological integrity, with regional applications. In *Assessing the Sustainability and Biological Integrity of Water Resources Using Fish Communities*; CRC Press: Boca Raton, FL, USA, 2020; pp. 301–337.
45. Kolpin, D.W.; Blazer, V.S.; Gray, J.L.; Focazio, M.J.; Young, J.A.; Alvarez, D.A.; Iwanowicz, L.R.; Foreman, W.T.; Furlong, E.T.; Speiran, G.K.; et al. Chemical contaminants in water and sediment near fish nesting sites in the Potomac River basin: Determining potential exposures to smallmouth bass (*Micropterus dolomieu*). *Sci. Total Environ.* **2013**, *443*, 700–716. [CrossRef]
46. Mofu, L.; Dalu, T.; Wasserman, R.J.; Woodford, D.J.; Khosa, D.; Weyl, O.L.F. Seasonal variation and drivers of zooplankton, macroinvertebrate and littoral fish communities from irrigation ponds in a semi-arid region in the Eastern Cape (South Africa). *Afr. J. Aquat. Sci.* **2021**, *46*, 452–463. [CrossRef]
47. Arkia, S.; Siahkalroudi, S.Y.; Kheradpir, N. Chironomidae (Insecta: Diptera) biodiversity at generic level in Lar river, Tehran province with introducing two new genera for Iranian fauna. *J. Wildl. Biodivers.* **2019**, *3*, 31–39.
48. Al-Shami, S.A.; Rawi, C.S.M.; HassanAhmad, A.; Nor, S.A.M. Distribution of Chironomidae (Insecta: Diptera) in polluted rivers of the Juru River Basin, Penang, Malaysia. *J. Environ. Sci.* **2010**, *22*, 1718–1727. [CrossRef] [PubMed]
49. Trottier, G.; Embke, H.; Turgeon, K.; Solomon, C.; Nozais, C.; Gregory-Eaves, I. Macroinvertebrate abundance is lower in temperate reservoirs with higher winter drawdown. *Hydrobiologia* **2019**, *834*, 199–211. [CrossRef]
50. Tickner, D.; Opperman, J.J.; Abell, R.; Acreman, M.; Arthington, A.H.; Bunn, S.E.; Cooke, S.J.; Dalton, J.; Darwall, W.; Edwards, G.; et al. Bending the curve of global freshwater biodiversity loss: An emergency recovery plan. *BioScience* **2020**, *70*, 330–342. [CrossRef] [PubMed]
51. Sanganyado, E.; Gwenzi, W. Antibiotic resistance in drinking water systems: Occurrence, removal, and human health risks. *Sci. Total Environ.* **2019**, *669*, 785–797. [CrossRef] [PubMed]
52. Häder, D.P.; Banaszak, A.T.; Villafañe, V.E.; Narvarte, M.A.; González, R.A.; Helbling, E.W. Anthropogenic pollution of aquatic ecosystems: Emerging problems with global implications. *Sci. Total Environ.* **2020**, *713*, 136586. [CrossRef]
53. Huang, W.; Song, B.; Liang, J.; Niu, Q.; Zeng, G.; Shen, M.; Deng, J.; Luo, Y.; Wen, X.; Zhang, Y. Microplastics and associated contaminants in the aquatic environment: A review on their ecotoxicological effects, trophic transfer, and potential impacts to human health. *J. Hazard. Mater.* **2021**, *405*, 124187. [CrossRef]
54. Odume, O.N.; Muller, W.J.; Arimoro, F.O.; Palmer, C.G. The impact of water quality deterioration on macroinvertebrate communities in the Swartkops River, South Africa: A multimetric approach. *Afr. J. Aquat. Sci.* **2012**, *37*, 191–200. [CrossRef]
55. Gething, K.J. Physicochemical drivers of managed river and agricultural drainage channel macroinvertebrate communities. *River Res. Appl.* **2021**, *37*, 675–680. [CrossRef]
56. Sylvain, F.É.; Holland, A.; Audet-Gilbert, É.; Luis Val, A.; Derome, N. Amazon fish bacterial communities show structural convergence along widespread hydrochemical gradients. *Mol. Ecol.* **2019**, *28*, 3612–3626. [CrossRef]

57. Zhang, J.; Li, X.; Guo, L.; Deng, Z.; Wang, D.; Liu, L. Assessment of heavy metal pollution and water quality characteristics of the reservoir control reaches in the middle Han River, China. *Sci. Total Environ.* **2021**, *799*, 149472. [CrossRef] [PubMed]
58. Tamiru, S.M. Impacts of human activities on selected physico-chemical parameters and macroinvertebretes of Lake Tana, Northwestern Ethiopia. *Ethiop. Renaiss. J. Social. Sci. Humanit.* **2021**, *8*, 146–159.

water

MDPI

Article

Anomalies of Sponge Spicules: Exploring Links to Environmental Pollution

Stefan Andjus [1,*], Bojana Tubić [1], Božica Vasiljević [1], Vera Nikolić [2] and Momir Paunović [1]

[1] Institute for Biological Research "Siniša Stanković"—National Institute of the Republic of Serbia, University of Belgrade, 11108 Belgrade, Serbia; bojana@ibiss.bg.ac.rs (B.T.); bozica@ibiss.bg.ac.rs (B.V.); mpaunovi@ibiss.bg.ac.rs (M.P.)
[2] Faculty of Biology, University of Belgrade, 11158 Belgrade, Serbia; vera@bio.bg.ac.rs
* Correspondence: stefan.andjus@ibiss.bg.ac.rs

Abstract: The aim of this study was to assess the frequency of spicule malformations in freshwater sponges in relation to selected environmental parameters of the streams and the presence of river pollutants. A total of 50 sponge samples were collected from ten rivers in Serbia. Selected parameters of the water varied considerably at every site where sponges were found. After spicule preparation, the samples were subjected to morphological analysis by light and scanning electron microscopy, and the number of anomalies were recorded (spicules with bulbous enlargements, sharply bent, bifurcated, scissor- and cross-like, and t-shaped). The frequencies and types of malformations within the analyzed specimens varied from 1 to 100 per 1000 spicules, with an average number of 12 per 1000. The main types of anomalies were single- and double-bent spicules. The highest number of anomalies was found in a specimen of *Eunapius fragilis* collected at Markovac (Velika Morava River), and the lowest number was found in a specimen of *Ephydatia fluviatilis* from Kanjiža (Tisa River). The sites with the lowest and the highest numbers of anomalies showed statistically significant differences in concentrations of ammonia, orthophosphates, sodium, chloride, manganese, and lead. This study indicates that several pollutants potentially affect the occurrence of spicule anomalies.

Keywords: Spongillidae; sponges; freshwater; spicule; anomalies; malformations; pollution; heavy metals; environmental parameters

Citation: Andjus, S.; Tubić, B.; Vasiljević, B.; Nikolić, V.; Paunović, M. Anomalies of Sponge Spicules: Exploring Links to Environmental Pollution. *Water* **2024**, *16*, 332. https://doi.org/10.3390/w16020332

Academic Editors: Eva Papastergiadou and Kostas Stefanidis

Received: 4 December 2023
Revised: 31 December 2023
Accepted: 1 January 2024
Published: 19 January 2024

1. Introduction

Freshwater sponges are a relatively neglected group of macroinvertebrates, and yet, they are omnipresent. Sponges are predominantly marine, with about 240 species, grouped in 48 genera and 6 families, and inhabit freshwater ecosystems all over the world [1]. Freshwater sponges (Phylum Porifera, family Spongillidae) are sessile organisms attached to hard substrates of all kinds, and similarly to their marine relatives, they produce many bioactive compounds. Freshwater sponges are widespread and often a common component of many ecosystems. They can be found in rivers, streams, springs, rapids, estuaries, lakes, caves, ponds, anthropogenic reservoirs, pools, etc. [1]. They are mainly attached to a solid substrate, which includes stones, rocks, submerged branches, roots, and vegetative organs of various aquatic plants. They can often be found on mollusk shells, as well as on live shellfish or snails. They also show great affinity for various types of anthropogenic substrates submerged in water, such as glass, cement, plastic, and metal objects [1]. Artificial bank fortifications are especially suitable for the colonization of sponges. Even so, their distribution is fairly irregular. In some habitats, they are sporadic, with very little share in the total biomass, while in others, they are very abundant [2–4].

Sponges spend their full vegetative life cycle in one place, exposed to the conditions of the water body they inhabit. Moreover, they are filter feeders with substantial filtering potential (often amounting to hundreds of liters per hour) and consequently, with a role in processes of water purification. This makes them a vital element of aquatic ecosystems and a potentially suitable water quality bioindicator. Freshwater sponges show a propensity towards calmer and cleaner waters, in which there is less possibility of silting up of the pores and obstruction of breathing and nutrition. However, they are often found in polluted waters as well, for example near industrial plant discharges or waste water collectors, where there is a higher level of organic input [5].

Freshwater sponges are frequently found in symbiosis with green algae, which affect host gemmule germination rates, enhance its growth rate due to net gain from photosynthesis, and provide fixed carbon for metabolism [6]. Due to their low nutritional value and high mineral content, sponges do not have many predators. However, some organisms do feed on them. Rare sponge predators include ducks [7], crayfish [8], some aquatic insects, and snails [9,10].

Aquatic ecosystems of southeast Europe have been poorly investigated in terms of freshwater sponge presence and their precise distribution. Though they represent an important component in many aquatic ecosystems, they are often overlooked, as they are rarely collected during standard sampling procedures. Previous investigations of Serbian waters revealed mainly *Spongilla lacustris* (Linnaeus, 1759) species, sporadically found in the benthic macroinvertebrate communities of large rivers (Globaqua, JDS3 report). Recent studies on freshwater sponges of the western Balkans and some parts of the Pannonian Plain by Andjus et al. [11,12] confirmed the presence of five species in different water bodies.

In recent years, data regarding Spongillidae morphology, distribution, phylogeny, etc. are increasing. Spicules, as structural elements of the skeleton of most sponges, have been the subject of numerous research efforts. Interestingly, one aspect of freshwater sponge biology remains rather uninvestigated. Namely, different studies have mentioned the existence of spicule malformations, yet very few have elaborated on this topic. Earlier studies have shown that environmental factors have a fundamental impact on Spongillidae spicule size and shape [13,14]. Among the most important factors in the synthesis of these structural elements are silica concentration, water temperature, and depth [15–19]. The importance of these parameters has been assessed both in field and in laboratory experiments. Lowering concentrations of silica in experimental conditions inhibits spicule synthesis. Also, most continental freshwater sponges prefer summer temperatures when they grow, while they are receding, and in dormant gemmule form during cold periods [20]. The toxicity of copper and zinc rises with water "softness" (lower mineral content), contributing to altered spiculogenesis as well [21]. For other benthic organisms, particularly diatoms, numerous studies have shown a connection between morphological changes and environmental stressors, mostly trace metals and multiple stressors [22,23]. It was noted that some of the genera from this algae group tend to develop deviations from the normal shape or ornamentations of the cell wall, which reflect sub-lethal responses to contaminants [24].

Although molecular genetics offers great possibilities in the sponge identification process, morphological analysis of the sponge mineral skeleton remains an unavoidable tool in Porifera taxonomy. However, it is not rare that ecomorphs are mistakenly registered as a new species due to aberrant spicules, and thus, additional information on sponge skeletal malformations is needed. The aims of the present study were to describe and quantify the most frequently encountered spicule anomalies in different sponge species collected in Serbian rivers and, if possible, to relate their presence to the streams' physico-chemical characteristics and the pollutants' presence.

2. Materials and Methods

2.1. Study Area

For the present study, ten rivers of the Danube River Basin in Serbia were investigated as major representatives of the basin: Velika Morava, Zapadna Morava, Južna Morava, Tisa, Kolubara, Porečka River, Mlava, Beli Timok, Crni Timok, and Nišava. On average, three to five localities in the upper, middle, and lower sections of the rivers were inspected in the period from August–November 2017 (Figure 1).

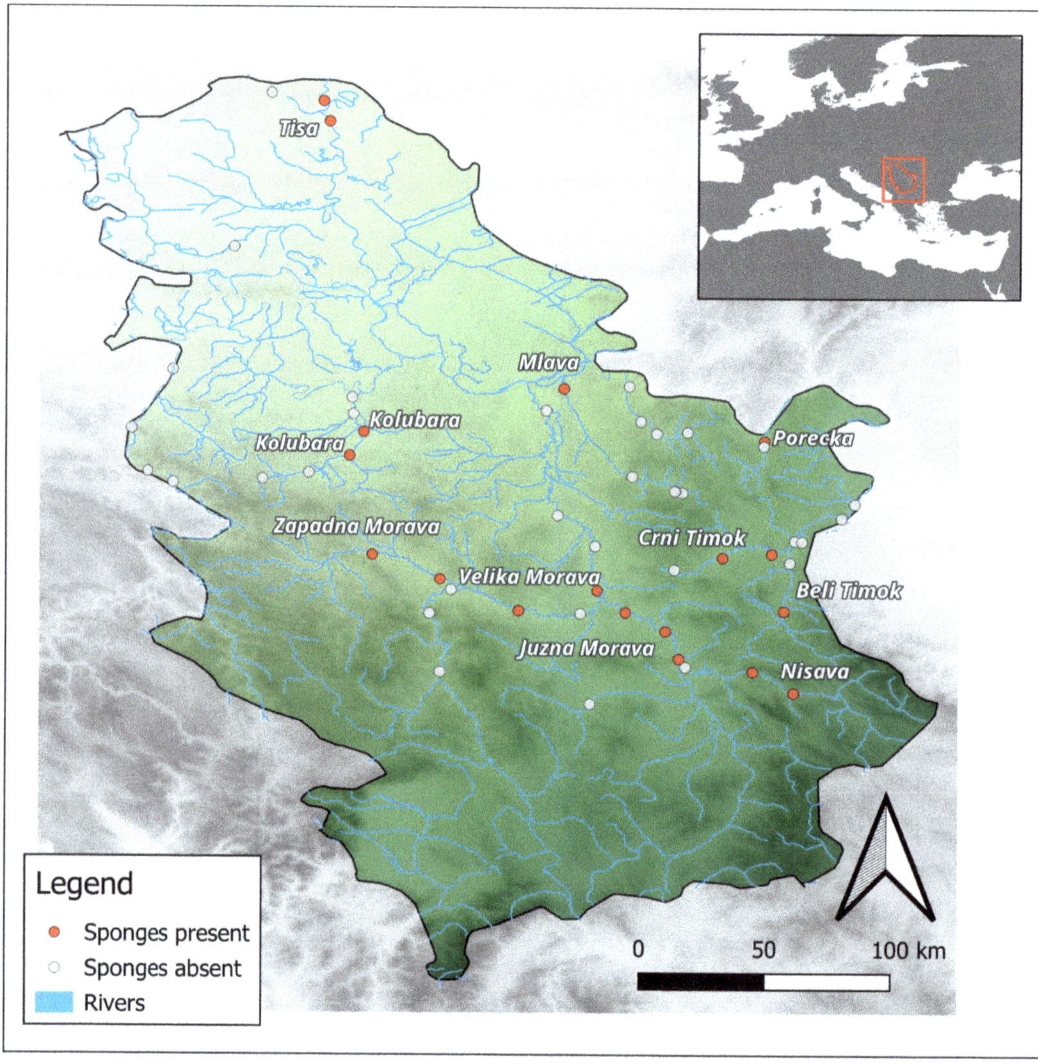

Figure 1. Map of localities searched during the study.

2.2. Sponge Sampling

Wadeable rivers were searched in their entire width, and the larger ones were in the coastal zone. They were examined on selected localities within 100 m stretches of wadeable areas and at depths usually between 0 to 1.5 m. The main sponge substrates were medium-sized rocks, wood debris, and submerged tree roots, but some specimens were also found attached to barrels and concrete underwater constructions, like piers. A total of 50 sponge specimens were collected and kept in 96% ethanol until spicule preparation.

2.3. Analysis of Selected Environmental Parameters of the Rivers

Physico-chemical parameters considered relevant for sponge ecology based on literature data [25] were included in the statistical analysis: water temperature, pH, electrical conductivity, suspended solids, dissolved oxygen, oxygen saturation, total water hardness, dissolved carbon dioxide, bicarbonates, total dissolved salts, silicates, and calcium. Temperature, pH, conductivity, and dissolved oxygen were measured in situ, using the WTW probe model Multi 3630 IDS SET G.

For other selected parameters, such as heavy metal concentrations, nitrates, etc., data were obtained from the Agency for Environmental Protection of Serbia [26]. The Environmental Protection Agency measures a vast number of physical and chemical parameters on a monthly basis, and average values for the period of sponge collection were calculated from the available data.

2.4. Spicule Preparation and Light Microscopy

The nitric acid technique, as described by Manconi and Pronzato [27], was used to dissolve sponge tissue and prepare spicules for light microscope analysis. Sponge fragments of approximately 2–5 mm were washed with ethanol, dried, and put into glass tubes. Then, they were covered with 2–5 mL of concentrated nitric acid and left to decompose for 24 h. The acid was removed with a pipette, and the spicule pellet was washed repeatedly with distilled water. Finally, the spicules were rinsed with and re-suspended in 96% ethanol. A drop of suspension was placed on a cover slip, and after the alcohol dried, the cover slip was placed over the microscope slides with a drop of Canada balsam. Slides were analyzed under magnifications of 20× and 40× on a light microscope (model ZEISS AXIO Lab.A1), and spicule malformations were scored. Spicule sizes were measured as well.

2.5. Quantification and Classification of Spicule Malformations

For the quantification of anomalies, light microscopy was used, and all spicules without apparent malformations and within normal size ranges [28] in five randomly chosen fields of view (FOV) were counted. Then, the average number of spicules per FOV was multiplied by the total number of FOV on the microscopic slide. The anomalies were counted on the entire slide, and the results were expressed in the number of anomalies per thousand spicules (1×10^3). All detected anomalies were sorted into main groups and named. The classification of known types of anomalies was proposed (Section 3.4).

2.6. Scanning Electron Microscopy

To analyze the sponge spicules more closely, specimens were prepared for SEM analysis. Drops of spicule suspension in ethanol were placed on specimen holders and coated with gold in a gold sputter at 18 mA for 1 min. Elements of the mineral skeleton were analyzed and photographed in a VEGA TS 5133MM Scanning Electron Microscope (SEM) in high-vacuum mode using the SE detector with accelerating voltage. Measurements of spicules (widths and lengths) were taken, and a more accurate observation of anomalies was performed.

2.7. Statistical Analysis

Differences in selected parameters between the rivers containing sponges were analyzed. A comparison was made between the measured values of parameters at sites with the highest number of sponge spicule anomalies and values at sites with the lowest number of anomalies. Statistical analyses were performed using IBM SPSS Statistics for Windows Software (Version 22.0; IBM Corp, Armonk, NY, USA). The statistical significance between numerical data was determined by Student's t-test. Assumptions of normality were verified using Kolmogorov–Smirnov and Shapiro–Wilks tests. All p-values less than 0.05 were considered significant.

3. Results

3.1. Sponge Distribution

In the ten investigated rivers, 51 localities were selected and analyzed. Three to five locations in the upper, middle, and lower sections of each river were chosen as sampling sites. Sponges were found at 18 out of the 51 investigated localities (Figure 1), and 50 samples were collected in total. Among all examined samples, five species of sponges were recorded.

The detected species, from the most to the least frequent in the samples, were the following: *Ephydatia fluviatilis* (Linnaeus, 1759) (24 samples), *Ephydatia muelleri* (Lieberkühn, 1856) (8 samples), *Spongilla lacustris* (Linnaeus, 1759) (9 samples), *Trochospongilla horrida* Weltner, 1893 (5 samples), and *Eunapius fragilis* (Leidy, 1851) (4 samples). It must be emphasized that not every species was detected at each of the 18 locations, indicating diverse community compositions. Interestingly, the presence of one of the species, characteristic of this part of Europe, *Heteromeyenia stepanowii* (Dybowski, 1884), was not confirmed.

3.2. Sponge Characteristics

Briefly, the appearance of *E. fluviatilis* was highly variable, from a relatively thin lichen-like layer covering the substrate to massive capped or ridged forms. The color of the body also varied greatly from whitish to brown, and due to symbiosis with algae, green specimens could be found as well (Figure 2).

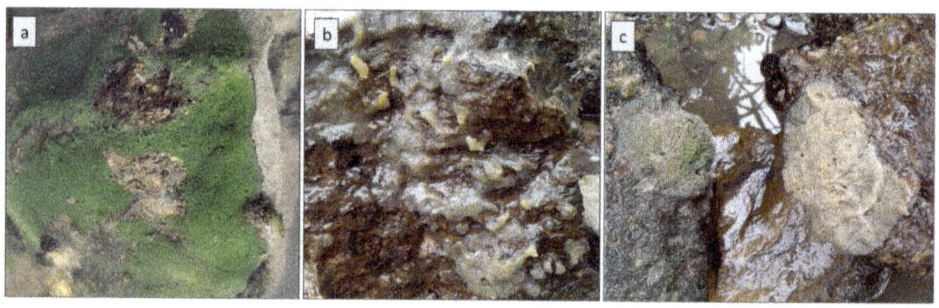

Figure 2. In situ appearances of three of the five sponges encountered on typical rocky substrates: *S. lacustris* (green color indicating symbiosis with algae)—(**a**); *E. fragilis*—(**b**); *E. fluviatilis*—(**c**).

E. muelleri was phenotypically very similar to *E. fluviatilis*. *E. muelleri* was found in various shades of brown, gray, and yellow, but as a result of symbiosis with algae, some green specimens were also seen. The shape of the body was irregular, and the surface was often papillar.

The vegetative body of *S. lacustris* was characterized by cylindrical, finger-like outgrowths that extended and branched from the axis of the body attached to the substrate. Thanks to these outgrowths, *S. lacustris* could be fairly accurately distinguished from other species in situ. Due to the presence of symbiotic algae, the body color was usually green.

E. fragilis was also often in symbiosis with algae, which turned its color green. Otherwise, it was grayish to brown. After growing a couple of centimeters wide, it had a cushion-like form, with visible ostia and sometimes bud-like protuberances.

The body of *T. horrida* was mostly flat and covered the substrate in the form of a thin layer with an irregular border. The color ranged from yellowish to dark brown.

3.3. Spicules Characteristics

Before analyzing spicule anomalies, normal spicules of the five sponge species were examined, and their morphology and size were recorded. The appearance of the spicules was generally in agreement with previous reports. The mineral skeleton of *E. fluviatilis* is characterized by two types of spicules: megascleres and gemmuloscleres. There are no microscleres in this species. The sizes of the megascleres (oxa) are in the 250–380 µm range in length and the 12–19 µm range in width. They are needle-shaped, slightly curved, and rarely straight. The surface of the megascleres may be smooth or covered with micro-thorns (Figure 3B). Gemmuloscleres are in the form of birotules, with a smooth or thorny axis, flat rotules of the same diameter, and micro-thorns and irregularly incised edges, with 15–20 teeth. The dimensions of the gemmuloscleres range from 21 to 25 µm in length, while the shafts are 2–4 µm wide. The diameters of the rotules are 18–23 µm (Figure 3B). *E. muelleri* is characterized by the presence of two variants of megascleres. They are usually densely covered with micro-thorns (except at the ends), but they can be completely smooth. The megascleres measure 170–320 µm in length. This species does not have microscleres. The microskeleton of *S. lacustris* differs from the previously mentioned species in that it has another class of spicules—microscleres. These spicules are located between megascleres and give the sponge extra support. Megascleres are in the shape of smooth amphioxea, 160–350 µm long, and 5–17 µm wide. The microscleres are also of the amphioxea type, slightly to strongly curved, and densely covered with minuscule thorns. The microscleres measure 35–95 µm in length. Gemmuloscleres are slightly to strongly curved and covered with microspines. The gemmuloscleres range from 18 to 70 µm in length and 3 to 10 µm in width. *E. fragilis* is characterized by amphioxa-type megascleres that are smooth, 165–261 µm long, and 10–14 µm wide. This type of sponge does not have microscleres. Gemmuloscleres belong to the amphistrongile type. They have blunt ends and are covered with thorns, mostly concentrated toward the tips (Figure 3A). The length ranges from 50 to 110 µm, and the width ranges from 4 to 9 µm.

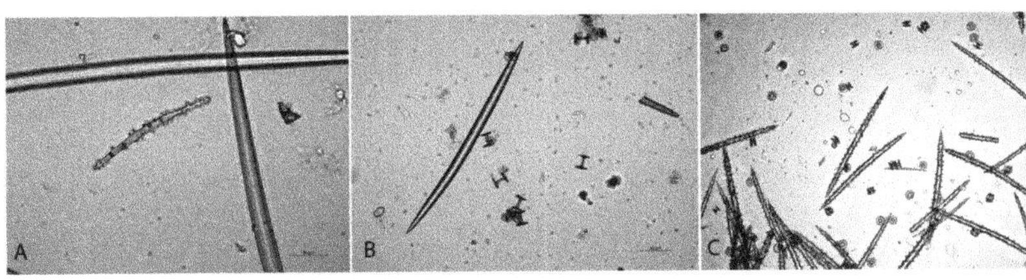

Figure 3. Representative spicules of *E. fragilis* (megascleres detail and gemmulosclere)—(**A**); *E. fluviatilis* (megasclere and gemmuloscleres–birotules)—(**B**); and *T. horrida* (numerous megascleres and birotules)—(**C**).

Megascleres of *T. horrida* species are of the amphioxea type and densely covered with blunt, thorny outgrowths. Dimensions range from 165 to 245 µm. Microscleres do not exist. Gemmuloscleres are in the form of birotules, with a short, smooth axis (the length of the axis is shorter than the diameter of the rotule) and rotules with smooth edges (Figure 3C). One is usually slightly larger in diameter than the other. The ranges of different spicule dimensions are shown in Table 1, while several representative skeletal elements of different Spongillida species are displayed in Figure 3.

Table 1. Spicule types and size ranges for species collected during the study.

Species/Spicule Sizes (µm)	Megasclere		Gemulosclere			Microsclere	
	Length	Width	Length	Width	Rotule Diameter	Length	Width
E. fluviatilis	250–380	12–19	21–25	2–4	18–23	/	/
E. muelleri	170–320	12–21	8–24	2–4	9–25	/	/
S. lacustris	160–350	5–17	18–70	3–10	/	35–95	4–7
E. fragilis	165–261	10–14	50–110	5–9	/	/	/
T. horrida	165–245	8–12	8–10	2–5	13–15	/	/

3.4. Spicule Anomalies

The types and incidences of spicule malformations varied considerably within the analyzed specimens. The frequency of anomalies ranged from 1×10^{-3} to 97×10^{-3}, with an average number of 12×10^{-3}. Among different spicule types, megascleres were the most affected by malformations, followed by microscleres. Gemmulosclere anomalies were seldom detected. The most frequent anomalies were spicules sharply bent at different angles near one end, both ends, or medially (Figure 4A–C). Spicules with split ends (bifurcated) were also common in different species (Figure 4D–F). Scissor-like or cross-like spicules (Figure 4G–I) and T-shaped spicules were recurrent as well (Figure 4J–L).

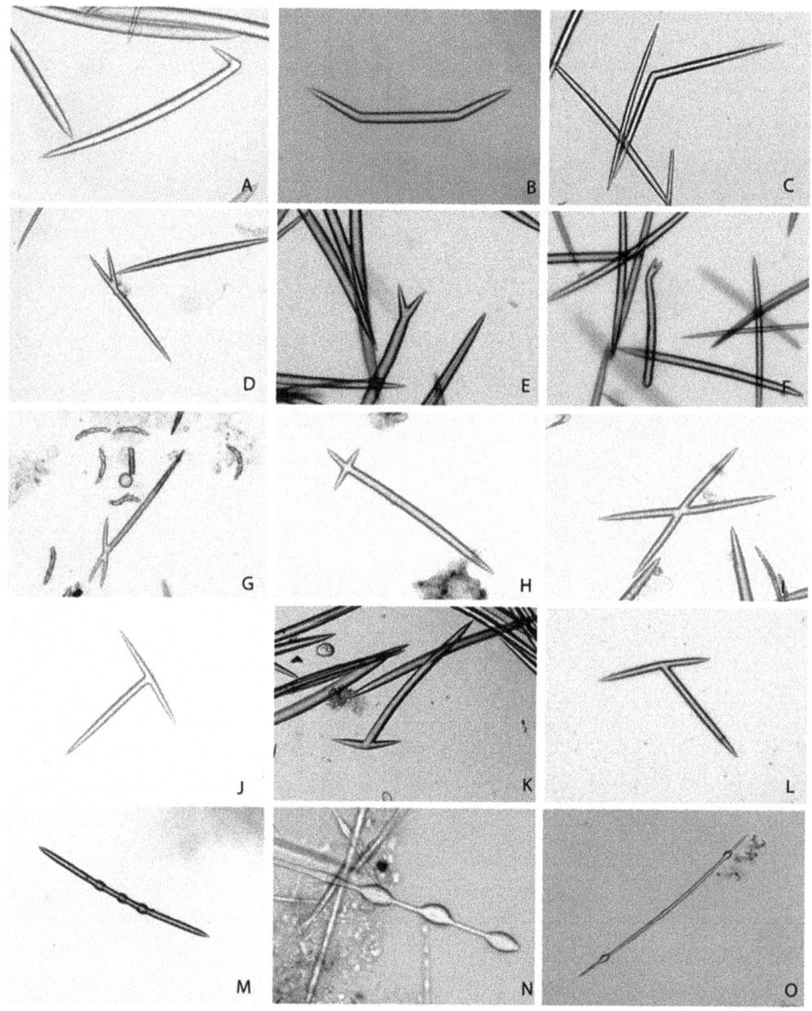

Figure 4. Compilation of encountered forms of spicule anomalies seen in specimens of freshwater sponges (light microscopy): megasclere bent near one end, both ends, and medially (**A–C**); bifurcated (split-end) spicules (**D–F**); scissor-like or cross-like spicule formations (**G–I**); T-shaped spicules (**J–L**); and spicules with bulbous, spherical enlargements (**M–O**).

As an example of a peculiar anomaly—spicules with bulbous, spherical enlargements along the axis were found at high frequency in specimens of *E. fragilis* but could be seen in *E. fluviatilis* and *E. muelleri* as well. Single or multiple bulbous enlargements were recorded on both megascleres and microscleres (Figure 4M–O). These were the most often encountered anomalies. A few anomalies appeared extremely rarely in our samples, such as gemmuloscleres forming loops, polyaxial spicules resembling skeletal structures in marine sponges, and spicules with complex malformation (Figure 5A–C).

Some of the most frequent malformations were observed with SEM as well (Figure 6).

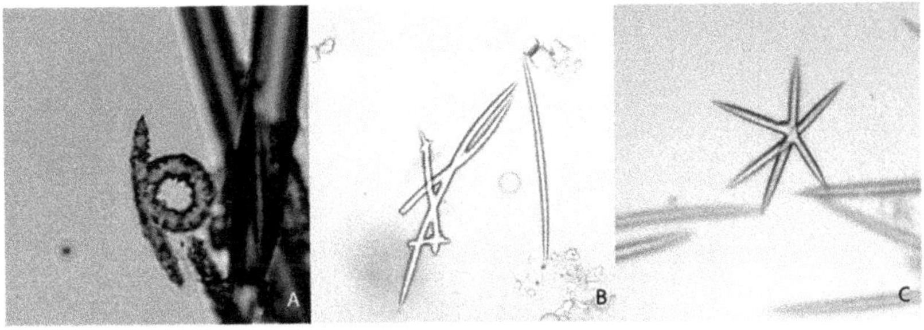

Figure 5. Rare types of spicule anomalies: gemmulosclere of *S. lacustris*, usually straight or slightly curved, forming a circle here (**A**); spicule with multiple anomalies (**B**); polyaxial spicule aberration, resembling spicules from marine species (**C**).

3.5. Spicule Anomalies in Relation to Environmental Parameters and Pollutants

Considerable variations among the ten surveyed rivers were noted regarding the main parameters (oxygen saturation, pH, temperature, and conductivity), levels of metals (iron, copper, led, etc.), and other pollutants, such as nitrates, orthophosphates, sodium, chlorides, etc. However, those levels did not exceed limits defined by legislation [29]. In order to uncover potential factors conditioning the appearance of spicule anomalies, we compared the rivers that harbored sponges with extreme values of anomaly frequency (lowest and highest). The two rivers, Tisa and Velika Morava, showed statistically significant differences in the following factors: pH, ammonia, orthophosphates, sodium, chloride, manganese, and lead (Table 2).

Table 2. Physico-chemical parameters and pollutants on sites with the least detected anomalies (Tisa, Martonoš) versus the most detected spicule anomalies (Velika Morava, Bagrdan); ns—non significant.

Parameter	**Tisa, Martonoš**	**V. Morava Bagrdan**	***t*-Test**	
	Mean ± SD	**Mean ± SD**	**df**	**Sign.**
Watertemperature (°C)	13.6 ± 9.5	14.6 ± 9.1	20	ns
Turbidity (NTU)	23.4 ± 17.6	58.3 ± 10.8	20	ns
pH	8.00 ± 0.13	8.21 ± 0.16	20	0.010
Oxygensaturation (%)	88.6 ± 4.7	95.9 ± 10.0	20	ns
Conductivity (µS/cm)	427.4 ± 71.9	402.9 ± 41.7	20	ns
Ammonia(NH_4-N)	0.067 ± 0.069	0.207 ± 0.228	20	0.048
Nitrates(N-NO_3) mg/L	0.675 ± 0.314	1.200 ± 0.455	20	ns
Orthophosphate(O-PO_4) mg/L	0.045 ± 0.012	0.083 ± 0.031	20	0.006
Sodium(Na) mg/L	32.5 ± 8.4	13.1 ± 4.5	20	0.000
Chloride(Cl) mg/L	38.3 ± 14.5	12.9 ± 3.6	20	0.000
Manganese(Mn) µg/L	61.6 ± 35.9	139.2 ± 73.8	20	0.028
Zink(Zn) µg/L	32.6 ± 25.2	13.6 ± 7.9	20	ns
Copper(Cu) µg/L	5.8 ± 1.0	7.0 ± 3.6	20	ns
Led(Pb) µg/L	2.2 ± 1.1	9.5 ± 8.7	20	0.046
Aluminum(Al) µg/L	599.3 ± 452.6	1086.8 ± 1049.8	20	ns

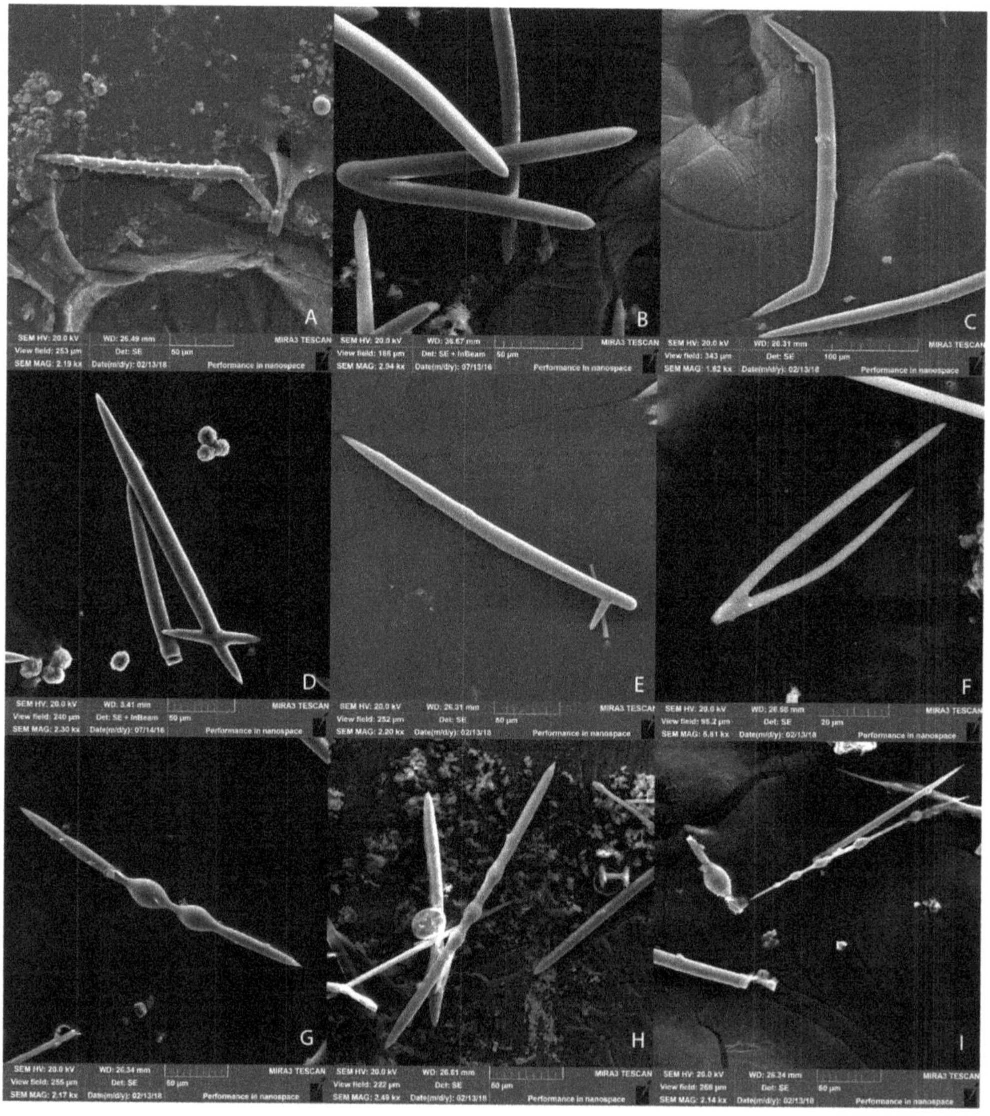

Figure 6. Collection of spicule anomalies analyzed with SEM: a thorny spicule of *T. horrida* sharply bent near one end (**A**); spicule sharply bent medially (**B**); spicule symmetrically bent on both ends (**C**); cross-like spicule (**D**); spicule with a thorn (**E**); bifurcating spicule (**F**); spicule anomaly in the shape of bulbous enlargements on megascleres (**G**,**H**); and microscleres (**I**).

The lowest number of anomalies was found in a specimen of *E. fluviatilis* (1×10^{-3}) from Kanjiža near Martonoš (Tisa River), and the main type of anomaly was the bent megasclere (single distal bend). The highest number of anomalies was found in a specimen of *E. fragilis* (97×10^{-3}) collected at Markovac near Bagrdan (Velika Morava River), and the predominant anomalies were bulbous megascleres. The orthophosphates concentration was lower in Tisa than in V. Morava, while sodium and chloride were both present at higher concentrations on the site that harbored sponges with less anomalies.

4. Discussion

According to current as well as our previous fieldwork, freshwater sponges were not frequent organisms in the studied water bodies. The number of sites where sponges were recorded represents a smaller part of the total number of searched sites. Their abundance, with a few exceptions, was low as well, since only a small number of specimens were collected from the sites where their presence was established. They were observed to display a tendency towards coastal zones and areas with slower water currents. However, in shallow rivers, they were discovered in locations further away from the banks, even in the midst of the stream. Sponges were mostly detected and collected at depths ranging from 0.5 to 1.5 m, and can be found on both natural and artificial substrates. Furthermore, they exhibited a distinct affinity for slightly alkaline and well-oxygenated water, as well as higher temperatures and conductivity.

In the present study, we tried to make a catalogue of the most frequent spicule anomalies encountered in Porifera from Serbian rivers and look for a relationship, if any, between the occurrence of anomalies and selected environmental parameters. Quite remarkably, a number of studies have dealt with the process of spiculogenesis and its regulation in freshwater and marine sponges, both in vitro and in vivo [30–34], and yet, information regarding spicule anomalies is very limited. Holwoet and Van der Vyer, for instance, demonstrated the substantial impact of silicate concentration on spicule formation in *Ephydatia fluviatilis* in terms of their quantity. The number of spicules was inversely related to silicate concentrations, both when sponges were grown under standard conditions in mineral medium, as well as when the differentiation of the aquiferous system was inhibited (hydroxyurea added to the medium), or the integration of spicules into a tridimensional network was disrupted (puromycine added to the medium). Yet, there was no comment about the possible impact on spicule morphology [35]. In the study by Nakayama et al., the complex steps of spicule formation and positioning, which include the activities of different classes of cells, were described in great detail in *E. fluviatilis*, but again, only normal morphogenesis was considered [34].

Interestingly, novel data on the specific topic of spicule malformations are missing. Several important historical papers have raised the question of spicule anomalies and their relation to water pollution. For instance, malformations were observed in spicules of *Trochospongilla leidyi* growing in iron water pipes [36] and megascleres of *E. fluviatilis* growing in water polluted by industrial wastes [37]. Some field observations have suggested that sponges are sensitive to chemical pollution [5], while other studies have been unable to find a clear relationship between spicule malformations and specific causal agents [3].

Sponge skeletal elements in the present study were observed using both light microscopy and scanning electron microscopy, and the morphologies of normal spicules belonging to the five recorded species, as well as their dimensions, corresponded with literature data [38,39]. On the other hand, the encountered anomalies varied greatly in terms of both morphology and frequency; the frequencies of spicules with structural anomalies varied practically two orders of magnitude between the localities and between the species.

The highest percentage of anomalies was found in *E. fragilis* (almost 10% of spicules were aberrant), while the lowest was in *E. fluviatilis* (only 1‰). The two "antipode" rivers, i.e., the rivers with sponges that had the biggest (Velika Morava) and smallest (Tisa) numbers of anomalies, showed statistically significant differences in a number of pollutants. Hence, we can speculate that they might have impacted anomaly occurrence. For instance, the concentrations of ammonia, manganese, and lead were 3, 2.3, and 4.3 times, higher in Velika Morava, respectively, compared to Tisa and could potentially affect spiculogenesis. This finding is in line with the study of Richelle et al., who demonstrated that *E. fluviatilis* was affected by Cd and Hg and that relatively low metal concentrations could cause spicule malformations [13]. However, in the mentioned study, the sponges were artificially grown in selected portions of rivers characterized by different types of pollution. On the other hand, it seems that the scarcity of some elements, such as sodium and chloride (2.5 and

3 times, respectively, lower concentrations in Velika Morava than in Tisa) might play a role in the development of malformations, given the statistically highly significant difference.

Interestingly some authors have pointed to the influence of copper and zinc on spiculogenesis [21]. In the present study, these metals did not seem to influence the appearance of anomalies.

The fact that *E. fragilis* had 10 times more aberrations than recorded on average, in the absence of extreme environmental conditions (none of the pollutant's concentrations exceeded limits defined by legislation), should be emphasized. It suggests that *E. fragilis* might be more sensitive to different pollutants than other species. This finding is in agreement with a report on sponges from Lake Ilopango, El Salvador [14]. Namely, a very high frequency of anomalies (bulbous structures) was registered in *Spongilla alba*, and their occurrence was related to elevated concentrations of arsenic, while at the same time, spicules of *E. fluviatilis* from this lake were without malformations, pointing again to differences in sensitivity, depending on the species.

Freshwater sponges were poorly represented in the surveyed localities in Serbia; a small number of sponges was collected, i.e., their abundance was low, which represents a limitation in the present study. In light of the scarcity of sponge samples, our focus in the analysis was on exploring potential trends and possibilities rather than establishing definitive correlations. We believe that even with a limited sample size, our study provides valuable insights into the interplay between physical and chemical parameters and sponge anomalies. The second limitation of sponge paucity was, consequently, the fact that not all sponge species were present at the same localities, preventing the comparison of species-specific propensities to aberrant spiculogenesis in a given environment.

5. Conclusions

The present study describes the pattern of sponge spicule malformations in different rivers from inland Serbia. Several factors can explain the observed pattern, ranging from environmental parameters to aberrant spiculogenesis, the origin of which is still unclear. The use of this invertebrate taxon as a bioindicator could be significantly increased if studies confirm a relationship between the changes in the chemical and physical parameters of streams and the frequency of anomalies found in spicules. Future investigations on larger samples of Porifera are needed.

Author Contributions: Conceptualization, S.A.; methodology, S.A.; validation, M.P. and V.N.; investigation, S.A. and B.T.; data curation, S.A., B.T. and B.V.; writing—original draft preparation, S.A.; writing—review and editing, S.A., B.V., B.T., M.P. and V.N.; supervision, M.P. and V.N. All authors have read and agreed to the published version of the manuscript.

Funding: This research was funded by the Ministry of Science, Technological Development, and Innovation of The Republic of Serbia, contract No. 451-03-47/2023-01/200007.

Data Availability Statement: Data are available upon request to corresponding author.

Conflicts of Interest: The authors declare no conflict of interest.

References

1. Pronzato, R.; Manconi, R. Atlas of European Freshwater Sponges. *Ann. Mus. Civ. Stor. Nat. Ferrara* **2001**, *4*, 3–64.
2. Frost, T.M. Freshwater Sponges. In *Tracking Environmental Change Using Lake Sediments*; Springer: Dordrecht, The Netherlands, 2002; pp. 253–263.
3. Penney, J.T.; Racek, A.A. *Comprehensive Revision of a Worldwide Collection of Freshwater Sponges (Porifera, Spongillidae)*, 1st ed.; Smithsonian Institution Press: Washington, DC, USA, 1968; p. 184. [CrossRef]
4. Ricciardi, A.; Reiswig, H.M. Freshwater Sponges (Porifera, Spongillidae) of Eastern Canada: Taxonomy, Distribution, and Ecology. *Can. J. Zool.* **1993**, *71*, 665–682. [CrossRef]
5. Harrison, F.W. Sponges (Porifera: Spongillidae). In *Pollution Ecology of Freshwater Invertebrates*; Academic Press: New York, NY, USA, 1974; pp. 29–66.
6. Hustus, K.; Díez-Vives, C.; Mitsi, K.; Nutakki, J.; Kering, V.; Nguyen, I.T.; Spencer, M.G.; Leys, S.P.; Hill, M.S.; Riesgo, A. Algal Symbionts of the Freshwater Sponge *Ephydatia muelleri*. *Symbiosis* **2023**, *90*, 259–273. [CrossRef]

7. McAuley, D.G.; Longcore, J.R. Survival of Juvenile Ring-Necked Ducks on Wetlands of Different PH. *J. Wildl. Manag.* **1988**, *52*, 169–176. [CrossRef]
8. Williamson, C.E. Crayfish Predation on Freshwater Sponges. *Am. Midl. Nat.* **1979**, *101*, 245–246. [CrossRef]
9. Frost, T.M. Porifera. In *Ecology and Classification of North American Freshwater Invertebrates*; Thorp, J.H., Covich, A.P., Eds.; Academic Press: New York, NY, USA, 1991.
10. Resh, V.H. Life Cycles of Invertebrate Predators of Freshwater Sponge. In *Aspects of Sponge Biology*; Elsevier: Amsterdam, The Netherlands, 1976; pp. 299–314.
11. Andjus, S.; Lazović, V.; Nikolić, N.; Tubić, B.; Nikolić, V.; Paunović, M. Distribution of Freshwater Sponges in Serbia. *Fundam. Appl. Limnol. Hydrobiol.* **2020**, *76*, 48–51. [CrossRef]
12. Andjus, S.; Nikolic, N.; Marjanovic, A.; Brankovic, M.; Lazovic, V.; Tubić, B.; Čanak Atlagić, J.; Nikolić, V.; Paunović, M. First Record of Freshwater Sponge *Trochospongilla horrida* Weltner, 1893 in Serbia—A Morphological and Genetic Study. *Limnologica* **2019**, *76*, 48–51. [CrossRef]
13. Richelle, E.; Degoudenne, Y.; Dejonghe, L.; de Vyver, G. Experimental and Field Studies on the Effect of Selected Heavy Metals on Three Freshwater Sponge Species: *Ephydatia fluviatilis, Ephydatia muelleri* and *Spongilla lacustris. Arch. Hydrobiol. Stuttg.* **1995**, *135*, 209–231. [CrossRef]
14. Poirrier, M.A.; Trabanino, S. Freshwater Sponges (Porifera: Spongillidae) from Lake Ilopango, El Salvador, with Observations on Spicule Malformation in *Spongilla alba. Trans. Am. Microsc. Soc.* **1989**, *108*, 211–214. [CrossRef]
15. Jørgensen, C.B. *On the Spicule-Formation of* Spongilla lacustris *(L.)*; I Komission hos Munksgaard: Copenhagen, Demark, 1944.
16. Elvin, D. Growth Rates of the Siliceous Spicules of the Fresh-Water Sponge *Ephydatia muelleri* (Lieberkühn). *Trans. Am. Microsc. Soc.* **1971**, *90*, 219–224. [CrossRef]
17. Pé, J. Étude Quantitative de La Régulation Du Squelette Chez Une Éponge d'eau Douce. *Arch. Biol.* **1973**, *84*, 147–173.
18. Bavestrello, G.; Bonito, M.; Sarà, M. Influence of Depth on the Size of Sponge Spicules. *Sci. Mar.* **1993**, *57*, 415–420.
19. Subagio, I.B.; Setiawan, E.; Hariyanto, S.; Irawan, B. Spicule Size Variation in *Xestospongia testudinaria* Lamarck, 1815 at Probolinggo-Situbondo Coastal. *AIP Conf. Proc.* **2017**, *1854*, 20034.
20. Erpenbeck, D.; Galitz, A.; Wörheide, G.; Albrecht, C.; Pronzato, R.; Manconi, R. Having the Balls to Colonize—The *Ephydatia fluviatilis* Group and the Origin of (Ancient) Lake "Endemic" Sponge Lineages. *J. Great Lakes Res.* **2020**, *46*, 1140–1145. [CrossRef]
21. Francis, J.C.; Harrison, F.W. Copper and Zinc Toxicity in *Ephydatia fluviatilis* (Porifera: Spongillidae). *Trans. Am. Microsc. Soc.* **1988**, *107*, 67–78. [CrossRef]
22. Falasco, E.; Ector, L.; Wetzel, C.E.; Badino, G.; Bona, F. Looking Back, Looking Forward: A Review of the New Literature on Diatom Teratological Forms (2010–2020). *Hydrobiologia* **2021**, *848*, 1675–1753. [CrossRef]
23. Vasiljević, B.; Simić, S.B.; Paunović, M.; Zuliani, T.; Krizmanić, J.; Marković, V.; Tomović, J. Contribution to the Improvement of Diatom-Based Assessments of the Ecological Status of Large Rivers—The Sava River Case Study. *Sci. Total Environ.* **2017**, *605–606*, 874–883. [CrossRef] [PubMed]
24. Lavoie, I.; Hamilton, P.B.; Morin, S.; Kim Tiam, S.; Kahlert, M.; Gonçalves, S.; Falasco, E.; Fortin, C.; Gontero, B.; Heudre, D.; et al. Diatom Teratologies as Biomarkers of Contamination: Are All Deformities Ecologically Meaningful? *Ecol. Indic.* **2017**, *82*, 539–550. [CrossRef]
25. Richelle-Maurer, E.; Degoudenne, Y.; de Vyver, G.; Dejonghe, L. Some Aspects of the Ecology of Belgian Freshwater Sponges. In *Sponges in Time and Space*; Balkema: Rotterdam, The Netherlands, 1994; pp. 341–350.
26. Agency for Environmental Protection. *Inland and Groundwater Quality Report—2017*; Agency for Environmental Protection: Belgrade, Serbia, 2018.
27. Manconi, R.; Pronzato, R. Phylum Porifera. In *Thorp and Covich's Freshwater Invertebrates*; James, T.H., Christopher, R.D., Eds.; Elsevier: London, UK, 2015; Volume 1, pp. 133–157. ISBN 9780123850270.
28. Lucey, J.; Cocchiglia, L. Cocchiglia Distribution of Sponges (Porifera: Spongillidae) in Southern Irish Rivers and Streams. *Biol. Environ. Proc. R. Ir. Acad.* **2014**, *114B*, 89–100. [CrossRef]
29. Official Gazette of RS 50/2012. *Regulation on Limit Values of Pollutants in Surface and Ground Waters, Sediment, and Deadlines for Reaching Them*; Ministry Agriculture, Forestry and Water Management: Belgrade, Serbia, 2012.
30. Imsiecke, G.; Steffen, R.; Custodio, M.; Borojevic, R.; Müller, W.E.G. Formation of Spicules by Sclerocytes from the Freshwater Sponge *Ephydatia muelleri* in Short-Term Cultures In Vitro. *Vitr. Cell. Dev. Biol.* **1995**, *31*, 528–535. [CrossRef]
31. Cao, X.; Fu, W.; Yu, X.; Zhang, W. Dynamics of Spicule Production in the Marine Sponge *Hymeniacidon perlevis* during in Vitro Cell Culture and Seasonal Development in the Field. *Cell Tissue Res.* **2007**, *329*, 595–608. [CrossRef]
32. Müller, W.E.G.; Wang, X.; Burghard, Z.; Bill, J.; Krasko, A.; Boreiko, A.; Schloßmacher, U.; Schröder, H.C.; Wiens, M. Bio-Sintering Processes in Hexactinellid Sponges: Fusion of Bio-Silica in Giant Basal Spicules from *Monorhaphis chuni. J. Struct. Biol.* **2009**, *168*, 548–561. [CrossRef]
33. Wang, X.; Gan, L.; Jochum, K.P.; Schröder, H.C.; Müller, W.E.G. The Largest Bio-Silica Structure on Earth: The Giant Basal Spicule from the Deep-Sea Glass Sponge *Monorhaphis chuni. Evid.-Based Complement. Altern. Med.* **2011**, *2011*, 540987. [CrossRef]
34. Nakayama, S.; Arima, K.; Kawai, K.; Mohri, K.; Inui, C.; Sugano, W.; Koba, H.; Tamada, K.; Nakata, Y.J.; Kishimoto, K. Dynamic Transport and Cementation of Skeletal Elements Build up the Pole-and-Beam Structured Skeleton of Sponges. *Curr. Biol.* **2015**, *25*, 2549–2554. [CrossRef]

35. Van De Vyver, G.; Holvoet, S.; Dewint, P. Variability of the Immune Response in Freshwater Sponges. *J. Exp. Zool.* **1990**, *254*, 215–227. [CrossRef]
36. Potts, E. Contributions Towards a Synopsis of the American Forms of Fresh Water Sponges with Descriptions of Those Named by Other Authors and from All Parts of the World on JSTOR. Available online: https://www.jstor.org/stable/4061447?seq=1#metadata_info_tab_contents (accessed on 1 November 2019).
37. Old, M.C. Environmental Selection of the Fresh-Water Sponges (Spongillidae) of Michigan. *Trans. Am. Microsc. Soc.* **1932**, *51*, 129–136. [CrossRef]
38. Hooper, J.N.A.; Van Soest, R.W.M. Systema Porifera. A Guide to the Classification of Sponges. In *Systema Porifera: A Guide to the Classification of Sponges*; Hooper, J.N.A., Van Soest, R.W.M., Willenz, P., Eds.; Springer: Boston, MA, USA, 2002; pp. 1–7. ISBN 978-1-4615-0747-5.
39. Manconi, R.; Pronzato, R. Suborder Spongillina Subord. Nov.: Freshwater Sponges. In *Systema Porifera: A Guide to the Classification of Sponges*; Hooper, J.N.A., Van Soest, R.W.M., Willenz, P., Eds.; Springer: Boston, MA, USA, 2002; pp. 921–1019. ISBN 978-1-4615-0747-5.

MDPI AG
Grosspeteranlage 5
4052 Basel
Switzerland
Tel.: +41 61 683 77 34

Water Editorial Office
E-mail: water@mdpi.com
www.mdpi.com/journal/water

www.ingramcontent.com/pod-product-compliance
Lightning Source LLC
Chambersburg PA
CBHW061453210926
51850CB00008B/1706

* 9 7 8 3 7 2 5 8 1 7 2 4 5 *